Foundations of Developmental Genetics

Foundations of
Developmental Genetics

D. J. Pritchard

Department of Human Genetics
University of Newcastle upon Tyne, England

Taylor & Francis
London and Philadelphia
1986

UK	Taylor & Francis Ltd, 4 John St., London WC1N 2ET
USA	Taylor & Francis Inc., 242 Cherry St., Philadelphia, PA 19106–1906

British Library Cataloguing in Publication Data

Pritchard, D. J.
    Foundations of developmental genetics.
    1. Genetics
    I. Title
    575.1   QH430

    ISBN 0-85066-356-3
    ISBN 0-85066-287-7 Pbk

Library of Congress Cataloging in Publication Data

Pritchard, D. J. (Dorian J.)
    Foundations of developmental genetics.

    Bibliography: p.
    Includes index.
    1. Developmental genetics.   I. Title.
    QH453.P75   1986   591.3   86-5884
    ISBN 0-85066-356-3
    ISBN 0-85066-287-7 (pbk.)

Illustrated by D. J. Pritchard
Typeset by Mathematical Composition Setters Ltd, Salisbury
Printed by Taylor & Francis (Printers) Ltd, Rankine Road, Basingstoke, Hants.

Preface

In the 'Aphorismen' of Georg C. Lichtenberg is the following quotation: "The thing that astonished him was that cats should have two holes cut in their coat exactly at the place where their eyes are." The individual to whom this refers might well also have wondered why, when he looked in the mirror, the right side of his face was so very much like the left, but very different from everyone else's; or when he found a centipede or millipede in the garden, whether he might one day find an 'infinipede', with an unlimited number of segments and legs. It is for persons with just this kind of ingenious curiosity about biology, albeit tempered by the maturity of wider experience, that this book is written. It has grown from a course in developmental genetics that I teach to second-year science students at the University of Newcastle upon Tyne, and while its direction has derived very largely from the questions research workers are currently asking about how animals develop, the emphasis placed on the various topics has been influenced to a considerable extent by the interests and background of my students. For example, I have assumed all readers to have a basic knowledge of transmission genetics, biochemistry and biology, and I have felt it necessary to emphasize at the outset the enormous gulf between eukaryotes and prokaryotes. I have dealt briefly with invertebrate embryogenesis, but have paid a lot of attention to the details of molecular biology. At all times I have tried to relate the happenings among molecules to those that proceed on grosser levels. My main aim, however, has been to strengthen and establish bridges between embryology and molecular biology, to facilitate mutual understanding and traffic between the two fields. Scattered throughout the text are pointers to and suggestions for research projects which should reinforce these links.

Although primarily intended for undergraduates, I have tried to develop the subject matter sufficiently to interest postgraduate students and research workers in other subjects, and where they have presented themselves I have deliberately chosen intellectually attractive examples, that I believe inspire new interest in both young heads and old. Somewhat to my surprise my analysis has led to what amounts to a new theory relating development, adaptation and evolution, although unbeknown to me evolution theorists have been following a convergent course for some considerable time.

The genotypes and phenotypes of animals in general are already quite well

v

described elsewhere, but between the two lies another realm, the 'epigenetic space', into which numerous exploratory sorties have been made, but which still remains a largely uncharted region. This book is intended as a guide to this area, to the epigenetic phenomena by which during development genotype becomes converted into phenotype. In this area we must tackle the literature of many disciplines: anatomy, biochemistry, cell biology, embryology, evolution theory, genetics, molecular biology, physiology and zoology. But the developmental biologist is somewhat like the architect who respects and incorporates the skills of many other trades yet adds something else of his own. Epigenetics thus provides a very special kind of intellectual challenge.

Of all scientific disciplines this must be the most difficult in which to make generalizations. One of the tenets of the scientific approach is that if we dig deep enough we will eventually discover common elements in various systems, which when identified and described will allow laws to be written and predictions made. However, in developmental biology we are bedevilled by the problems of the levels of organization at which we should expect to find similarities and where we can safely draw boundaries around groups of phenomena with common elements. A powerful camp currently supports the idea that body patterning in vertebrates and insects is probably defined in a similar fashion. To my mind this is a highly dubious notion. At the molecular level there are still hopes that punctuation of the genetic message will prove to be carried out in similar fashions in bacteria and higher organisms, but this cherished belief must probably soon be abandoned. Where experimental work is concerned it is very difficult to be sure which of the results one obtains in modified situations can be extrapolated to intact organisms. We still do not know what kinds of experiments can validly be carried out with cells taken out of the body and cultured in man-made vessels.

While writing this book I have at times felt the need to adopt a determined stance in order to correct what seems to me a misunderstanding, or bias of emphasis in recent literature. For example, one currently popular idea is that a species' DNA encodes *all* the instructions necessary to bring about development of its members. This is in fact untrue. Not only are many other components of the fertilized egg essential precursors of elements of body cells, but also, in order to form a normal viable animal, the information encoded in its DNA must be interpreted in the correct molecular environmental context. This is perhaps best illustrated by an analogy. For example, imagine that the orders issued to a soldier on manoeuvres include the instruction: "Using locally available materials build a shelter to protect yourself from the prevailing elements". Depending on circumstances this could be interpreted as a direction to construct a windbreak of snowblocks, a canopy of dried leaves, or a platform of mangrove roots. An alternative instruction might say "Using the kit provided assemble the building shown on the enclosed plan", in which case only one outcome would be correct. Interpretation by a developing embryo of the message contained in its DNA falls somewhere between these situations. Individuals of similar genotypes will utilize essentially the same raw materials, but may incorporate them in rather different proportions, or, taking cues from their surroundings, they may over- or under-emphasize particular bodily features. The resultant derived phenotypes may therefore differ considerably even though encoded by similar genotypes. The plumage of flamingos is typically a brilliant pink, but they develop this colour only if their food contains a high concentration of carotenoids.

Conversely, similar phenotypes can be derived from radically different genotypes since the genes do not contain a complete programme for development in the way that concept is generally understood. Development occurs in a predictable fashion largely because the particular set of conditions present at the outset naturally gives rise to a specific consequence. An analogy for this principle would be the natural colonization of a newly formed island. The different plant and animal species become established in a sequence defined by natural laws, but without external direction. Similarly, a fertilized egg begins its ontogenetic journey along certain developmental pathways because there is only a limited number of open alternatives. It proceeds towards a defined phenotypic outcome by a series of cause-and-effect events, only some of which are actually dictated by the genome. In elucidating this progression genetic mutants are of particular value since they allow us to distinguish the genetically coded aspects from those that are merely the automatic consequences of preceeding situations.

A more fundamental problem faced by the developmental biologist relates to the very basis of what is widely accepted as 'the scientific approach'. Many investigators operate on the principle that analysis of natural phenomena must necessarily involve subdividing the object of investigation into ever smaller elements and describing those elements in the minutest detail, on the assumption that if this kind of analysis is taken far enough, all there is to know will eventually be revealed. This is equivalent to pulling a bicycle pump to pieces and examining the composition and structure of its parts in order to find out how it performs its major function of pumping air. Clearly all the answers are not to be found in that way, but the exercise of examining the relationships between the diverse parts of living organisms and the impact of external non-biological forces upon them is something many of us are not trained to do.

Throughout most of this book I also have followed the reductionist approach, first examining aspects of gross phenotype, then moving down through cell physiology and the properties of proteins, through translation and RNA, to the DNA and transcription. However, I have then tried to pull together some of the threads revealed in previous chapters, in order to elucidate the principles by which animal development is achieved, and to look at whole organisms in a developmental and evolutionary context. It is at this stage more than any other that the text departs from the less imaginative style of reading matter often recommended to undergraduates. It is hoped that the latter chapters particularly may contribute something to the more theoretical approach to living systems which has traditionally played such an important part in the development of our subject.

One topic I have not covered, but which promises to contribute valuable ideas for the future, is the study of viruses. As yet it is difficult to discern to what extent viruses are prokaryote infiltrators, fragments of eukaryotes that have broken loose, something in between, or something different altogether. It is therefore at present impossible to know whether the rules that govern expression of viral genes integrated into eukaryote DNA should also apply to the rest of the genome, or vice versa. Inter-species transfer of genetic material as viral particles has apparently occurred on many occasions and when we know the full extent of this exchange we may well need to revise some of our best-loved theories.

Each of the disciplines I have entered has its own standards of excellence and its own style of communication, but I have tried where possible to unify the different approaches. In the interest of encouraging the flow of positive thinking so essential in undergraduate teaching I have avoided the critical appraisal of all observations that would be essential in a more advanced text. For the same reason, references and the names of individual workers are rarely included in the text, although all relevant sources, plus fuel for arguments to counter those I advance, should be accessible by an intelligent perusal of the reading lists.

In attempting to cover such a wide range of subject matter errors and misunderstandings are bound to arise, and I am very grateful to colleagues more knowledgeable than myself who have taken the trouble to read and criticize sections of the manuscript. In this respect I owe particular thanks to Professors Ken Burton and Stuart Glover and to Doctors Donald Ede, Tim Horder, Monica Hughes, Alec Panchen, Surinder Papiha, Tony Samson, Clarke Slater and Robert Whittle. However, I must admit I have not always taken their advice and must accept full responsibility for those errors that remain. I wish to express my grateful thanks to Mrs Valerie Webb and Mr Ian Munro for the diagram annotation and to our secretarial staff, especially Mrs Tessa Havelock and Miss Debbie Hayles. My family, Penny, Ceri and Hamish earn my sincere gratitude for their interest and support and for foregoing trips and holidays they might otherwise have had. I gratefully acknowledge permission granted by the McGraw-Hill Book Company Ltd to reproduce Figures 2.11 and 4.7, and the Company of Biologists Ltd to reproduce a diagram from one of my own papers as Figure 7.9.

This book could not have come into existence without the additional stimulus afforded by the thought-provoking words of many other authors, in particular those of my late Professor, C. H. Waddington, and by discussions with many people, including numerous students, my colleagues in Newcastle, and my former teachers and colleagues at the Department of Genetics in Edinburgh. I owe a special debt of gratitude to Professor Derek Roberts for his constant guidance and enthusiastic support and for initiating the idea for this work.

Dorian Pritchard
Newcastle upon Tyne
January 1986

Contents

Dedication

To my parents, who gave me not only life itself, but also a love of it in its many aspects and who, despite extreme provocation, never discouraged me from asking questions. Also to Penny, for providing some of the most important answers.

Chapter 1 Prokaryotes and the origins of eukaryotes

1.1. Genotype and phenotype

Most of the readers of this book will already be aware of Darwin's theory of evolution through natural selection. From the point of view of the evolution of genetic material, an important consideration is that although mutational forces introduce variation into the genes, the forces of selection do not act upon the genes themselves, but rather upon their products: they act upon the living organism, the 'survival machine', that carries those genes. The genetic information which is inherited by an organism from its parents becomes translated into physiological, anatomical and behavioural properties as it develops, and it is these properties which may ensure survival of the organism to the point where it can reproduce and pass its genes on to the next generation.

The identifiable or measurable properties of an organism are what we call its phenotype. Development involves the translation of the genetic blueprint, or genotype, into phenotype, which can be considered to be the product of interaction between genes and environment:

$$\text{Phenotype} = \text{Genotype} \times \text{Environment}$$

This book is about the nature of that interaction, the way in which an organism is produced from its genes.

Classical genetic theory deals with many aspects of the structure of the genetic material, its replication, its mutation and its distribution between progeny, or between members of a population. But what classical genetics has largely avoided is the way in which genes are selectively expressed at different times in development, or in different parts of the body. For example, if a *Drosphila* fruit fly carries a gene for scarlet eye colour, why is this gene not expressed in all the cells of the body? Why is a fly with a scarlet gene not scarlet all over? Why is the larva not abnormally pigmented also? All over the world many laboratories are tackling just this type of problem, using a wide variety of living materials and some of the most advanced molecular genetic techniques. Many of the observations of the early embryologists, which at the time seemed inexplicable, are now falling into place in the minds of present-day developmental biologists. It therefore seems an opportune time to draw together some of the early observations, and to try to see

1

them in the context of the new knowledge acquired from the very high-powered techniques which have arisen in the field of molecular biology, founded upon the simpler systems of bacteria.

As a starting point to our investigation, we will consider a major type of genetic control which operates in bacteria.

1.2. *Prokaryote control systems*

The living world is divided into two major groups: the eukaryotes, those with a 'true nucleus', which includes all the plants, animals, fungi and protozoa, and the prokaryotes (literally 'before a nucleus') — the blue-green algae (more correctly called Cyanobacteria), and the true bacteria. (The viruses do not easily fall into either major category and their origin is uncertain.)

In the prokaryotes there are various systems for the control of gene expression. We are not concerned here with the intricacies of prokaryote genetics and for our purpose it will be sufficient to consider a simplified treatment of the negative and positive control of transcription. In both cases there are induction or inducible systems, and repression, or repressible systems. Inducible systems characteristically cope with the situation where the organism needs to respond appropriately to the occasional presence of a particular biochemical in the medium, such as a food substance like lactose. In contrast, repressible systems come into operation to cope with scarcity of an essential substance, such as tryptophan, which is normally present in adequate concentration in the environment.

Figure 1.1 illustrates the two types of negative control. In both there is a linked array of structural genes of related function. The term structural gene, or cistron, describes a gene which codes for an enzyme or structural protein utilized by an organism in carrying out its normal life processes. In each system three cistrons are illustrated, labelled A, B and C, and D, E and F. Close to the 'upstream' or 3′ end of the A gene is what is called an operator site (O) and upstream from this a promoter site (P). The promoter site is the part of the DNA to which the RNA polymerase (the enzyme which carries out transcription) first becomes attached. The cistrons in each group have related functions and this group of linked genes, with associated operator and promoter sites, is termed an operon. In both the inducible and repressible systems there is also a regulator gene (R) which may or may not be structurally linked to the operon. The regulator gene codes for a repressor protein, that controls the transcription of the structural genes.

In a negative inducible system, the repressor protein, in its native state, actively binds to the operator site. If RNA polymerase becomes attached to the promoter site, its passage along the DNA is then blocked by the repressor protein in its path. However, the repressor protein has the capacity to change its shape and its binding properties should it become associated with a specific regulatory metabolite. This property of a protein which allows it to change shape is known as allostery, and the protein is said to be an allosteric protein.

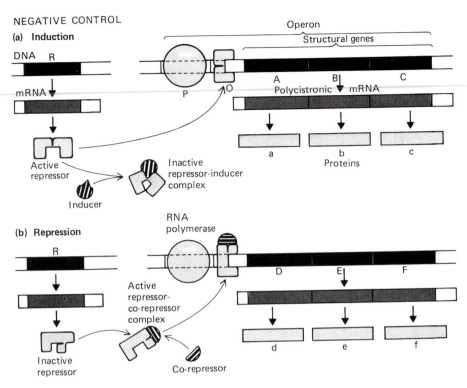

Figure 1.1. Negative control of transcription in prokaryotes.
(a) Induction. A regulator gene, R, codes for a repressor protein, that binds directly on to the operator site, O, upstream of the three structural genes, A, B and C. RNA polymerase attaches to the promotor site, P, but its transcription of the structural genes is blocked by bound repressor. If inducer molecules are present in the cell, they bind to the repressor and reduce its binding affinities for O, so that A, B and C can be transcribed. One polycistronic messenger RNA molecule is produced and translated into three enzymes with related functions. P, O, A, B and C constitute an operon. (b) Repression. In this case the repressor protein is inactive in its native state, so that the structural genes D, E and F are transcribed constitutively. The repressor becomes active in the presence of co-repressor. It then binds to O, blocks the passage of RNA polymerase from P and so prevents the expression of D, E and F.

The operon model was proposed by Jacob and Monod as an explanation of some features of what has become considered to be the classic inducible system, the lac operon. This is concerned with the uptake and metabolism of the sugar lactose. In this system allolactose is the regulatory metabolite, which is formed in the cell when lactose is present. The lac repressor protein responds to allolactose, changing its shape so that it can no longer bind to the lac operator site (Figure 1.1(a)). RNA polymerase which binds to the promoter, P, now passes down the DNA and transcribes A, B and C as a single long molecule of RNA—a polycistronic message

— which is translated into three enzymically active proteins in the ribosomes. When the cytoplasmic concentration of lactose diminishes, due to the action of the enzymes which are produced by the operon, insufficient allolactose is available to inactivate the repressor protein. The repressor therefore binds back on to the operator site and blocks further transcription, pending a later increase in lactose.

In negatively repressible systems (Figure 1.1(b)) the regulator gene codes for a repressor protein which in its native state is *inactive*. RNA polymerase that binds to the promoter site can thus transcribe D, E and F without inhibition. The repressor protein becomes activated by combining with a specific co-repressor, which in some cases is the end-product of the synthetic pathway. The repressor/co-repressor complex binds on to the operator site and blocks further transcription. When cytoplasmic levels of the end-product (i.e., co-repressor) diminish, the complex dissociates, the repressor becomes inactive and transcription of D, E and F proceeds once more.

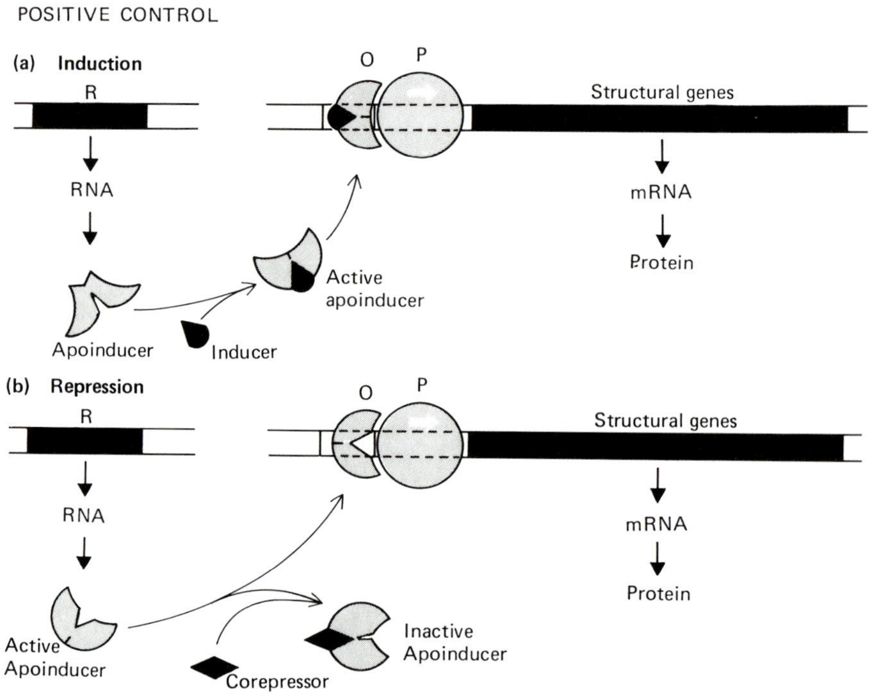

POSITIVE CONTROL

Figure 1.2. Positive control of transcription in prokaryotes.
(a) Induction. An inactive apoinducer protein is produced by expression of a regulatory gene sequence. This becomes active on combination with an inducer molecule and the complex binds to an operator site beside the structural gene(s). This facilitates binding of RNA polymerase to the promoter, allowing transcription to take place. (b) Repression. In this case the apoinducer is active in its native state, but will not bind to the operator site if it becomes complexed with a co-repressor molecule.

There is a variety of positive controls in bacteria. Those that act at the initiation of transcription are the exact counterpart of negatively controlled systems, except that the regulator protein is *essential* for transcription instead of blocking it. Such proteins have been called apoinducers. Figure 1.2 summarizes the role of such proteins in controlling transcription. In inductive positively controlled systems the apoinducer is inactive until it becomes complexed with an inducer molecule. The converse occurs in repressive positively controlled systems, the apoinducer being active in its native state, but becoming inactivated by combination with a co-repressor molecule.

In all types of operon, one of the most important characterisitics is *the co-ordination of expression of several genes of related function through production of a polycistronic messenger RNA molecule.* Apart from the expression of the genes carried by the organelles (see below and Chapter 12), this seems to be a unique feature of the prokaryote operon, but it is interesting that genes which are organized into operons in some bacteria are not necessarily organized in the same way in others. Additional prokaryote positive controls involve interference with the termination of RNA transcription (cf. synthesis of immunoglobulin heavy chains, page 203) and the substitution of a variant protein subunit into the RNA polymerase molecule, causing it to attach to a different set of promotor sequences.

1.3. Eukaryote control systems

In studying the control of gene expression in eukaryotes there is a major problem, since it is carried out in many different ways: we cannot yet put forward a simple all-embracing model like the operon. However, until very recently the majority of authors of general textbooks would sidestep these problems, by suggesting one or other of the superficially reasonable models derived directly from the bacterial operon. We now know that although there are very many important similarities between bacteria and the eukaryotes, the organization of their DNA is quite different and there are good reasons why we should not expect operons to exist in quite the same form in higher organisms. In fact, in spite of hundreds, perhaps thousands, of man-years spent in searching for eukaryote operons, not one really convincing example has been reported. However, some features of the operon system are present in eukaryotes and it is beginning to look as if a transcriptional control system of a rather similar basic type may exist as just one element in a vastly more complex hierarchy of controls. We will return to these transcriptional controls in Chapter 12.

1.4 Differences between prokaryotes and eukaryotes

Life arose on this planet nearly 4 000 000 000 years ago. The first life-forms lacked cell nuclei so would fall into the prokaryote category. Cells with nuclei arose nearly

2 000 000 000 years ago. If we look at present-day prokaryotes, we find they are highly specialized for rapid responses to changes in their environment; they are streamlined for efficiency. In contrast the eukaryotes have specialized in stability of gene expression, and the more complex eukaryotes have developed highly refined mechanisms for insulating them from the effects of environmental fluctuations. The operon is a highly efficient system for rapid response, so except in situations where higher organisms are required to respond rapidly to new stimuli, there is an *a priori* expectation that eukaryotes would have little requirement for operons. Indeed, since the two lines diverged so long ago, it is quite possible that operons evolved in prokaryotes after the eukaryotes had branched off on their own evolutionary course.

There are also important differences between eukaryote genomes and those of prokaryotes. (1) Eukaryotes have an enormous amount of DNA, perhaps 400–1000 times the estimated amount required to command all necessary processes. In our inquiry we will need to account for this apparently surplus DNA. (2) Whereas in prokaryotes practically all genes are present only as single copies, in some eukaryotes as much as 30% of the genome is composed of sequences repeated up to 1 000 000 times or more which are not translated into protein. As we will see, there are good reasons to suppose that some of this repetitive DNA is required for the controlled expression of the single-copy sequences which are translated. (3) Eukaryote DNA is complexed with protein and RNA in a very organized fashion, forming a substance called chromatin, the structure of which changes when transcription takes place. This is a very important feature that has no counterpart in prokaryotes. Eukaryote chromatin is also typically divided into several units called chromosomes.

The major component of the chromosomal proteins is a class of basic proteins called histones which have a composition that has been very strictly conserved throughout evolution. For example, the amino acid sequence of histone H4 is identical in the cow, the pig and the rat and there are only two amino acid differences in the pea! Histone H4 has thus evolved at something like $1/200$ of the rate of most proteins. Obviously the structures of the histones must be of extreme importance for eukaryotes. The properties and functions of the histones are described in Chapter 10.

Another characteristic of the eukaryote genome is the separation of DNA from cytoplasm by the nuclear membrane. In prokaryotes translation of messenger RNA into protein frequently begins before transcription of the messenger is completed. In eukaryotes there is a considerable time-gap between RNA synthesis in the nucleus and its translation in the cytoplasm.

In the prokaryotes the gene is co-linear with messenger RNA and polypeptide — the order of the bases in the DNA is precisely the same as the order of bases in the messenger and the order of amino acids in the polypeptide (or protein) for which it codes. In contrast, in eukaryotes many structural genes are 'split', that is they contain 'introns', or sequences of bases which have no counterpart in the polypeptide or protein product. These introns are represented in the intital RNA transcript, but are removed in the production of messenger at a processing stage before the RNA leaves the nucleus (see Chapter 9). As we shall see in Chapter 11, sections of eukaryote genes can even become rearranged to produce entirely novel combinations. Some of the more important differences between the prokaryotes and eukaryotes are summarized in Table 1.1 and Figure 1.3.

Table 1.1. Differences between prokaryotic and eukaryotic cells.

Feature	Prokaryotes	Eukaryotes
Organisms represented	Bacteria and 'blue-green algae'	Protozoa, fungi, plants and animals
Time of origin	3.5×10^9 years ago	1.5×10^9 years ago
Size	Usually $1-10$ μm diameter	Somatic cells usually $10-100$ μm diameter, Protozoa $0.2-1500$ μm
Cellular organization	Mainly unicellular	Mainly multicellular, with differentiation and mutual interdependence of cell types
Cell division	By simple binary fission	By mitosis or meiosis, using centrioles and spindle apparatus
Nucleus	Absent, but DNA and RNA synthesis located in specific areas	Present, bounded with a membrane perforated by pores
Nucleolus	Absent	At least one present per nucleus
Cell wall	Present, may contain muramic acid	Present in plants but does not contain muramic acid
Internal membranes	If present, simple and often transient	Complex, permanent structures such as endoplasmic reticulum and Golgi apparatus
Organelles	Absent	Mitochondria, plastids and centrioles present
Elecron-transport system	Localized on cell membrane	Localized on inner membranes of mitochondria and chloroplasts
Metabolite pools	Very small or absent	Frequently very large
Motility	Simple flagella of flagellin, if present	If present, cilia and flagella composed of tubulin and based on a 9-outer-and-2-central microtubular structure
Genetic material		
Quantity of DNA	Very small, e.g., 0.0014 units/cell in *E.coli*	Very large, e.g., 1 unit†/cell in the mouse
Organization of DNA	DNA in form of simple duplex. Linkage group single and circular	DNA mainly as several complex chromosomes. Linkage groups linear and multiple
DNA-associated protein	Very little, limited to repressor proteins and enzymes involved in DNA and RNA synthesis	Large quantities present, especially histones, conferring nucleosomal structure
Sequence complexity	All sequences unique	Many repetitious sequences present
Structural differentiation	None	Presence of heterochromatin and other regions with differential staining properties

(Continued)

Table 1.1. (*Continued*)

Feature	Prokaryotes	Eukaryotes
Structure of gene sequences	Colinearity of DNA and polypeptide, no introns	DNA and polypeptide not colinear, due to presence of introns
RNA and protein synthesis		
Polycistronic mRNA	Present	Absent or very rare
RNA processing	No extensive processing of initial RNA transcript; translation commences immediately	Extensive processing of initial DNA transcript, including excision-splicing and polyadenylation. Translation delayed while RNA is transported to cytoplasm
Longevity of mRNA	Average about 3 min in *E.coli*	On average close to one day in cultured cells, sometimes many years in seeds
Ribosomes	70s (subunits 30s + 50s)	80s (subunits 60s + 40s)
Rate of protein synthesis	Very rapid, addition of amino acids per ribosome approx. 20 times that in mammals	Relatively slow

† The conventional unit of DNA is the quantity present in the diploid genome of the mouse.

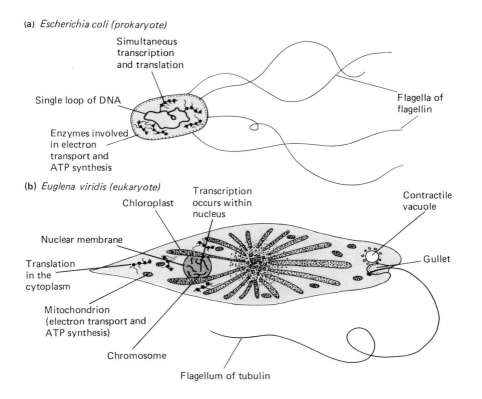

(a) *Escherichia coli (prokaryote)*

Simultaneous transcription and translation

Single loop of DNA

Enzymes involved in electron transport and ATP synthesis

Flagella of flagellin

(b) *Euglena viridis (eukaryote)*

Chloroplast

Transcription occurs within nucleus

Contractile vacuole

Nuclear membrane

Translation in the cytoplasm

Gullet

Mitochondrion (electron transport and ATP synthesis)

Chromosome

Flagellum of tubulin

1.5. The origins of the eukaryote cell

Eukaryotes cannot be traced back to a single prokaryote species, they appear instead to have properties similar to those of several prokaryotes and, although it seems quite an astonishing idea, there is a lot of evidence to suggest that our ancestors actually originated as an association of several simple prokaryotes. One of the most plausible explanations of how this came about is what is known as the serial endosymbiotic theory (SET) (Figure 1.4) proposed by Margulis. According to this theory, the ancestral eukaryote, or 'protoeukaryote', was a phagocytic amoeboid cell, perhaps with a primitive nucleus, which respired anaerobically. This protoeukaryote is thought to have ingested a species of bacterium which obtained its energy in a rather different fashion, by operation of the tricarboxylic acid (TCA) cycle and the cytochrome system: these ingested bacteria respired aerobically. Instead of digesting the aerobic bacteria which had been taken in, it is thought that the protoeukaryote may have retained them within its cytoplasm, where they developed a symbiotic relationship based upon the metabolism of oxygen. Oxygen is toxic to anaerobic organisms and the aerobic bacteria may have been tolerated within the protoeukaryote because they provided a detoxification mechanism. They would also have provided a valuable and abundant supply of adenosine triphosphate (ATP) which can be considered as a form of universal energy 'currency' for all living organisms. According to the SET, symbiotic bacteria are the ancestors of the mitochondria which now perform a similar function in our own bodies.

The next stage in eukaryote evolution is thought to have been the association of this, now aerobic, amoeboid cell with a species of the motile bacteria of the group known as Spirochaetes. Their motility depends on a specialized arrangement of microtubules composed of the protein tubulin. Present-day Spirochaetes possess these structures in common with eukaryote cilia, flagella, their basal bodies and the centrioles involved in the separation of chromosomes during cell division. Symbiotic Spirochaete bacteria are thought to be the ancestors of all these eukaryote subcellular structures.

The resultant 'organism', really an association between three prokaryotes, it is suggested, was the ancestor of all the animals, the protozoa and the fungi.

Figure 1.3. Prokaryote and eukaryote organization.
(a) *Escherichia coli*. Characteristic features are the single loop of naked DNA, absence of a nuclear membrane and simultaneous translation and transcription. The flagella are of flagellin and the enzymes of the electron-transport chain are situated just below the cell membrane. (b) *Euglena viridis*. This primitive eukaryote has a nucleus in which there are several chromosomes containing a large amount of protein and RNA, as well as DNA. Translation occurs only in the cytoplasm, where the enzymes of the electron-transport chain are carried by the mitochondria. Chloropolasts are also present and the flagellum is of tubulin.

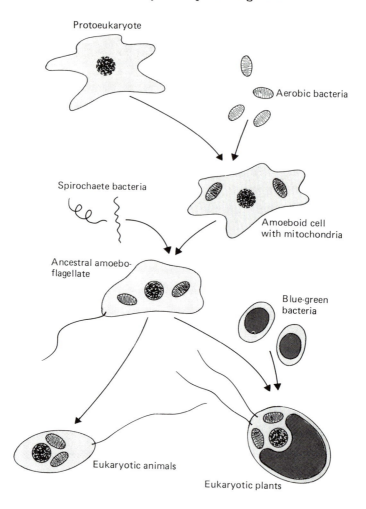

Figure 1.4. The early evolution of eukaryotic cells, according to the serial endosymbiotic theory.

The phagocytic amoeboid protoeukaryote is thought to have incorporated aerobic bacteria which evolved into mitochondria. This now aerobically respiring organism incorporated Spirochaete bacteria to produce the ancestral amoeboflagellate, from which the eukaryote metazoan animals, fungi and protozoa are believed to have arisen. Incorporation of blue-green bacteria led to the evolution of the eukaryotic plants. (After Tribe *et al.* 1981.)

According to the SET, green plants arose following a further step which involved the incorporation of Cyanobacteria. These are green bacteria possessing chlorophyll and capable of photosynthesis — the elaboration of molecules containing high-energy phosphate bonds, using sunlight as an energy source.

1.6. *The significance of a multi-ancestral origin for the eukaryotes*

There are several lines of evidence in favour of the serial endosymbiotic theory, and several against it. Among those in support are the presence of DNA in mitochondria and chloroplasts, and the fact that these organelles replicate independently of the nucleus and can be produced only from pre-existing organelles. The ribosomes which operate on mitochondrial and chloroplast RNA transcripts are also characteristically different from those that translate nuclear transcripts and actually resemble those of modern prokaryotes.

There are also present-day 'organisms' which are really symbiotic associations similar to those postulated by the SET. Notable among these is *Myxotricha paradoxa*, an inhibitant of the guts of termites. This organism propels itself through the termite gut contents with what appear to be cilia, but which are really Spirochaete bacteria attached to the surface of the organism by another species of bacterium. A third bacterial species occupies the cytoplasm. *Myxotricha* is not thought to be on the main line of eukaryote evolution, but its existence shows that the idea of symbiotic associations between simple organisms should not be lightly dismissed.

Despite being an attractive theory, the SET is not accepted by all authorities. They point out that the mitochondrial genome is more the size of that of a small plasmid, or virus, than that of a bacterium. The mechanism of genetic recombination which can occur in mitochondria, so far as it is understood, seems more like that of viruses than of bacteria. The sensitivity of mitochondrial DNA to some mutagens is also reminiscent more of plasmid than of bacterial DNA. On the other hand, some features, like the high poly-A content of their RNA transcripts, are characteristic of eukaryote systems (see Chapter 9). The lack of a 5S component in their ribosomal RNA is not shared with prokaryotes or eukaryotes.

Whether or not the SET does describe what really happened, it has drawn attention to some very important features of the cytoplasm of eukaryotes. Neither the nucleus nor the organelles of eukaryote cells can exist independently of one another, so there must be interactions between them. By considering the possible origins of eukaryote cells we are therefore forced towards the realization that the cytoplasm, traditionally regarded as a subordinate product of the nucleus, may have a greater significance in controlling nuclear activity. The existence of cytoplasmic DNA in chloroplasts and mitochondria also illustrates that not all inherited information is nuclear. Sperm and pollen do not normally carry chloroplasts, or mitochondria, or DNA coding for them; these organelles are usually all inherited from the maternal parent. A father is therefore normally incapable of transmitting a mitochondrial mutation to his son or daughter, although there is evidence of it occurring occasionally.

To return to the problems of development, the fact that mitochondria are transmitted through the maternal cytoplasm has an important bearing on the type of metabolism which takes place in the early embryo before its nuclear genes come into operation.

1.7. *Multicellularity and the principles of cytodifferentiation*

Whereas the typical prokaryote is a single cell, the typical present-day eukaryote is a multicellular system of different cell types. The generation of differences between cells is known as cytodifferentiation. Some of the important principles that apply to the differentiation of eukaryote cells can be revealed by considering very simple multicellular organisms such as the sponges.

The basic structure of a typical sponge is rather like an elongated bag, as shown in Figure 1.5. The outer surface of the bag consists of a single layer of flattened epithelial cells, forming a waterproof casing. The inner surface is covered with cells called choanocytes, or collar cells, each of which has a flagellum. There are pores in the wall of the sponge, guarded by long cells called porocytes, and the form of the organism is maintained by skeletal elements called spicules, secreted by mesenchyme cells within the wall. Water is drawn through the pores and into the central cavity by the action of the flagella on the choanocytes, and this water is then forced out through the mouth at the top of the sponge. Food materials and oxygen are extracted from this water within the central cavity.

Some of the principles of cytodifferentiation illustrated by the sponge are:

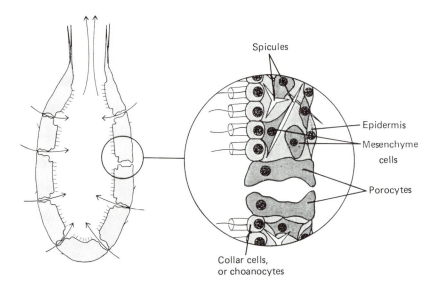

Figure 1.5 The integration of cell functions illustrated by a sponge.
This primitive metazoan animal feeds on particles drawn into its body cavity through pores in the wall. The direction of water flow is indicated. Collar cells, or choanocytes, create the water flow by movements of their flagella and the pores are regulated by porocytes. The outer surface of the body is rendered waterproof by a sheet of epithelium and the body is supported by spicules secreted by mesenchyme cells.

1. The cells display different shapes and different types of activity which are related to their function.

2. The function of each cell type is related to the organization of the whole organism.

3. Specialization by the cells involves deficiencies in some functions, compensated by mutual dependence between cells of different functions: there is division of labour and co-operation between cell types.

If we were to take the investigation further we would find that differentiation with respect to function also involves differences in ultrastructure, enzyme activities, structural proteins, numbers of mitochondria and many other features.

Sponges can be disaggregated into single cells by forcing them through a fine sieve. If the cells are then placed together in sea-water they will re-aggregate into one or more sponges. This takes place by sorting out of the different cell types, combined with their movement into appropriate positions, rather then by radical changes in the phenotypes of the cells (see also Chapter 13). This illustrates another principle: that the differentiated states of cells are relatively stable.

4. Differentiation involves stabilization of cell phenotype and patterns of gene expression.

Many species undergo what is known as an 'alternation of generations' between two different forms of organization. One of the best-known examples is in the ferns, where one generation is diploid and the other haploid. In jelly fish, tapeworms and hydroids, both generations are diploid. When one form of organization of a species is more complex than the other, it tends to be the complex form that occupies the greater part of the life-cycle. This suggests that the development of a complex system, with differentiation of cell types and division of labour between them, is in evolutionary terms a successful strategy.

1.8. *Summary and conclusions*

Eukaryotes are the organisms with nuclei. They represent quite a different form of life from the prokaryotes and possibly evolved from associations of prokaryotes. By analogy with human society, the prokaryote is a 'Jack of all trades' working alone, whereas the multicellular eukaryote is a community of specialists joining forces for the common good. Unlike prokaryotic DNA, that of eukaryotes is complexed with proteins and contains many repeated sequences, and there are also non-coding stretches, called introns, within structural gene sequences. The overriding tendencies in the evolution of the two types of genome seem to have been: in the prokaryotes, a facility for rapid replication and rapid transcriptional response to external influences; in the eukaryotes, more stable patterns of transcription and different kinds of response by the different cell types.

Cytodifferentiation necessitates selective expression of appropriate genes in each cell type. As we shall in later chapters, this is achieved through exposure of the progenitors of each cell lineage to a unique series of environmental stimuli. These stimuli indirectly promote or obstruct access to the DNA by RNA polymerase, with the result that quantitative and qualitative differences in RNA synthesis become established in different parts of the embryo. The RNA transcripts and their translated products are then further subjected to a battery of influences, which ensure that a unique selection of genes is finally expressed as functional protein in each cell type. In the next chapter we will investigate the ways in which metabolic distinctions between the different parts of early embryos first become generated.

Bibliography

Alberts, B., Bray, D., Lewis, J., Raff, M., Roberts, K. and Watson, J., *Molecular Biology of the Cell*. Garland, New York/London (1983).

Beale, G. and Knowles, J., *Extranuclear Genetics*. Edward Arnold, London (1978).

Folsome, C. E., *The Origin of Life. A Warm Little Pond*. Freeman, Oxford (1979).

Goldberger, R. F., Strategies of genetic regulation in prokaryotes. In *Biological Regulation and Development*, Vol. 1, *Gene Expression*, edited by R. F. Goldberger. Plenum, London (1979).

Hyman, L., *The Invertebrates, Protozoa through Ctenophora*. McGraw-Hill, London (1940).

Lewin, B., *Genes*. John Wiley, Chichester (1983).

Margulis, L., *Symbiosis in Cell Evolution*. W. H. Freeman, Oxford (1981).

Reid, R. A. and Leech, R, M., *Biochemistry and Structure of Cell Organelles*. Blackie, Glasgow (1980).

Schoff, J. W., The evolution of the earliest cells. *Sci. Am.*, **239**: 85 (1978).

Tribe, M. A., Morgan, A. J. and Whittaker, P. A., *The Evolution of Eukaryotic Cells*. Edward Arnold, London (1981).

Wheatley, D. N., *The Centriole: A Central Enigma of Cell Biology*. Elsevier, Amsterdam (1982).

Chapter 2 The initiation of cytodifferentiation

2.1. Overview of animal development

Most multicellular eukaryotes go through a standard sequence of ontogenic stages during their life-cycle. The first of these, gametogenesis, occurs in the bodies of their parents and involves formation of the gametes. This is followed by fertilization, when the fusion of gametes from two usually genetically distinct parents produces what is, in all probability, an entirely unique and novel combination of genes in the zygote. In the third stage the zygote nucleus replicates repeatedly to form a mass of cells, which in most species contain identical genetic information. In most animal groups this occurs by cleavage of one whole cell into two, two into four, and so on, but in insects a similar result is achieved by multiplication of the nuclei within an initially undivided cytoplasm. The result of either process is usually a hollow, often spherical body, the blastula, composed of a wall of cells, the blastoderm, surrounding a central cavity.

During the fourth phase, gastrulation, the blastoderm gives rise to two or more layers of cells known as the germinal layers. These are the ectoderm externally, the mesoderm next to this and the endoderm on the inside. The rearrangements by which the germinal layers come to occupy these positions vary considerably between animal groups. The ectoderm is the origin of the epidermis and the nervous system, the mesoderm is the source of the muscles, the circulatory system, the lining of the body cavities and the sex organs. In many species, including our own, the excretory system and most of the skeleton are also derived from mesoderm. The endoderm forms the gut and its associated digestive glands and a variety of other organs is formed by various combinations of the primitive germinal layers.

Following gastrulation the basic body plan becomes laid out (see Chapter 5). The most conspicuous change at this stage involves formation of the major elements of the future nervous system by the process of neurulation. In vertebrates this is the spinal cord, which extends along the entire dorsal surface of the elongating embryo. The germinal layers may partially dissociate at this stage and reform in mixed clusters of cells derived from more than one layer. These are the organ rudiments, the modelling of which occupies the following stage of organogenesis. Internal organs are mainly derived from endoderm together with mesoderm, while ectoderm usually combines with mesoderm to form peripheral structures (see Chapter 3).

Organogenesis is followed by a period of growth and cytodifferentiation, during which the cells of the organs acquire the unique structures and properties appropriate to their role in the larva or young adult. When this has been achieved the animal is ready to embark upon an independent existence and procure its food from the outside world. If it possesses special organs necessary for its early existence, which are absent from the adult, it is called a larva. In this form it may exploit a food source different from that its parents use, and perhaps, unlike them, be capable of active movement and of being dispersed widely.

Eventually the larva undergoes the process of metamorphosis during which it becomes transformed into an adult. During metamorphosis many radical organizational changes take place. In insects, for example, many structures break down to be replaced by the growth of adult organs from new rudiments (imaginal discs). During metamorphosis, or during a more extended period of adult growth, the gonads mature and the animal becomes capable of reproduction. In some species growth continues indefinitely, depending on availability of food, but eventually body processes begin to deteriorate and the animal enters a period of senescence, followed ultimately by death.

2.2. Developmental strategies

Classical embryological theory conceived two basic strategies by which animal development is achieved. These were called the mosaic and the regulative plans, these concepts being based on some of the very early experiments with simple marine organisms. Subsequent research has shown these early ideas to be oversimplistic, but the experiments on which they were based provide a useful starting point for our discussion.

The Mosaic Theory

The concept of mosaic development is illustrated by the common limpet (*Patella*). The first stages in development of the limpet involve cleavage of the fertilized egg into two cells. These then divide into four, the four into eight and then 16 and so on. In the early days of this century E. B. Wilson removed cells from early limpet embryos. He found that if one cell is taken away from an eight-cell embryo, that single cell will develop into only a part of a limpet, while the remainder of the embryo forms an incomplete animal (Figure 2.1). The loss of a single cell from the early embryo thus causes irreparable damage, as each cell can develop only into that part of the animal which it was predestined to become.

The Mosaic Theory of development, derived from such observations, made three postulates:

   1. The membranes or cytoplasm of the egg contain different specific instruc-

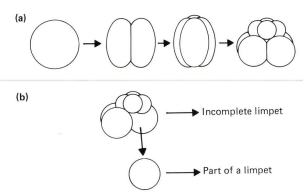

Figure 2.1. Mosaic development illustrated by the limpet.
(a) The early development of the limpet embryo. (b) If one cell is removed from an embryo at the eight-cell stage, the organism lacks the capacity to regulate its development to make good the loss. (Based on the work of E. B. Wilson.)

tions which become packaged around the individual nuclei. The nuclei then express only those genes dictated by the instructions with which they are associated.

2. These instructions are initially located at precise positions in the egg.

3. Each cell develops autonomously, i.e., without requiring instructions from its neighbours.

Regulative development

When the same kind of experiment was performed on early embryos of Echinoderms such as sea urchins rather different results were obtained. In Echinoderms the first two divisions result in cells of equal size, but the third division, in the horizontal plane, produces an embryo with small cells on the dorsal surface, in what is known as the animal hemisphere, and large cells ventrally, in what is called the vegetal hemisphere (Figure 2.2). At the eight-cell stage the embryo is thus asymmetric in a vertical sense. The cells continue to divide to produce a hollow ball covered with cilia, the blastula, which develops into a ciliated pluteus larva, with a rudimentary gut and skeleton as shown in Figure 2.2(a).

If the embryo is cut horizontally at the eight-cell stage, so that the animal and vegetal cells are separated, the two halves can undergo further development, but they produce rather different structures. The animal cells produce a blastula-like form with long cilia, then cease development. The cells of the vegetal hemisphere will continue to develop for a longer period, but the organism which is formed lacks cilia and is poorly formed, with a distorted skeleton, very large gut and no mouth (Figure 2.2(c)). Dividing the eight-cell embryo about the horizontal plane thus produces two morphologically distinct but incomplete forms, in accordance with the Mosaic Theory. However, if the plane of division is made vertical instead

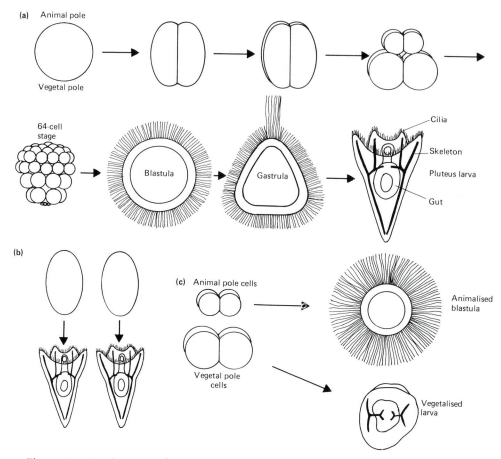

Figure 2.2. Development of the sea urchin.
(a) Some critical stages in early development. (b) Division of the embryo at the two-cell stage leads to formation of two small, but perfect embryos. (c) Separation of animal hemisphere from vegetal hemisphere cells at the eight-cell stage leads to two types of abnormal embryo, an abnormal blastula from the animal pole cells and a grossly distorted pluteus larva lacking some elements. (Based on Hörstadius, 1973.)

of horizontal, to produce two mirror-image halves, for example, at the two-cell stage, each half will develop into a small but otherwise normal larva (Figure 2.2(b)). An embryo which has been damaged in the latter way can thus regulate its development and adjust it for the loss of half its cells.

The capacity of sea urchin embryos to carry out developmental readjustment thus seems to depend on the plane about which it is divided. If the embryo is divided with the plane of division in the horizontal dimension, the result agrees with the Mosaic Theory, but when in the vertical dimension, subsequent development cannot be interpreted in these terms.

These were just a few of many observations that revealed a capacity for organisms to undergo a certain degree of developmental regulation. One idea that was put forward to explain it was based upon the Reference Points Theory. This theory suggests that gradients of biochemical morphogens are established within the egg cytoplasm with respect to reference points, such as the animal and vegetal poles. The activity of nuclei is considered to be dictated by the ratio of the influences of different morphogens in the cytoplasm which becomes associated with each nucleus.

The best evidence for the existence of morphogen gradients in the sea urchin comes from an experiment of Hörstadius, in which cells from the vegetal pole were 'titrated' against different parts of the rest of the embryo at the 64-cell stage. Tiny cells called micromeres divide off at the vegetal pole and it is these that seem to be the source of a vegetal morphogen. The 64-cell embryo was divided into four horizontal slices, which we can call An 1, An 2, Veg 1 and Veg 2, from the animal to the vegetal pole (Figure 2.3). If An 1 is combined with micromeres it can form a small but otherwise normal embryo, but four micromeres are required. If less than four are used an 'animalized' embryo results, resembling a blastula with long cilia. However, a normal embryo can be produced from the An 2 slice combined with only two micromeres. 'Vegetalized' embryos are produced from combinations of the vegetal slices with micromeres, but the most normal of these develops

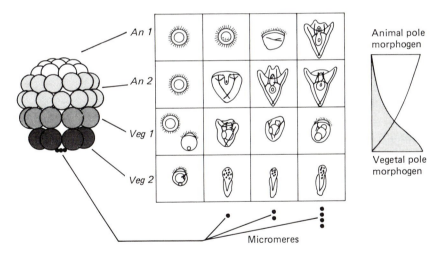

Figure 2.3. Development of sea urchin embryos from reconstituted fragments. Tiers of cells from 64-cell stage embryos were combined with different numbers of micromeres and their development examined. Normal larvae result from the combination of the most dorsal tier, An 1 with four micromeres, or a combination of the second tier, An 2 with only two micromeres. The most normal of the embryos produced from vegetal slices results from combination of Veg 1 with one micromere. The postulated gradients of morphogen are illustrated. (Based on Hörstadius, 1973.)

from Veg 1 combined with only one micromere (Figure 2.3). Although other explanations are possible, the simplest is that there are two major gradients of
morphogen, each part of the embryo being defined by the concentration ratio.
Similar influences are believed to control morphogenesis in the majority of
eukaryote species.

According to the Reference Points Theory, new reference points, which lay down
further gradients of influence, arise later in development. Loss of cells is not
critical, provided that representatives of important tissues, which include the
sources of the morphogens, are still present and responsive tissues retain the
capacity for appropriate differentiation under their influence.

The main points of difference between the mosaic and regulative theories are:

1. In the Mosaic Theory a large number of different kinds of instructions are
postulated; in the Reference Points Theory only a few are necessary, but the
capacity of cells to discriminate this information is more important.

2. In the mosaic situation determination of the fates of cells is rapid and
relatively inflexible, whereas in the regulative situation determination is progressive and cell phenotype is more dependent on environmental factors.

According to current opinion, neither theory adequately describes the controls
upon the fates of animal cells. The development of most species seems to involve
both mosaic and regulative aspects, as well as other features, such as extracellular
matrices and hormones (see Chapter 4), which were not considered by the authors
of either theory. An important distinction between mosaic and regulative aspects
of development seems to reside in the partitioning of mitochondria which are
divided unequally between the cells of mosaic embryos. For example, in *Ciona
intestinalis*, which undergoes mosaic development, the first cleavage division
segregates the vast majority of the mitochondria into the posterior blastomere
which gives rise to neural and somitic structures, while the yolk-enriched
mitochondria-poor cells derived from the anterior blastomere develop into
endoderm. In contrast, mitochondria are fairly evenly divided between the
blastomeres of regulative eggs such as those of *Xenopus*. Mitochondria obviously
play a very important role in the differentiative aspects of development (see
Chapters 1 and 4).

2.3. The early development of insects

Insect development occurs within a rigid egg case and follows quite a different
course from that of the species we have considered so far. The two haploid
pronuclei from the male and female gametes fuse and undergo division, resulting
in a large number of diploid nuclei. These migrate outwards and come to rest in
the layer of cytoplasm around the periphery of the egg, just inside the outer case,
or chorion. Cell walls then develop around these nuclei, forming an epithelial
layer, the blastoderm. A series of morphogenetic movements and other changes

then occurs and the head, thorax and abdomen of the embryo develop in appropriate positions, as shown in Figure 2.4(a).

Some of the most interesting work on insect development was carried out by Sander with the leaf-hopper *Euscelis*. This insect is unusual in that each egg contains a ball of symbiotic bacteria near the posterior pole. Sander studied the development of *Euscelis* by tying ligatures around the egg at different stages during development, in order effectively to isolate the anterior and posterior parts (Figure 2.4).

When the ligature was tied around the middle of an embryo at the late blastoderm stage the anterior section developed head structures and the posterior section parts corresponding to thorax and abdomen, all in the correct relative positions, showing that the cells were already committed to their future fate by this stage. If the ligature was tied in the same position, but at the earlier nuclear migration stage, no thoracic structures were formed, just a head in the anterior section and an abdomen in the posterior (Figure 2.4(c)). Sander reasoned that tying the ligature before the blastoderm stage had obstructed the free diffusion of morphogens through the egg cytoplasm from the anterior and posterior poles. He suggested that a high concentration of morphogen from the anterior end of the egg determines head structures, a high concentration of morphogen from the posterior end dictates abdominal structures, while a mixture of the two provides information necessary for the formation of the thorax. These gradients must presumably have become established by the blastoderm stage. The gradients are thus thought to be acting and having their effects just before the blastoderm stage, when the nuclei finally acquire specific types of cytoplasm in association with the formation of plasma membranes around them.

Sander later carried out experiments which showed that the source of the posterior morphogen is associated with the site occupied by the ball of symbiotic bacteria (Figure 2.4(d),(e)). If this ball is pushed mid-way along the embryo before formation of the blastoderm and a ligature tied behind it, head, thoracic and abdominal structures form faithfully in the anterior section, but the posterior half forms only abdomen. If the bacteria are moved forward, but the ligature is tied anterior to them no recognizable structures arise in the anterior section, while a grossly abnormal embryo develops in the posterior part. This unusual embryo has an abdomen at both the posterior and anterior ends and a thorax in the centre!

These experiments are remarkable in two respects: they provide good evidence for the existence of morphogenetic gradients within insects eggs and they locate the source of one of the morphogens.

The possession of symbiotic bacteria as a possible source of morphogen is an unusual feature of *Euscelis*. Reference points are thought to be established in most insects by the products of the nurse cells of the maternal ovary, during the synthesis of the egg. There is a mutation in *Drosophila melanogaster*, the common fruit fly, called *dicephalic* in which embryos are formed with two heads, one at each end (Figure 2.5(b)). This is due to the eggs being formed with a micropyle, or entrance tube, at each end instead of just the anterior. The nurse

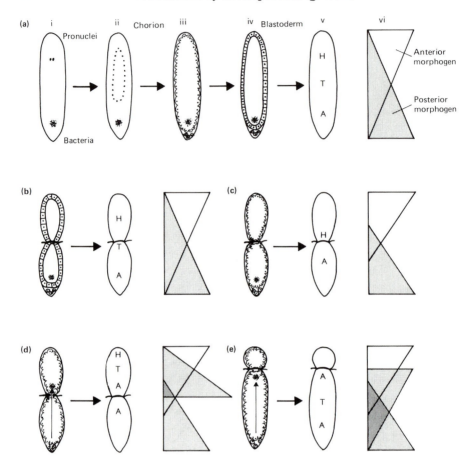

i: fertilized egg;   ii: nuclear division stage;   iii: nuclear migration stage;   iv: blastoderm stage;
v: regions of embryo;   vi: theoretical gradients
H: head;   T: thorax;   A: abdomen

Figure 2.4. Evidence for the morphogen double-gradient model in insects.
(a) The normal sequence of events during the early development of *Euscelis*. The position
of the ball of symbiotic bacteria and the theoretical gradients of morphogen originating
at the anterior and posterior ends are indicated. (b) Tying a ligature around the middle
of the embryo at the blastoderm stage does not affect the formation, or positions, of the
major subdivisions of the embryo. (c) Tying a ligature around the embryo at the nuclear
migration stage inhibits the establishment of morphogen gradients, resulting in non-
development of a thorax. (d) If the embryo is ligatured before the blastoderm stage and
the ball of bacteria is moved into the rear part of the anterior section, all the major body
parts develop in the anterior part, but an abdomen only forms in the posterior section. (e)
If the bacteria are moved just posterior to a ligature tied before the blastoderm stage, an
embryo with two abdomens and no head forms in the posterior section. (Based on experiments of Sander.)

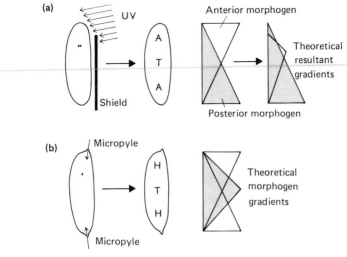

H: head;   T: thorax;   A: abdomen

Figure 2.5. Observations on the sources and nature of insect morphogens.
(a) Irradiation of the anterior region of a *Smittia* egg with u.v. light causes destruction of the head-forming morphogen. (Based on experiments of Kalthoff.) (b) The *dicephalic* mutant of *Drosophila melanogaster* lays eggs with a micropyle at each end instead of just at the anterior. The direction of access of anterior morphogen into the egg is indicated.

cells pump material into the egg through the micropyle, and in the *dicephalic* mutant material enters from both ends and establishes two gradients of the same morphogen. Another mutant gene, *bicaudal*, causes the reverse effect, embryos being formed with abdomens at both ends. Like the dicephalic mutation this operates through the maternal system and presumably either involves a defect in the gradient of the anterior morphogen or a failure in the response of the cells to that gradient.

One set of experiments suggests that in the Chironomid midge, *Smittia*, the anterior morphogen is composed of RNA. Working with this species, Kalthoff irradiated the anterior pole of the egg with ultraviolet light before the nuclear migration stage (Figure 2.5(a)). the resultant embryo had an abdomen at both the anterior and the posterior end, with a thorax in between. When u.v.-irradiated eggs were exposed to visible light, whatever damage had been caused was repaired and normal embryos were produced. Nucleic acids are damaged by u.v. light, but can be repaired by enzymes which are activated by visible light. This suggests that the molecules which carry the anterior-specific information and which suffered the u.v. damage are either DNA or RNA. Its identity as RNA was demonstrated by injecting ribonuclease into a non-irradiated egg. This is an enzyme with the property of specifically cleaving RNA but not DNA. The ribonuclease produced effects similar to u.v. light. This lends considerable support to the theory that anterior structures in *Smittia* embryos are formed through the mediation of RNA molecules derived from the mother and laid down in the egg cytoplasm during oogenesis.

The germplasm

The theory that the fate of nuclei is determined initially by the interaction of morphogenetic gradients arising from reference points holds for most cells in insects, but not for the germ cells — the cells which give rise to the ova and sperm. In this case there seems to be a special signal which is irrespective of the gradients and is associated with a particular type of cytoplasm, known as the germplasm or poleplasm, at the posterior pole of the egg. A recessive gene called *grandchildless* (*gs*) occurs in a species of fruit fly called *Drosophila subobscura*. Females homozygous for this mutation are fertile, but their offspring are sterile as the eggs have defective germplasm.

The suspected role of the germplasm was confirmed in a very elegant experiment carried out by Ilmensee and Mahowald, using *Drosophila melanogaster* (Figure 2.6). They removed germplasm from the posterior pole of one egg and transferred it to the anterior pole of a recipient egg (Recipient 1) very early in development, before nuclear migration. The recipient egg was allowed to develop to the blastoderm stage, then its anterior cells were removed and placed at the posterior end of a second recipient egg (Recipient 2), adjacent to the germ cells. The second recipient was then allowed to develop to maturity and used for breeding experiments, in order to determine whether or not nuclei which had migrated to the anterior end of the first recipient had become germ cells by virtue of their association with the transplanted germplasm.

In order that the origins of the germ cells could be distinguished they were marked with mutant genes. The donor of the germplasm was a homozygote for the dominant gene *Bar Eye* (*Bar/Bar*), but had wild-type alleles at all other loci. The first recipient was homozygous for the recessive alleles *multiple wing hairs* (*mwh*) and *ebony body* (*e*), all other alleles being of wild type. Recipient 2 was homozygous for three recessive alleles, *yellow body* (*y*), *white eye* (*w*) and *singed bristles* (*sn*).

The mating partner of Recipient 2 was homozygous for *yellow*, *white* and *singed* and the expectation was that all progeny should be homozygous at the *yellow*, *white* and *singed* loci and would show that phenotype if the germplasm transplant had not succeeded. On the other hand, if any of the germ cells carried by Recipient 2 were derived from Recipient 1, some progeny should be heterozygous for *y*, *w* and *sn* and also for *mwh* and *e*. Since these genes are all recessive, offspring resulting from transplanted cells would have the wild-type phenotype. In fact, 10–20% of the progeny did have the wild-type phenotype, showing that the transplant had been successful.

Recipient 2 was also mated with flies which were similar to Recipient 1. If the

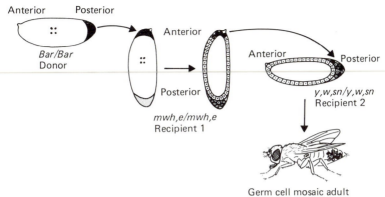

Adults test-mated with:

(a) *y,w,sn/y,w,sn* flies. Some wild-type progeny produced.

(b) *mwh,e/mwh,e* flies. Some *mwh,e/mwh,e* progeny produced.

None of the progeny had *Bar* eyes.

Figure 2.6. Experiment to test the role of the germplasm in determination of the germ cells in *Drosophila melanogaster*.
Cytoplasm was removed from the posterior poles of homozygous dominant bar-eyed eggs (*Bar/Bar*) and transferred to the anterior poles of fertilized eggs homozygous for the recessive alleles multiple wing hairs (*mwh*) and ebony body (*e*). These embryos were allowed to develop to the blastoderm stage and their anterior cells transferred to the posterior region of blastoderm-stage embryos homozygous for the recessive alleles yellow body (*y*), white eye (*w*) and singed bristles (*sn*). Mating experiments indicated that inclusion of posterior poleplasm into the cytoplasm of anterior cells caused them to develop the properties of germ cells. (Based on Illmensee and Mahowald, 1974.)

transplant had not succeeded, all the progeny of this mating should be heterozygous for each of the recessive marker alleles and should have wild-type phenotype. On the other hand, if any of the cells of Recipient 1 had become germ cells some progeny should be homozygous for *mwh* and *e* and would have multiple wing hairs and ebony bodies. Some of the progeny did in fact have these features, confirming the deduction made from the first mating.

If the germplasm taken from the original *Bar-eyed* homozygous donor had contained nuclear material, it would be expected that some of the progeny in either of these matings might have *Bar* eyes, but no *Bar-eyed* progeny were observed.

The experiment therefore demonstrates that the cells at the posterior pole of a normal insect egg, which become germ cells, do so by virtue of their association with germplasm. In insects therefore, it seems that the major body plan is most easily understood in terms of gradients of morphogens, in accordance with the Reference Points Theory, while the special properties of the germ cells are acquired by association with specific determinants which are precisely located, as postulated by the Mosaic Theory.

2.4. The early development of amphibians

Fertilization and cleavage

The chorion of the insect egg protects the early embryo from predators, infection and environmental fluctuations, but it also performs the important function of acting as a reference frame, stabilizing biochemical gradients and maintaining the relative positions of the different parts of the body. The eggs of amphibians, such as frogs, toads and newts, have much softer surfaces. With amphibians, as with most animals, an increase in cell numbers always involves cleavage of cells, instead of subdivision of the cytoplasm as in insects. Protection of the embryo is afforded by a mucilaginous jelly which surrounds it, and reference points are provided by gravity, the light and heavy components of the egg cytoplasm separating out before cleavage begins (Figure 2.7).

As in insects, the germ cells of amphibians are set apart at an early stage. A histologically recognizable region of the cytoplasm, at the vegetal pole of the egg, becomes associated with a small number of early cleavage nuclei and it is these which become the germ cells and pass their genetic information to the next generation.

If the vegetal pole of an amphibian egg is irradiated with u.v. light at the wavelengths absorbed maximally by nucleic acid (see Chapter 11) the adult which develops from that egg will be sterile. If, following u.v. irradiation, the germplasm is replaced with non-irradiated germplasm from another egg, the adult which develops is (usually) fertile. The inference is that the germplasm may owe its special properties to polynucleotides (see above), but the obvious experiments to test this hypothesis have not been carried out.

The remainder of the cells of the amphibian body are determined progressively, as a result of a series of environments to which they are subjected. The basic organization of the animal develops with reference to the animal and vegetal poles, at the extremes of the vertical axis, and, in the frog, a third reference point, the point of entry of the sperm. The vertical axis is established by gravitational settling out of the heavy yolk into the vegetal half and positioning of the nucleus in the animal half (Figure 2.7). At fertilization the cytoplasm rapidly becomes reorganized and the second polar body (a waste product of the meiotic events which gave rise to the haploid female nucleus) is ejected at the animal pole, so reinforcing the uniqueness of this site.

The outer coat of the amphibian egg is composed of a membrane lined with a layer of dense cytoplasm, the double layer being known as the egg cortex. At fertilization a change takes place in the membrane rendering it impermeable to further sperm. The cortex rotates over the deeper cytoplasm, setting up a zone of interaction between the cortex of the animal hemisphere and the yolk of the vegetal hemisphere (Figure 2.7). This zone is at its broadest diametrically opposite the point of entry of the sperm and becomes visible on the exterior as a paler crescent-shaped belt around the equator of the egg, known as the grey crescent.

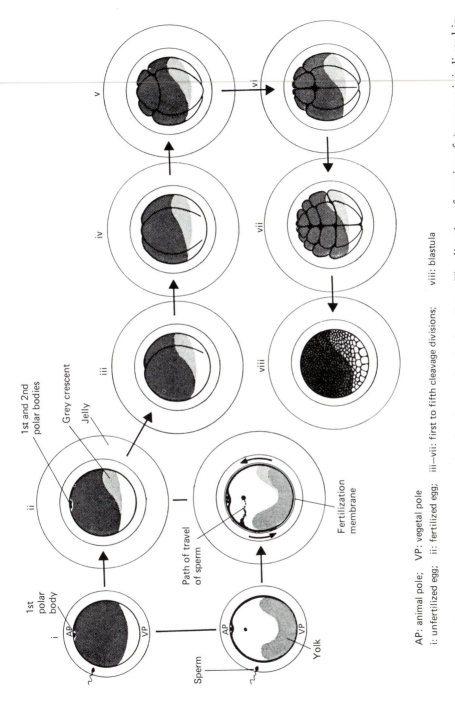

AP: animal pole;   VP: vegetal pole

i: unfertilized egg;   ii: fertilized egg;   iii–vii: first to fifth cleavage divisions;   viii: blastula

Figure 2.7. Fertilization and cleavage up to the blastula stage in the frog, *Rana*. The direction of rotation of the cortex is indicated in vertical section. (After Balinsky and Fabian, 1981; Grant, 1978.)

The grey crescent becomes the future dorsal (back) side of the embryo. The point of entry of the sperm dictates not only the position of the grey crescent, but also the plane of the first cleavage, which is the future plane of bilateral symmetry of the adult animal (see Chapter 5). The grey crescent and its derivative tissue, the dorsal lip of the blastopore, become what is known as the primary organizer, from a developmental point of view the most important region of the whole embryo (see below and Chapter 3).

Following fertilization, the egg divides in the vertical plane, through the point of entry of the sperm and the centre of the grey crescent. This is followed by a second vertical division at right angles to the first and a third horizontally. Further divisions follow until a hollow ball of cells, the blastula, is formed (Figure 2.7).

If a frog's egg is allowed to divide naturally up to the four-cell stage and then these cells are separated, each one can develop into a completely normal tadpole. If the embryo is disaggregated between the four- and 16-cell stages, (some of) the individual cells can still develop into tadpoles, but these will be smaller than normal. Compared with the limpet and sea urchin, the individual cells of early amphibians show quite a striking difference in potency. After the 16-cell stage, tadpoles produced from separated cells are defective.

If cells from an embryo after the 16-cell stage are transplanted to other sites in another embryo of the same species they merge with the host cells and become indistinguishable from them. Therefore at this stage although the cells are no longer capable of forming a whole animal, they are still totipotent.

If the grey crescent area is destroyed, development stops at the blastula stage. Also, if the fertilized egg is ligatured and separated into equal halves before the first cleavage, a half will develop beyond the blastula only if it contains a sizeable portion of grey crescent. The critical importance of the grey crescent is demonstrated if a piece is removed from one egg and transplanted into the opposite side of another fertilized egg. In this case a double embryo forms (Figure 2.10(a)). The grey crescent is thus of extreme importance in directing development beyond the blastula stage.

Up to the end of the blastula stage, most of the protein synthesis which has occurred in the embryo has utilized messenger RNA deposited in the egg by the mother, but at gastrulation the embryo accelerates synthesis of RNA and protein from its own DNA. At this stage the cells lose a great deal of their earlier developmental plasticity: they are becoming determined. The terms 'determination' and 'commitment' are used by embryologists to describe the restriction of the possible future pathways of cytodifferentiation to only one.

Gastrulation

It is possible to stain a live frog blastula with 'vital' stains which do not interfere with development, but which mark the cells so that their fate can be followed. By this means it is possible to identify which areas will form the germinal layers and future organ rudiments (Figure 2.8)

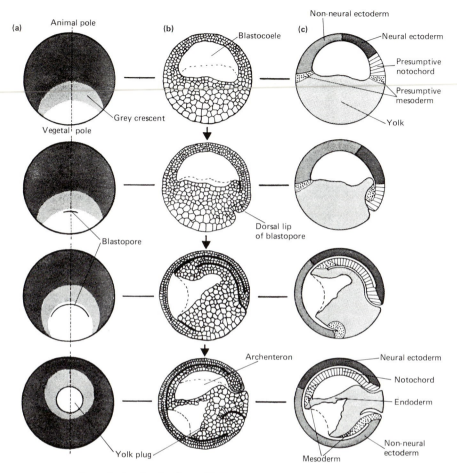

Figure 2.8. Gastrulation in the frog, *Rana*.
(a) Surface views. (b) Sectional view along the line indicated in (a) to show the size and numbers of cells. (c) Diagram of prospective fate of embryonic cell types. (From Balinsky and Fabian, 1981.)

Gastrulation begins with the appearance of a small intucking in the grey crescent region of the yolk tissue, just ventral to the presumptive mesoderm. This invagination, the blastopore, gradually extends into the central cavity or blastocoele, gradually obliterating it, with formation of a second cavity, the archenteron, that later develops into the gut (Figure 2.8). During these events physiological changes take place in cells in the different parts of the embryo, which initiate patterns of gene expression that cause them to begin their differentiation into the 200 or so cell types distinguishable in the adult (see Figure 2.11). The distribution of the different presumptive tissues at the end of gastrulation is shown in Figure 2.8.

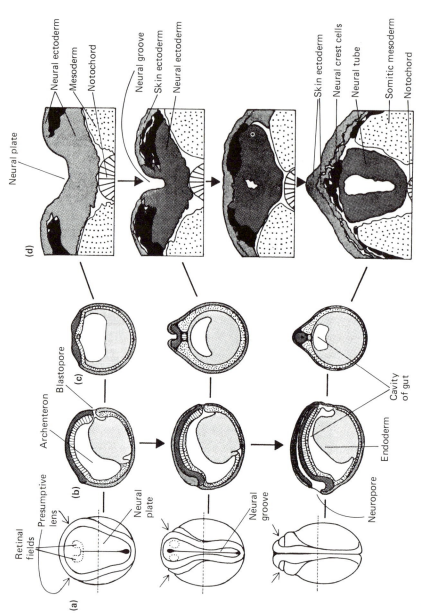

Figure 2.9. Neurulation in the frog, *Rana*.
(a) Surface views showing the positions of the presumptive retinas and lenses (arrowed; see Chapter 3). (b) Sectional views at right angles to the lines indicated in (a). (c) Sectional views along the lines indicated in (a). (d) Detail of formation of neural tube and origin of the neural crest cells. Tissue type coding as shown in Figure 2.8.

Neurulation

The dorsal ectoderm, neural ectoderm or neurectoderm becomes thickened at the end of gastrulation and forms a recognizable area called the neural plate. The lateral edges of this plate rise up like symmetrical waves and approach one another, eventually fusing in the midline to form a tube, the neural tube, covered by another layer of non-neural ectoderm that later becomes skin (Figure 2.9). The neural tube is the origin of the spinal cord and brain. The process of formation of the neural tube is called neurulation and the embryo at this stage is called a neurula.

A group of cells known as the neural crest separate off between the neural tube and the skin and later migrate out ventrally (Figure 2.9(d)). These have a fascinating fate, forming a variety of very different tissues, but we will leave a discussion of the ways in which they do this to the next chapter.

Primary induction

Although the grey crescent is visible from the egg up to the blastula stage, its role seems to have been accomplished by the eight-cell stage, since its removal then does no great harm to the embryo. Implantation of grey crescent material from a fertilized egg into an eight-cell embryo also has no major effect upon the development of the embryo.

The blastopore develops at the centre of the grey crescent, and at gastrulation its dorsal lip performs an essential function in directing the fates of the tissues which become positioned closely beside it. The crucial importance of the dorsal lip of the blastopore as the primary organizer of the embryo was demonstrated by Spemann and Mangold in a classic series of transplantation experiments in the 1920s and 30s.

If a piece of prospective neurectoderm from the dorsal surface of an *early* gastrula is transplanted to the presumptive skin ectoderm on the ventral side of another embryo of a similar stage, as the latter develops, the transplant will merge with the recipient tissues and become indistinguishable from them (Figure 2.10(c)). This is because, at these early stages of gastrulation, the ectoderm has been determined only as ectoderm, not specifically as skin or neural tissue. If a similar experiment is performed with a piece of ectoderm taken from the dorsal surface of a *late* gastrula (i.e., after it has been exposed to dorsal lip tissue moving underneath) and implanted in the ventral surface of a host late gastrula, it does not merge with the cells in its new position, but develops as neurectoderm instead (Figure 2.10(e)). Having been exposed to the influence of the blastopore dorsal lip it has become determined as neurectoderm and stays as such. This capacity of one tissue to direct the fate of neighbouring tissues is known as embryonic induction, the action of the primary organizer in this respect being called primary induction. Induction is of extreme importance in the development of multicellular species, but even now we really have little understanding of how it actually works (see Chapters 3, 4, 10, 11).

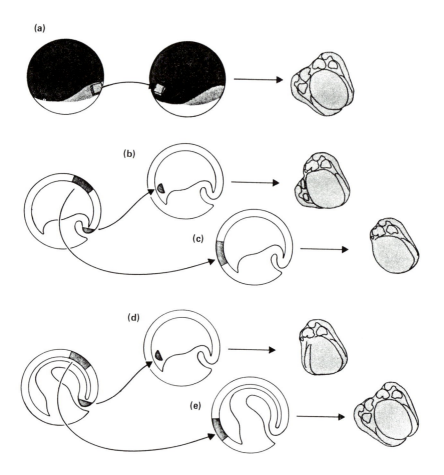

Figure 2.10. The primary organizer in the frog, *Rana*.
(a) Implantation of a piece of grey crescent material taken from a fertilized frog egg into the opposite side of another fertilized egg causes the latter to develop into a double embryo. (b) Implantation of the dorsal lip of the blastopore from an early gastrula beneath the non-neural ectoderm of a recipient early gastrula causes induction of an extra set of head structures in the ectoderm. (c) Presumptive neural ectoderm from an early gastrula implanted into the non-neural ectoderm of another early gastrula soon becomes indistinguishable from the surrounding tissues. (d) Implantation of dorsal lip tissue from a late gastrula beneath the non-neural ectoderm of a recipient early gastrula causes induction of an extra set of tail structures (cf. (b) above), showing that the properties of the dorsal lip change during gastrulation. (e) Neural ectoderm taken from a late gastrula and implanted among the non-neural ectoderm of another late gastrula develops into a second neural plate (cf. (c) above), showing that it has been induced during gastrulation. (Based on Curtis, 1960; Spemann and Mangold, 1924; Spemann, 1931, 1938.)

During gastrulation the inductive properties of the dorsal lip change, as can be demonstrated by transplanting dorsal lip tissue of different stages beneath the ventral surface of early gastrulas. If the dorsal lip tissue is taken from an *early* gastrula it induces head structures in the overlying ectoderm (Figure 2.10(b)). If taken from a *late* gastrula it induces tail structures instead (Figure 2.10(d)). One way of viewing this change is as a gradient of properties in relation to time, as distinct from the gradients in space that characterize the invertebrate embryos we have considered (see also Chapter 6).

2.5. Summary and conclusions

One of the widespread principles of embryonic development is that nuclei or cells which are initially totipotent become progressively restricted in developmental potential — they become determined. This acquisition of a programme of determination depends to some degree upon the association of the nucleus with specialized cytoplasm derived from the mother, which may contain specific concentrations of particular determinants. Determination also occurs through the influence of morphogens, which arise from organizers, or points of reference in the egg and later the embryo (see Chapter 5). The early distinction between 'mosaic' embryos, in which cells are more-or-less irreversibly determined at an early stage, and 'regulative' embryos, in which cells acquire less rigidly defined phenotypes over a more extended time period, is not now considered to be valid. In the sea urchin, for example, experiments in which the embryo is divided into halves lead to contradictory results, depending on the plane of division. The distribution of mitochondria is a major factor in experiments of this sort. Some of the determinative factors which are present in the egg and early embryo are almost certainly RNA, or proteins derived from maternal RNA, but there may be other kinds of molecules as yet unidentified. At gastrulation, the embryo's own genes become expressed to a significant extent and the patterns of gene expression in the tissue types differ, emphasizing the differentiative characteristics which have already begun to emerge.

The development of a cell is said to be canalized along defined pathways towards particular end-points. For example, nuclear DNA, which will later be found in skin cells gradually acquires the programme of synthetic capacities appropriate to skin, to the exclusion of other properties. This concept of the canalization of development has been illustrated by Waddington with his 'Epigenetic Landscape', a modified version of which is shown in Figure 2.12 (cf., Figure 2.11). In this diagram, the fate of the lineage of vertebrate cells derived from a single cell at the cleavage stage is represented by the pathway followed by a sphere rolling from the summit of a mountain peak. The further it rolls, the more restricted becomes the choice of gulleys, or valleys, into which it can move. The valleys represent the different cytodifferentiative pathways which terminate in end-points such as skin, muscle and nerve. A variety of experiments suggests that the state

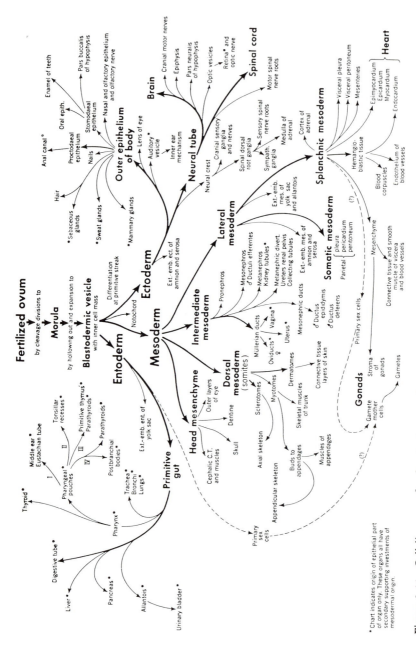

Figure 2.11. Cell lineages of a developing vertebrate.
(Reprinted from Patten and Carlson, 1974 with permission.)

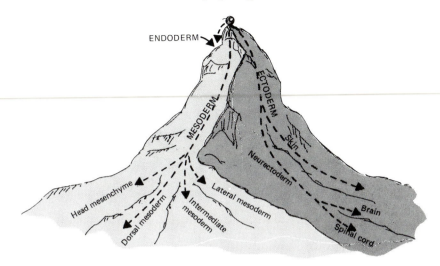

Figure 2.12. The canalization of development illustrated by a version of the Epigenetic Landscape.
The path of travel of a ball pushed off the peak represents the course of differentiation of an embryonic cell lineage. The further the ball travels, the more difficult it becomes to deflect it into an alternative pathway. (Based on the concept proposed by Waddington, 1956.)

of determination of cells is retained more by the cytoplasm than the nucleus, and in a later chapter we will explore the role of the cytoplasm in this respect.

The key features of embryonic development described in this chapter are embryonic induction and determination. What these expressions mean in molecular terms has largely been ignored by present-day molecular biologists. In the next chapter we will explore these concepts in more detail.

Bibliography

Baker, P. C. and Schroeder, T. E., Cytoplasmic filaments and morphogenetic movements in the amphibian neural tube. *Dev. Biol.*, **15**: 432 (1967).

Balinsky, B. I. and Fabian, B. C., *An Introduction to Embryology*, 5th edn. Saunders College Publishing, Philadelphia (1981).

Billett, F. S., *Egg Structure and Animal Development*. Edward Arnold, London (1985).

Curtis, A. S. G., Cortical grafting in *Xenopus laevis. J. Embryol. Exp. Morphol.*, **8**: 163 (1960).

Ede, D. A., *An Introduction to Developmental Biology*. Blackie, Glasgow (1978).

Gardner, R., Control of early embryonic development. (In '*Reproduction*', edited by R. V. Short.) *Br. Med. Bull.*, **35**: no. 2.

Grant, P., *Biology of Developing Systems*. Holt, Rinehart and Winston, New York (1978).

Hopper, A. F. and Hart, N. H., *Foundations of Animal Development*. Oxford University Press, Oxford (1980).

Hörstadius, S., *Experimental Embryology of Echinoderms*. Clarendon Press, Oxford (1973).

Ilmensee, K. and Mahowald, A. P., Transplantation of posterior pole plasm in *Drosophila*. Induction of germ cells in the anterior pole of the egg. *Proc. Natl Acad. Sci. USA*, **71**: 1016 (1974).

Kalthoff, K., Specification of the antero-posterior body pattern in insect eggs. In *Insect Development*, edited by P. A. Lawrence. Blackwell Scientific, Oxford, pp. 53–75 (1975).

Mangold, O., Autonome und kompletementäre Induktionen bei Amphibien. *Naturwissenschaften*, **20**: 371 (1932).

Patten, B. M. and Carlson, B. M., *Foundations of Embryology*, 3rd edn. McGraw-Hill, San Francisco (1974).

Pollak, J. K. and Sutton, R., The differentation of animal mitochondria during development. *Trends Biochem. Sci.*, **5**: 23 (1980).

Sander, K., Pattern specification in the insect embryo. In *Cell Patterning*, CIBA Symposia, Vol. 29 (new series), pp. 241–263.

Sander, K., The evolution of patterning mechanisms: gleanings from insect embryogenesis and spermatogenesis. In *Development and Evolution*, edited by B. C. Goodwin, N. J. Hodder and C. G. P. Wylie. Cambridge University Press, Cambridge, pp. 137–159 (1983).

Saxen, L. and Toivonen, S., *Primary Embryonic Induction*. Logos Press, London (1962).

Spemann, H., Ueber den Anteil von Implantat und Wirtskeim an der Orientierung und Beschaffenheit der induzierten Embryonalanlage. *Wilhelm Roux Arch. Entw. mech. Org.*, **123**: 389 (1931).

Spemann, H., *Embryonic Development and Induction*. Yale University Press, New Haven (1938).

Spemann, H. and Mangold, H., Uber Induktion von Embryonalanlagen durch Implantation artfremder Organisatoren. *Wilhelm Roux Arch. Entw. mech. Org.*, **100**: 599 (1924).

Waddington, C. H., *Principles of Embryology*. Allen and Unwin, London (1956).

Waddington, C. H., *Principles of Development and Differentiation*. Macmillan, New York; Collier-Macmillan, London (1966).

Wilson, E. B., Experimental studies on germinal localization. I. The germ regions in the egg of *Dentalium*. II. Experiments on the cleavage-mosaic in *Patella* and *Dentalium*. *J. Exp. Zool.*, **1**: 1–72.

Chapter 3 Embryonic induction

The conversion of the structureless mass of cytoplasm contained in a hen's egg into an active and sensitive chicken is just one of the everyday extraordinary miracles of life. If you cut a window in the egg shell, build up its sides with wax and glaze it with a glass coverslip, it becomes possible to view this remarkable process taking place. Within 24 hours from the beginning of incubation, a disc-shaped mass on the side of the yolk, known as the blastodisc, already reveals the main axis of the future chick's body (Figure 3.11). The neural folds almost meet in the brain region and a spattering of red shows that haemoglobin is being synthesized in the surrounding blood islands. A day later the primitive brain and optic vesicles are visible. Within another 24 hours the leg and wing buds are apparent, melanin pigment is accumulating in the eyes, and a tiny heart is pulsing rhythmically in a diminutive blood system. After three weeks, a sticky, vociferous chicken breaks open the shell and struggles out, already equipped with a personality and a capacity to learn whatever is necessary to stay alive, find a mate and raise its own young.

This extraordinarily complex biological entity that we know as a chicken achieves this state of organization in a mere 21 days, a fact at which we know biologists and philosophers have marvelled since the days of Aristotle, and no doubt since long before that also. We now know that the majority of relevant information is encoded in the animal's DNA, but the means by which this information is brought to bear is another matter. The key to understanding this subject is the knowledge that gene switching in any one cell type is controlled by interactions with neighbouring cells of dissimilar types, by the processes of embryonic induction, introduced in the last chapter. This represents a refinement of the universal principle of phenotype expression stated on the opening page of Chapter 1: acquisition of a *specific* cell phenotype depends on the interaction between genotype and a *specific* cellular environment.

In Chapter 2 we considered some of the factors that initiate programmes of development in early embryos. These include molecules distributed in the egg cytoplasm and environmental cues, such as gravity, operating from outside the egg. But if these were the only forces guiding gene expression in the nuclei the organism would remain a very simple one. As the embryo becomes more complex its internal 'environment' becomes ever more heterogeneous and this generates new points of reference. For example, tissues buried deep inside the body must

37

either become reliant on oxygen transported there, mainly by the blood, or else develop an anaerobic form of metabolism. Waste products accumulate in some cells, but are easily discharged from others. Diverse conditions such as these cause the tissues to diversify further from one another and proximity to tissues of dissimilar type brought about by tissue movements, introduces yet more diversity into the equation. During the gradual evolution of species, these internal metabolic features have become adopted as cues for directing gene expression into what have become accepted by that species as the pathways of normal development.

What is of particular interest, especially in inter-species comparative studies, is that although the overall strategy of development may be similar in diverse organisms, the 'tactics', the means by which an end is achieved, are often rather different. In other words the sequence of events, or the type of molecules which guide a particular course of development in one organism, may be dissimilar from those which perform the same task in other species. For example, development of skin pigmentation is triggered in some species by light, in others by temperature, in yet others by ionic environment. It is just this circumstance which bedevils the molecular study of embryology and frequently prevents formulation of general rules (see Chapter 13).

The concept of embryonic induction is one of the most important ideas in developmental biology. It is defined as 'communication between cells required for their cell-type-specific differentiation, morphogenesis and maintenance', and is considered traditionally under two headings, primary and secondary. (Some authors recognize a class of tertiary inductions also, but this classification becomes unwieldy.) Primary induction occurs at the beginning of gastrulation. It involves establishment of the main axis of the body and the initiation of changes leading to the development of the central nervous system. As we saw in the last chapter, in amphibians the primary inducer, or primary organizer, is the dorsal lip of the blastopore. We will consider secondary induction to include all later inductive interactions, which are the main topic of this chapter. Strictly, secondary induction can be considered to include the action of hormones, which however, will be described under another heading in Chapter 4.

The vertebrate eye is a particularly valuable organ for studying many aspects of development and will be used to illustrate some of the features of secondary induction.

3.1. Development of the vertebrate eye

Early development of the eye

The part of the blastula which will become the retina, the 'presumptive retina', can be identified even at this stage by use of vital stains. It originates as a patch in the presumptive neurectoderm, which divides into two retinal fields in the

neural plate (Figure 2.9). These sites initiate eye development by inductive interactions with the ectoderm that comes to overlie them after neurulation. If the
presumptive retina from an early gastrula is cut out and placed alone in tissue
culture medium, it will develop only into unspecialized ectoderm (Figure 3.1). If
it is transplanted to other regions of the body it performs inductions there, but
the tissues which form are characteristic of the region to which it has been
transplanted (Figure 3.1).

If the presumptive retina is taken from an embryo at the neurula stage its
properties are quite different. It develops into retina in the flank, or tail, or
anywhere else near the ectoderm that it is placed and around it forms a small eye,

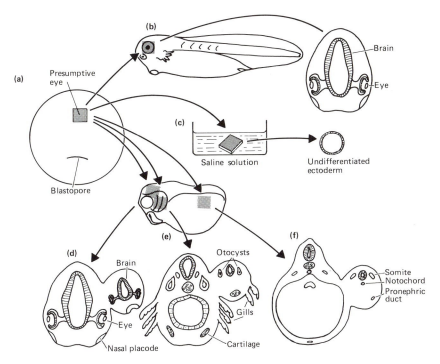

Figure 3.1. Acquisition of inductive capacity by presumptive retina.
The region shaded in (a) in the amphibian gastrula is the presumptive retina, which
normally initiates development of an eye as in (b) (lateral view and transverse section). If
this region is excised and cultured in a saline solution (c) it develops into a ball of undifferentiated ectoderm. If presumptive retina from the early gastrula is transplanted to extraneous sites in a neurula, it induces structures similar to those that would normally
develop close to that site: (d) at sites near the eye the implant differentiates as brain and
retina which induces a lens in the overlying ectoderm; (e) in the ear region otocysts and
gills develop; (f) implants in the flank induce somites, notochord and pronephric ducts:
this indicates that the inductive properties of the presumptive retina are not fully developed
at the early gastrula stage. (From Barth, 1949.)

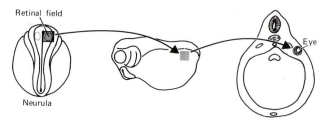

Figure 3.2. Induction of eye structures in extraneous sites.
A retinal field taken from an amphibian neurula embryo and implanted beneath the ectoderm at an unusual site causes formation of a whole eye at that site. (From Barth, 1949.)

gazing out from this unusual position (Figure 3.2). This experiment demonstrates that the tissue which will become the retina acquires this precise instruction during neurulation and thereafter remains determined as such.

This determined state is imposed upon the neural tissue by the underlying tissues that roof the archenteron (see Figure 2.9). If, at the gastrula stage, the presumptive retina is rotated or damaged, this has little effect on eye development, but if the underlying archenteron roof is destroyed, an embryo with a single central eye results, a condition known as cyclopea. Not only does the retinal specification depend on proximity of the archenteron roof, but division of the single retinal rudiment into two halves also seems to depend on the influence arising from the archenteron roof below.

After induction by the archenteron roof, the dorsal ectoderm, now presumptive neural tissue, itself becomes an inducer. The retinal fields occupy positions on each side of the fore-end of the neural tube (Figure 2.9). These become the tips of a pair of evaginations called optic vesicles, which grow out and eventually make contact with the inner surface of the overlying skin ectoderm (Figure 3.3). At each point of contact a layer of extra-cellular material composed largely of collagen (see Chapters 4 and 8) is laid down between the two surfaces and the ectoderm thickens to form the lens placodes. The tip of each optic vesicle then invaginates to form an optic cup, which later forms the back of the eye. As the optic cup forms, the lens placode also sinks in as the lens cup. This later becomes pinched off, forming a bag, the lens vesicle, that later develops into the lens.

The outer surface of the optic cup remains as a monolayer of epithelium which becomes pigmented, forming the pigment epithelium. Meanwhile its inner layer thickens, differentiates into a complex of several cell types, and becomes the sensory tissue of the eye, the neural retina.

The development of these new types of cell from undifferentiated neural ectoderm and the formation of a lens from non-neural ectoderm will take place correctly only if certain critical inductions are allowed to occur. Some idea of the complexity of the inductive interactions that take place during organogenesis can

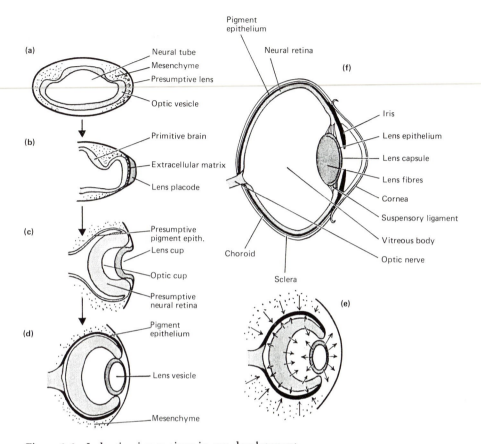

Figure 3.3. Inductive interactions in eye development.
(a) Transverse section through the eye region of a chicken embryo soon after closure of the neural tube. (b) Induction of the lens placode by the optic vesicle. Extra-cellular matrix material is laid down between the two tissues. Mesenchyme probably also plays a part in this interaction. (c) The optic cup and lens cup sink inwards. (d) The lens cup becomes pinched off as a lens vesicle and pigment is synthesized in the pigment epithelium. (e) Inductive influences (at stage (d)). Pigment epithelium receives inductive instructions from the surrounding mesenchyme and the underlying neural retina. The neural retina also emits inductive instructions for lens development. The lens emits influences which induce the overlying ectoderm to differentiate as cornea. (f) The mature eye (after Coulombre, 1965).

be gained from Figure 3.6, which shows those thought to take place during development of the eye.

Conditions for induction of lens and retina

Induction of lens tissue in non-neural ectoderm will occur only if the latter is in the correct stage of competence (see below). It also requires close proximity of the optic vesicle (see Figure 3.3). If the optic vesicle is surgically removed before it contacts the outer ectoderm, or if an impermeable barrier is interposed between them, then no lens will form. However, if a permeable barrier such as a piece of agar gel is placed between the two tissues, this does not prevent lens induction. It is at present uncertain whether under these circumstances contact is made between the two tissues by extension of long cell processes through the gel. Alternatively, the critical communication may occur through diffusion of small molecules, or by the accumulation of insoluble extra-cellular material between the interacting tissue rudiments.

Invagination of the tip of the optic vesicle requires even more critical conditions and will occur only after contact with the ectoderm of the presumptive lens. If contact is not achieved or maintained, the optic vesicle remains distended and its whole surface becomes a monolayer of pigment epithelium. Normal development of the multi-layered neural retina also requires the presence of the surrounding mesenchyme.

Development of the lens

The lens vesicle is initially a simple bag, its walls composed of a single layer of cells, but before long each cell on the internal aspect gradually elongates. These develop into long thin fibres as they become tightly packed and crowded by new cells pushing in from the sides. Eventually a tightly crammed mass of cells is formed, built up in concentric layers, until the collagenous membrane, called the capsule, which surrounds it is bulging tightly with turgid lens fibres. The lens capsule originates as the collagenous layer secreted between optic vesicle and lens placode, but is later synthesized by lens cells alone.

The external wall of the lens vesicle remains as a monolayer of epithelium and new cells are produced by mitosis around the equator. Some of these contribute to the increase in area of the lens epithelium, but the majority move towards the inner side of the lens, elongate, shed their mitochondria, endoplasmic reticulum and other organelles and even their chromosomes, until they eventually become just translucent elongated fibres packed tightly with characteristic proteins known as crystallins. Although these proteins do not crystallize in the chemical sense, they interdigitate with one another to produce the densest protein complex in the body, with the characteristic high refractive index associated with all lenses. Despite their very unusual composition and the lack of normally important elements, the lens fibres remain alive throughout life, under the maintenance of the other eye tissues.

As the lens forms, it in turn induces the overlying skin ectoderm to become cornea. This involves the synthesis of a very tough multi-layered membrane composed to a large extent of laminated sheets of collagen fibres, built up like multi-layered plywood. The cornea remains clear and translucent due to the exclusion of skin cells that would otherwise render it opaque.

Lens development is controlled by a balance between stimulatory factors which arise in the retina, and inhibitory factors produced by the lens itself. Fibres develop only on the side of the lens vesicle which faces the retina and this rule is obeyed in lens vesicles implanted in reverse orientation, so that the normal outer surface faces inward. Furthermore, any lens vesicle material placed within the plane of the edge of the optic cup develops into fibres, whereas external to this position it remains as epithelium. A variety of other experiments support the theory that one of the roles of the retina is to influence growth and differentiation of lens cells in such a way that the lens develops to precisely the right size, at precisely the right position for it to perform its visual function of focusing incident light on to the retina.

One of the best pieces of evidence for integrated control of growth and differentiation of retina and lens comes from experiments in which the optic cup of a large, rapidly growing species of salamander was grafted under the dermal ectoderm of a small, slow-growing species. In this and the reciprocal experiment, lenses of appropriate size were induced and harmoniously proportioned eyes of intermediate size were formed.

Development of the retina

For a time the presumptive neural retina can form pigment epithelium and vice versa, but gradually their phenotypes become stabilized. This requires prolonged contact between the two tissues. If they are separated, the pigment layer proliferates and generates a replacement layer of neural tissue at the position where neural retina should be. The reverse regeneration does not seem possible *in vivo* although similar changes will take place in cultured neural retina cells (see Chapter 8).

Another condition for pigment epithelium differentiation is physical tension within the cell sheet. If the outer wall of the optic cup is not kept under tension, pigment is not synthesized.

Once the neural retina has begun to grow, its increase in thickness and area seems to be independent of any specific influences of the surrounding ocular tissues, but the pigment epithelium services the neural retina with respect to supplies of nutrients and removal of waste products. Disorders of the pigment epithelium are therefore rapidly expressed as injury to the neural retina.

The role of the retina in controlling growth of the lens does not stop when the eye becomes functional, but continues throughout life. If the neural retina is excised, the lens will regress and retinal disease is very frequently associated with disorders of the lens.

Competence for lens induction

As described above, if presumptive retinal tissue from an amphibian is transplanted to a site under the ectoderm of another part of the body it will induce a lens and then a complete eye there, since all the skin ectoderm is initially competent to form lens. At the true lens site, however, the threshold of response is lower. The retention and increase in competence at the correct site is related to prior exposure of this ectoderm to two other embryonic tissues, the endoderm of the presumptive foregut and the mesoderm of the presumptive heart (see Figure 2.5). All three germinal layers — endoderm, mesoderm and ectoderm (optic vesicle and retina) — are thus responsible for inducing lens differentiation. The times at which these inductors act, in relation to the determinative response in the ectoderm in a species of salamander, is shown in Figure 3.4.

Although a similar sequence of inductive interactions probably occurs in most vertebrates, there are variations on this general theme. For example, in some species a lens will occur at more or less the correct site even if the optic vesicle is removed. In others, transplanted optic vesicles induce lenses in overlying ectoderm less effectively than when the ectoderm is transplanted to the eye region.

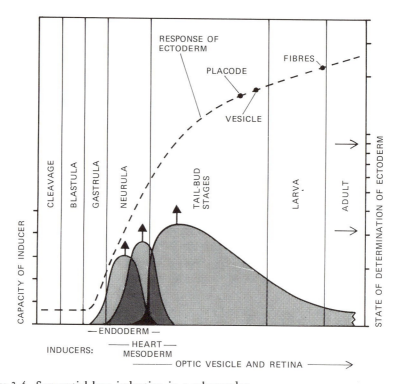

Figure 3.4. Sequential lens induction in a salamander.
The dashed line represents the response in the ectoderm to the normal lens inducers; foregut endoderm, presumptive heart mesoderm and optic vesicle or retina. The ordinate for this curve is logarithmic, the level of response being a function of the sum of all past inductions. This curve commences above the baseline, as the ectoderm has a certain degree of determination for lens differentiation before these inductions take place. The inductive capacity of the inductors through the different developmental stages is represented by the shaded curves, plotted on a linear scale. Head mesenchyme also probably plays a role in lens induction in some species, but is not included in this diagram. (From Jacobson, 1966.)

This suggests that the surrounding head mesenchyme also plays a role in lens induction.

As shown in Figure 3.4, even before exposure to foregut endoderm the presumptive lens ectoderm has a level of determination which is greater then zero. This could be due to segregation or selective expression of maternally derived egg components, as suggested in Chapters 4 and 9.

We do not know the nature of the inducing molecules, but their effects seem to be similar and additive, and can be mimicked merely by maintaining non-competent ectoderm in saline solution. In the experiment illustrated in Figure 3.5, presumptive neurectoderm from a newt early gastrula was wrapped around an excised retinal field taken from a neurula embryo. Under these conditions the inducer developed into brain and retina-like tissues, which induced a neural tube in the gastrula neurectoderm. However, when the gastrula ectoderm was previously incubated in saline solution for 36 hours it developed a lens instead, showing that its competence had changed. This could have been the result of ageing, or an effect of the salt solution, or be due to the absence of other inductive stimuli.

The word competence is used to describe the special responsiveness of a tissue to embryonic induction. A formal definition would be: 'the capacity of a multipotential embryonic tissue to respond to induction, by entering into certain pathways of differentiation which would not obtain in the absence of induction'. Judging by observations such as those described above, competence seems to be a property related to prior exposure to other inducers. It has been suggested that at a molecular level competence for lens induction may involve low-level production of mRNA characteristic of lens: this idea is discussed in Chapters 8 and

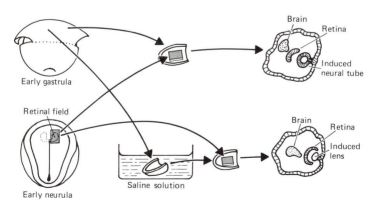

Figure 3.5. Acquisition by ectoderm of competence for lens induction.
Presumptive ectoderm taken from the animal hemisphere of an early newt gastrula is cultured with a retinal field from an early neurula. A neural tube is induced in the gastrula ectoderm. When the gastrula ectoderm is stored in saline solution for 36 hours before culturing with neurula retinal field, a lens is induced instead. (From Waddington, 1966.)

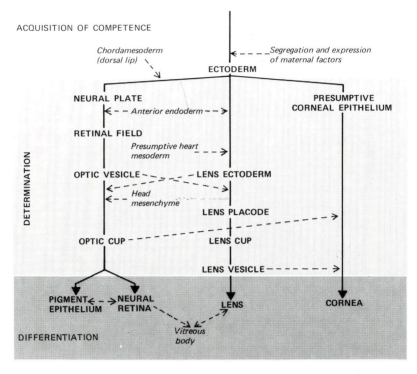

Figure 3.6. Flow sheet of inductive interactions during eye development.
Cell lineages are represented by heavy continuous lines leading from ectoderm to pigment epithelium, neural retina, lens and cornea. Inductive and other influences are represented by dashed arrows. The diagram illustrates the conventional concept that cytodifferentiation is preceded by gradual determination. In Chapter 7 the applicability of this concept to eye development is called into question. (Based on Coulombre, 1965.)

13. Another suggestion is that it could also involve specific patterns of methylation of cytosine residues in DNA (Chapter 11) and of modification of histone proteins associated with the DNA (Chapter 10). In Chapter 4, competence of chicken oviduct for induction by oestrogen is shown to involve association of a special combination of non-histone proteins with the DNA at specific sites, and the presence of a specific cytoplasmic receptor protein.

3.2. The fate of the neural crest

The neural crest cells are a particularly interesting group of cells which become detectable at neurulation (see Chapter 2). They arise from the neurectoderm and initially form a long band lying between the neural tube and the ectoderm. They constitute a unique population of migratory cells which give rise to a very wide range of tissues: the pigment cells of the skin, the epinephrine-secreting cells of the adrenal medulla, connective tissue cells of the dorsal fin in amphibia, some bones of the head, ganglia and nerves of the sympathetic and parasympathetic systems, the Schwann cells of the nervous system, and several others. A question that has long intrigued developmental biologists is whether these cells are programmed for differentiation into a particular cell type before they leave the neural crest, whether they acquire their instructions during migration, or whether they receive them at their final site. The most informative experiments on this topic were performed by Le Douarin, who used a combination of chicken and quail embryo tissues, taking advantage of the fact that quail and chick cells can be distinguished under the microscope.

Closure of the neural tube begins in the head region. The somites form as blocks of tissue on either side of the neural tube, and somite formation continues in a posterior direction in the wake of neural tube closure (see Chapter 6). Neural crest cells from the region between the first and seventh somites migrate out and give rise to the nerve ganglia of that part of the alimentary canal anterior to the umbilicus. The epinephrine-secreting cells of the adrenal medulla come from the neural crest in the region corresponding to somites 18–23. Le Douarin took neural crest tissue from beside somites 18–23 in a quail embryo and transplanted them

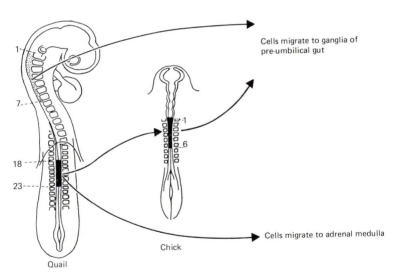

Figure 3.7. Differentiation of neural crest cells in relation to site of origin.
Neural crest cells in the region of somites 18–23 normally migrate to populate the adrenal medulla and differentiate as chromaffin cells secreting epinephrine. When cells from beside somites 18–23 are transferred to the region of somites 1–6, they follow the path of migration of neural crest cells that normally originate at this site and enter the ganglia of the pre-umbilical gut, where they differentiate as neuroblasts. In these ganglia quail cells are distinguished from chick cells by their staining properties. (Based on Le Douarin.)

to the 1–6 somite region in a chick embryo (Figure 3.7). The reciprocal experiment was also done, in which quail neural crest cells from the anterior region were transplanted to the posterior region of a recipient chicken embryo. It was found that cells taken from the hind-brain region (somites 1–6) and placed beside somites 18–23 will populate the adrenal gland instead of their normal sites in the anterior gut and will actively secrete epinephrine, instead of performing the functions of nerve ganglion cells. The reciprocal experiment yielded similar results, and it was concluded that the fate of the neural crest cells depends either on the tissues among which they migrate, or on their final site, but not on the position along the body axis from which they originated.

This deduction raises some further fascinating questions. How do migratory cells find their way? How do they recognize their destination? What experiences do they go through to obtain the information necessary to establish specific programmes of expression of their genes? No-one can yet answer these questions, but it is generally believed the mechanisms involve interactions between molecules carried on the neural crest cell surfaces and others displayed on the surfaces of the cells over which they move, or within the extra-cellular matrix that lies between them (see Chapter 4).

The biochemistry of neural crest cell differentiation

The differences between the neural crest derivatives may not be so radical as they appear at first. For example, the skin pigment cells, the ganglia of the sympathetic

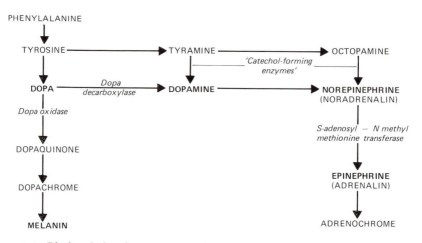

Figure 3.8. Biochemical pathways associated with the metabolism of DOPA in neural crest derivatives.

In melanocytes DOPA (3,4-dihydroxyphenylalanine) is acted upon by dopa oxidase, which leads to synthesis of the brown pigment melanin. In other neural crest derivatives dopa decarboxylase converts DOPA into dopamine, which, with norepinephrine and epinephrine, acts as a neurotransmitter in the ganglia of the sympathetic system. Epinephrine is produced in quantity in the adrenal medulla and secreted into the blood system.

system and the adrenal medulla cells all share a common biochemical pathway concerned with the synthesis of DOPA (3,4-dihydroxyphenylalanine) (Figure 3.8). In pigment cells DOPA is oxidized by the enzyme dopa oxidase to dopaquinone, which is converted through several steps to the brown pigment melanin. In the other cells dopa decarboxylase converts DOPA to dopamine, norepinephrine (or noradrenalin) and epinephrine (or adrenalin), which act as neurotransmitters in the sympathetic system (cf. acetyl choline in the parasympathetic system). In the adrenal glands the same synthesis takes place and epinephrine is secreted into the blood stream, where it acts as an endocrine hormone (see Chapter 4). The instructions received by the neural crest cells, which bring about their differentiation, may therefore merely regulate the relative activities of alternative enzymes, such as dopa oxidase and dopa decarboxylase, at major switch points in biochemical pathways.

3.3. Induction by mesenchyme

Many of the secrets of development of multicellular animals are hidden in the products of a mysterious tissue called mesenchyme. This is a migratory cell type which arises in the embryonic mesoderm and moves out, forming condensates adjacent to ectodermal and endodermal epithelia. It also gives rise to the blood cells, cartilage, bone, adipose and connective tissues. Virtually all morphogenetic processes and the origins of all organs can be accounted for by examining the reorganization, rearrangements and interactions of epithelia and mesenchyme. Epithelium of endodermal origin, together with mesodermal mesenchyme, form the glands associated with the gut, while peripheral body structures are formed from mesodermal mesenchyme acting in concert with ectodermal epithelium. The typical situation is that specialized differentiation of epithelium occurs only after, or during, induction by mesenchyme. In some cases a specific mesenchyme determines epithelial fate (see below). At the other extreme, a generalized type of mesenchyme merely enables epithelial cells to realize a predetermined potential.

Control of epidermal differentiation

The vertebrate epidermis is an example of an ectodermal epithelium, the fate of which is highly dependent upon the specificity of its adjacent mesenchymal derivative, the dermis (see Figure 4.12). These tissues jointly produce a variety of diverse structures, including hair, feathers, scales, skin glands, beaks, fingernails and teeth. Even our limbs originate in a variant of this interaction (see Chapter 6).

Although most of a chicken's body is covered in feathers, there are scales on its legs while some parts of its skin, called the brood patches, are bare. The form and texture of feathers also vary in different parts of the body. If mesenchyme is taken from beneath the ectoderm of the thigh of an early chicken embryo and transplanted beneath the ectoderm of the wing, at that site the wing will develop feathers of a type characteristic of the thigh. Mesenchyme from a feathered part

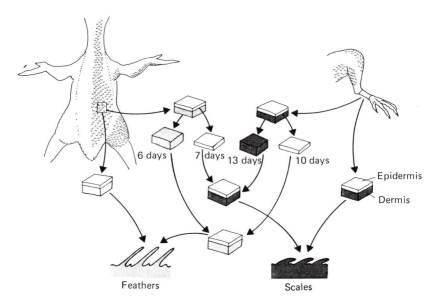

6 days   7 days   13 days   10 days

Epidermis

Dermis

Feathers   Scales

Figure 3.9. Interactions between dermis and epidermis and the differentiation of scales and feathers.
Skin from the mid-dorsal region of the trunk and the scaled region of the leg of chick embryos was separated into dermal and epidermal layers by treatment with 1% trypsin at 4°C. The tissues were recombined and cultured on the chorioallantoic membranes of 10-day-old embryos. The inductive capacity of dermis and the competence of epidermis varied with age, but in certain combinations mid-dorsal epidermis could be induced to form scales, while leg epidermis formed feathers. Note the hexagonal pattern of dermal papillae which is also induced in the epidermis by the underlying dermis. (Based on Rawles, 1963.)

of the body transplanted to regions where feathers do not normally grow causes feather-forming cells to differentiate there. If mesenchyme from the scaled region of the leg is transplanted into a region where feathers normally grow, that region will develop scales instead (Figure 3.9).

These experiments show that epidermis anywhere in the body has a variety of potentials, but the underlying dermis, derived from mesoderm, selects which potential will be realized. This is equivalent to the early widespread capacity of neurula skin ectoderm to respond to induction by presumptive retina, with formation of a lens (Figure 3.2). An interesting illustration of the developmental potential of skin ectoderm is provided by transplanting one of the mesenchymal condensates that normally induces mammalian tooth papillae into a position under the epidermis of the foot. Despite transplantation, the mesenchyme continues to do its normal work and induces a tooth in the skin of the foot!

Experiments in which tissues are transplanted between species confirm the expectation that the responding epithelium can react only within its own genetic limitations. For example, mouse mesenchyme transplanted into the anterior chamber of the eye of a chicken induces the cornea to grow feathers, but under no circumstances does chicken epidermis grow fur!

When dermis and epidermis from chickens and ducks are recombined, it is possible to show that some of the features which form are dictated by dermis, while others depend on the epidermis. Duck down feathers have a stem, or rachis, and minute barbules of a characteristic type, both of which are lacking in chicken down (Figure 3.10). When chick epidermis is transplanted on to duck dermis, the feathers which form have a duck-like rachis even though they are synthesized entirely by chick tissues. On the other hand, their barbules are of chicken type. When duck epidermis is combined with chicken dermis the feathers that form lack a rachis, but have duck barbules. Apparently the dermis issues instructions for the basic form of the feather, but the fine details, like the barbules (and the nature

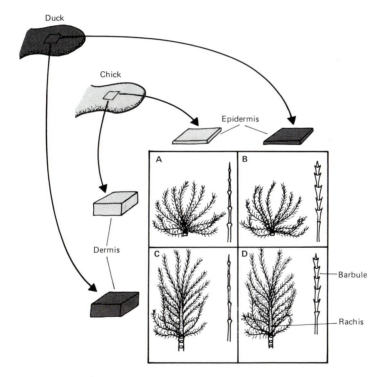

Figure 3.10. Inter-species interactions between ectoderm and mesoderm in feather development.
Pieces of skin were taken from limb buds of chicken and duck embryos, split into epidermis and dermis, recombined and cultured on chorioallantoic membranes. The presence (C,D) and absence (A,B) of a rachis depends on the influence of duck, as compared to chick, dermis. In contrast, the duck type of barbule arises only from duck epidermis (B,D) and the chick type only from chick epidermis (A,C). (Based on Sengel, 1976.)

of the proteins synthesized), are translated from genes in the epidermal cells and do not depend on detailed directions from the dermis.

Control of epidermal patterning

Not only is the form of the epidermal structures dictated by the underlying dermis, but so is the pattern of their distribution. Typically, scales, feathers and to a lesser extent hairs, are arranged in a hexagonal pattern which can be seen in chicken embryos in the distribution of the dorsal feather papillae (see Figure 3.11). The result is that the structures which eventually form produce a neat protective covering arranged like the slates on a roof. If a section of chick dermis is rotated through 90° or 180°, provided this is done before the overlying epidermis has received its instructions from the dermis, the orientation of the feather tracts becomes altered accordingly.

Directions for epidermal pattern will also cross the species barrier. For example, if dermis from the whiskered region of the snout of a mouse embryo is combined with embryonic chick epidermis, the latter forms feather buds arranged in the pattern typical of mouse whiskers. In all these cases, however, the dermis seems capable of emitting its correct inductive stimuli only during a restricted period of development. Likewise, the epidermis will respond appropriately only at certain periods when it is competent to do so.

3.4. Gene mutations affecting induction

During induction two classes of error can arise: failure in the transmission of information by the inductor, or defects in the capacity of potentially responsive cells to acquire or respond to that information. A dominant mutation called *Small-eye* (*Sey*), in the laboratory mouse, has eyeballs which are typically much reduced in size. There are several additional ocular defects, but the basic biochemical fault seems to be in the glycosylation of collagen, as revealed by the abnormal composition of the lens capsules. As pointed out above, the lens capsule originates in the extra-cellular layer secreted between optic vesicle and lens placode at lens induction. Although direct evidence is lacking, it seems likely that failure to develop eyes of normal size could derive from a deficiency in inductive stimuli given out by the presumptive lens, or lens placode, in the form of its secreted extra-cellular material. This observation also lends support to the idea that the inductive influence emitted by the lens placode on the optic vesicle involves the mediation of extra-cellular matrix (see above and Chapter 4).

Mutations are known which interfere with pigment production in neural crest-derived melanocytes without affecting the pigment cells of the ocular pigment epithelium which, as we have seen, has a different origin. Animals with these mutations develop pigment in their eyes, but in no other parts of their bodies. Two types of black-eyed white-coated mice fall into this category: the recessive *steel* (*st*) and *Dominant White Spotting* (*W*) mutants. In addition to their pigmentation defects, both have reduced fertility and *steel* mice also suffer from anaemia. If neural crest cells are taken from a normal wild-type (+ / +) mouse and transplanted into a *steel* homozygous embryo, the recipient mouse will remain defective, but if *st/st* neural crest cells are transplanted into a normal host, they can differentiate like normal cells into pigmented skin melanocytes. This suggests it is the tissue environment, rather than the neural crest cells themselves, which is defective in *steel* mice.

In the *Dominant White Spotted* mouse, the basic defect is in the neural crest cells. If *W* neural crest cells are transplanted into a normal host, they remain defective. On the other hand, normal or *st/st* cells transplanted into a *W* host respond to the inductive environment and become pigmented. Apparently neural crest cells which carry the *W* allele are unable to respond to the inductive influences which impinge upon them.

3.5. *The nature of inductive stimuli*

Primary induction

A great deal of attention was at one time paid to the chemical nature of the primary inducer. Many experiments were carried out on amphibians, but the more that were performed the more obscure became the problem. It was found that blastopore dorsal lip tissue will induce neural tube in competent ectoderm even if the dorsal lip has been crushed, frozen, heated or extracted with organic solvents. When protein, RNA or lipid were injected into amphibian blastulas, all were shown to be capable of inducing neurulation. Even inert chemicals such as silica and kaolin were found to have this property. It was eventually deduced that the nature of the stimulus was very much less important than the repertoire of possible responses by the induced tissue.

If gastrula ectoderm is dissected out and cultured in a hypertonic solution of sodium chloride, it will form a neural tube in isolation — it will autoneuralize. This observation led to the suggestion that the natural stimulus could be the accumulation of inorganic ions, such as Na^+, Mg^{2+} and Ca^{2+} between the closely apposed mesoderm and ectoderm. The latest theories postulate two inductive influences, possibly due to protein; one which induces head structures, acting early in gastrulation, the other inducing tail structures, acting late in gastrulation (cf. Figure 2.10(b),(d)). A mixture of the two is thought to induce mid-body structures.

This idea is reminiscent of the morphogenetic gradients thought to define the relative positions of the main body segments in insects (see Chapters 2 and 5), but whereas the insect gradients are considered to be gradients in space, the postulated inducer gradients in amphibians are laid down with respect to time.

In birds and mammals also, primary induction occurs progressively down the

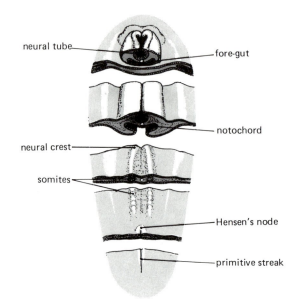

Figure 3.11. Structure of an early chick embryo about 24 hours after the start of incubation, at the beginning of somite formation. The embryo is situated at the interface between yolk and albumen.
(After Huettner, 1971, and Waddington, 1966.)

length of the body, as the primary organizer, Hensen's node, moves through the blastodisc, down the body axis (Figure 3.11). If a Hensen's node is transplanted from a duck or a rabbit to a chicken embryo (or vice versa), it will induce a secondary neural tube in the recipient embryo (Figure 3.12). This

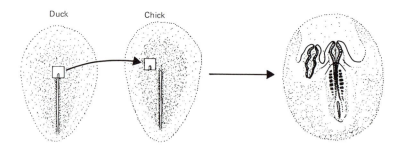

Figure 3.12. Induction of a second embryonic axis by implantation of a Hensen's node from another species.
A Hensen's node was excised from a duck embryo and transplanted into a chick host. The appearance of the double embryo that resulted 31.5 hours later is illustrated. (After Waddington and Schmidt, 1971.)

demonstrates that the inductive stimuli are similar, or at least have similar influences in different vertebrates, even when they are as evolutionarily distant as birds and mammals. A similar conclusion comes from the inter-species transplantation of dermis and epidermis (see below).

Induction in the eye

As suggested above, induction of neural retina, lens, or both, may require contact with extra-cellular matrix material laid down between the optic vesicle and lens placode during early eye development.

There is also evidence of physiological communication between optic vesicle and lens placode, since a radioactive label introduced into the optic vesicle in the form of the amino acid, phenylalanine, was picked up again in the lens placode after induction. However, the nature of the molecule that is transferred is unknown, nor is it known whether this transfer is in any way concerned with induction. A protein which stimulates mitosis of cultured lens epithelium has been isolated from retinas and vitreous humour of adult bovine eyes. This could be involved in growth and maturation of the lens, but as yet direct evidence is lacking.

Induction in the skin

The dermis and epidermis are normally separated by a collagenous extra-cellular membrane called the basal lamina (see Figure 4.12). It has been suggested that movement of epidermal cells over this layer and the points at which they aggregate to form dermal papillae are dictated by biochemical features of the lamina (see Chapter 4).

Contact with the basal lamina seems to maintain the capacity of the basal epidermal cells to divide while they receive instructions for differentiation. It is possibly hyaluronic acid in the basal lamina which promotes this cell division (see Chapter 4). Formation of feathers could be a response to mechanical pressure from below, as feather growth can be stimulated in the brood patches of chickens by insertion of a dense object beneath the skin. In other parts of the body this stimulus may be provided by packing of the underlying tissues, together with tension in the epidermis itself.

Instructions for differentiation may be carried by proteoglycans and laminin which coat the collagen of the basal lamina (see Chapter 4), as well as by biochemicals which diffuse across it. Hair, feathers and scales are composed largely of the protein keratin, but if epidermis is cultured in the presence of an excess of vitamin A it ceases keratin synthesis and secretes mucus instead. When vitamin A is withdrawn, keratin synthesis recommences (see Chapter 7). Mucus is secreted by amphibian epidermis, and it is interesting to speculate whether the evolution of reptiles from amphibians may have involved a change in the metabolism of vitamin A in the skin.

Inductive stimuli are therefore probably of many different types, but whatever their nature, the inductive response is determined very largely by the state of competence of the reactive tissue, rather than by the nature of the inducer. As we saw above, competence is a cumulative property, derived from the responding tissue's past history of inductive experience.

3.6. Summary and conclusions

The development of complex structures and the differentiation of the diverse tissue types they contain require initiation and maintenance through interaction with adjacent tissues. Such interactions come under the heading of embryonic induction. The initiation of gastrulation and neurulation is referred to as 'primary induction', all subsequent interactions of this type being classed as 'secondary'. Before a tissue will respond appropriately to inductive stimuli it must be in an appropriate state of competence. Competence is often initially widespread within a tissue, but becomes restricted and enhanced in the appropriate regions as a result of a series of pre-inductions. In later chapters it is suggested that this localization of competence may involve methylation of cytosine in the DNA, modification of histone proteins which are closely associated with it, or specific patterns of low-level synthesis of messenger RNA.

Some inductions are transitory, but others continue throughout life, the latter including the repression of regenerative changes in intact tissues (see Chapter 6). The eye may be exceptional in this respect, as its tissues show a quite remarkable capacity to undergo regenerative, or 'transdifferentiative', changes (see Chapter 7). Competence for induction of epidermis by dermis is restricted to certain short stages during development (Figure 3.9), but if feather or scale mesoderm is introduced into the anterior chamber of the eye i.e., between lens and cornea — after corneal differentiation has commenced, the cornea will switch developmental pathways and produce feathers, or scales, instead. This is one of several examples which suggest that the conventional concept that cell determination precedes cytodifferentiation (Chapter 2) may not be strictly applicable in all cases. This idea is explored further in Chapter 7.

During development, critical inductions occur between derivatives of all three germinal layers, but mesenchyme, derived from embryonic mesoderm, is one of the most potent inducers and plays a part in the formation of all the body organs. In general, inductive responses depend more on the state of competence of the induced tissue than on the chemistry of the inducer, which frustrates our attempts to investigate the biochemistry of inductive molecules. Despite these problems a great deal is known about how some secondary inductions take place and this topic is explored in the next chapter.

Bibliography

Ambrose, E. J. and Easty, D. M., *Cell Biology*, 2nd edn. Nelson, Middlesex (1977).

Balinsky, B. I. and Fabian, B. C., *An Introduction to Embryology*, 5th edn. Saunders, Philadelphia (1981).

Barth, L. G., *Embryology*, revised edn. Dryden Press, New York (1953).

Barth, L. G. and Barth, L. J., Ionic regulation of embryonic induction of cell differentiation in *Rana pipiens*. *Dev. Biol.*, **39**: 1 (1974).

Bourne, M. C. and Gruneberg, H., Degeneration of the retina and cataract. *J. Hered.*, **30**: 131 (1939).

Browder, D. W., *Developmental Biology*. Saunders, Philadelphia (1980).

Clayton, R. M., The molecular basis for competence, determination and transdifferentiation: a hypothesis. In *Stability and Switching in Cellular Differentiation*, edited by R. M. Clayton and D. E. S. Truman. Plenum, New York, pp. 23–38 (1982).

Coulombre, A. J., The eye. In *Organogenesis*, edited by R. L. de Haan and H. Ursprung. Holt, Rinehart and Winston, New York, pp. 219–251 (1965).

Coulombre, J. L. and Coulombre, A. J., Metaplastic induction of scales and feathers in the corneal anterior epithelium of the chick embryo. *Dev. Biol.*, **25**: 464 (1977).

Deuchar, E. M., *Cellular Interactions in Animal Development*. Chapman and Hall, London (1975).

Ede, D. A., *An Introduction to Developmental Biology*. Blackie, Glasgow (1978).

Grant, P., *Biology of Developing Systems*. Holt, Rinehart and Winston, New York (1978).

Hamburgh, M., *Theories of Differentiation*. Edward Arnold, London (1971).

Holtfreter, J., Uber die Verbreitung induzierender Substanzen und ihre Leistungen im Triton-keim. *Wilhelm Roux. Arch. Entw. Mech. Org.* **132**: 307 (1934).

Hopper, A. F. and Hart, N. H., *Foundations of Animal Development*. Oxford University Press, Oxford (1980).

Huettner, A. F., In *Early Embryology of the Chick*, edited by B. Patten, 5th edn. McGraw-Hill, New York (1971).

Jacobson, A. G. Inductive processes in embryonic development. *Science*, **152**: 25 (1966).

Le Douarin, N., Cell migration in early vertebrate development studied in interspecific chimeras. In *Embryogenesis in Mammals. Ciba Foundation Symposium*, Vol. 40 (new series). Elsevier/Excerpta/North-Holland, Amsterdam, pp. 71–101 (1976).

Pritchard, D. J. and Clayton, R. M., Abnormal lens capsule carbohydrate associated with the dominant gene 'Small-eye" in the mouse. *Exp. Eye Res.*, **19**: 335 (1974).

Rawles, M. E., Tissue interactions in scale and feather development as studied in dermal-epidermal recombinations. *J. Embryol. Exp. Morph.*, **11**: 765 (1963).

Rutter, W., In *Epithelial-Mesenchymal Interactions*, edited by R. Fleischmajor and R. E. Billington. Williams and Wilkins Co., Baltimore (1968).

Saxen, L. and Toivonen, S., *Primary Embryonic Induction*. Logos Press, London (1962).

Sengal, P., *Morphogenesis of Skin*. Cambridge University Press, Cambridge (1976).

Waddington, C. H., The origin of competence for lens formation in the amphibian. *J. Exp. Biol.*, **30**: 86 (1936).

Waddington, C. H., *Principles of Embryology*. Allen and Unwin, London (1956).

Waddington, C. H., *Principles of Development and Differentiation*. Macmillan, New York/Collier-Macmillan, London (1966).

Waddington, C. H. and Schmidt, In *Early Embryology of the Chick*, edited by B. M. Patten, 5th edn. McGraw-Hill, New York (1971).

Wessells, N. K., *Tissue Interactions and Development*. Benjamin, Menlo Park, California (1977).

Yamada, T., A chemical approach to the problem of the organiser. *Adv. Morphog.* **1**: 1 (1961).

Chapter 4 Cytoplasmic and extra-cellular controls

As has already been stressed, phenotype is the product of interaction between genotype and environment. At a cellular level, 'genotype' is usually regarded as the information encoded by the nuclear DNA, and it is often claimed that the nuclear DNA contains *all* the information required to produce a complete organism. However, this is not the case; some essential components of the body are derived from extra-nuclear factors that are present in the egg cytoplasm, while essential information is also provided by influences arising in other parts of the body and by physical and chemical features of the animal's surroundings. Organisms that are quite different in appearance can develop from similar eggs raised under slightly different conditions. A dramatic example is the minute parasitic wasp *Trichogramma semblidis*. This organism lays its eggs in those of other species of insects. If the unfortunate host egg is that of a butterfly or moth, the wasp that emerges has two pairs of wings and bushy antennae. On the other hand, its brothers and sisters raised in the eggs of the alder fly, *Sialis*, develop as wingless forms with smooth antennae (Figure 4.1).

PART I. EXTERNAL FACTORS

Some of the most important environmental influences are day length and temperature, and we will begin our investigation with a consideration of these factors.

4.1. Environmental physical factors

Day length

The butterfly *Araschnia* exists in two forms which at one time were believed to be distinct species: *A. levana*, a pale form that appears in spring, and a dark form, *A. prorsa*, that makes its appearance in late summer (Figure 4.2). Both are now known to be colour variants of the same species. The dark form emerges from pupae that develop in high summer, the light form from those that have over-wintered in diapause (a kind of hibernation). The trigger for entering diapause

butterfly host   alder fly host

Figure 4.1. Phenotype variation in *Trichogramma semblidis*.
(a) The winged form with bushy antennae develops parasitically in the eggs of Lepidoptera.
(b) The wingless form with modified antennae and legs is raised in the eggs of the alder fly, *Sialis* (Based on Wigglesworth, 1964.)

is the length of day experienced by the caterpillars before pupation. Short days result in the light form, long days in the dark form.

In the silk moth, *Bombyx*, day length regulates the rate of development of the next generation. Eggs laid in spring (short-day period) produce fertile adults that reproduce again the same summer. Those laid when the length of day is greater than 16 hours go into diapause and the animal overwinters in the egg. The diapause is broken by two to three months' chilling at 5 °C.

Day length affects reproduction in another way in greenflies and other aphids. When the days are long, greenfly reproduce parthenogenetically, giving birth to a tiny new daughter every two hours. These progeny are themselves pregnant at birth. Viviparous parthenogenetic reproduction changes in the autumn to the normal, sexual and egg-laying type, and this change is brought about by the shortening length of day. Aphids usually overwinter in the egg and on hatching

long day   short day

Figure 4.2. Phenotype variation in *Araschnia levana*.
The dark form (*A. levana*, form *prorsa*) develops from caterpillars that have grown during the long days of high summer. The light form appears in spring, the animals having over-wintered in the pupa after the caterpillars have grown during the short days of late summer. Male individuals are represented. (Redrawn from Higgins and Riley, 1975.)

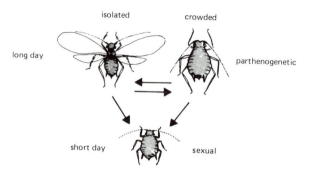

Figure 4.3. Phenotype variation in aphids.
When day length is short, aphids reproduce sexually, producing a standard form of offspring. In conditions of long day length females reproduce parthenogenetically, producing winged female offspring in crowded conditions and wingless females in uncrowded circumstances (After Wigglesworth, 1964.)

in the spring they resume parthenogenetic reproduction until after mid-summer, when the days again start to shorten (Figure 4.3; see below).

Temperature

Life as we know it can exist only within fairly narrow temperature limits, related to the stability of protein structure and the physical properties of water, the only widely available solvent on this planet. Temperature is important in controlling the overall rate of metabolism, but it also produces more specific effects. We have already seen that chilling is necessary for breaking the diapause of silkworm eggs. In the protozoan 'slipper animalcule' *Paramecium*, characteristic antigens are accumulated on the cell surface at specific extreme temperatures. Once a particular pattern of antigenic display has become established, this is maintained at less extreme temperatures in the same individual and its progeny, but can be modified again by changing the temperature to the other extreme. This seems to indicate the existence of gene 'switches' operated by temperature alone. Figure 4.4 shows an experiment which illustrates this effect. *Paramecium* of two strains known as 90 and 60 are maintained either at 25°C, when they express antigen G, or at 29°C, when they express antigen D. Thus, for example, strain 90 individuals maintained at 25°C have antigen 90G, while stain 60 maintained at 29°C display antigen 60D. When these animals are transferred to the intermediate temperature of 27°C, the strain 90 individuals continue to display 90G antigens and the other strain 60D. Both patterns of display are maintained through many rounds of binary fission.

Paramecium has two nuclei and during conjugation one of these, the micronucleus, is transferred to the co-conjugant. When 90G and 60D individuals conjugate at 27°C the pattern of gene expression in the transferred nucleus falls

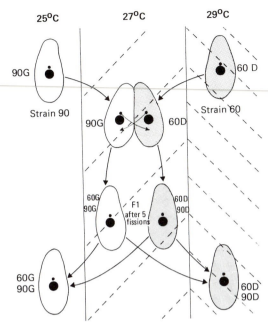

Figure 4.4. Temperature-induced antigenic variation in *Paramecium*.
Paramecium display surface antigens defined by their genotype and directed by environmental temperature. Thus individuals of strain 90 always display antigens characteristic of the strain, but which can be of type D or G depending on temperature. Strain 90 animals maintained at 25°C display antigen 90G and strain 60 animals at 29°C display antigen 60D. When moved to the intermediate temperature of 27°C they retain these antigenic displays. If they undergo conjugation at 27°C the micronuclei are transferred into cytoplasm which was conditioned at the other extreme temperature and this causes the micronuclei to change their patterns of synthesis. The new patterns are maintained through many binary fissions in the intermediate temperature. When transferred back to the extreme temperatures, they all adopt patterns of synthesis appropriate to those temperatures (Based on Beale, 1954.)

into line with that in the recipient cell. This occurs apparently because the temperature-conditioned cytoplasm exerts a gene-switching effect upon the new nucleus. Thus, for example, a 60D individual at 27°C, which receives a 90G nucleus, will give rise to F1 progeny expressing 60D and 90D antigens, while the 90G individual which receives a 60D nucleus will have F1 progeny expressing 90G and 60G antigens. If these F1 daughter cells are now transferred into the extreme temperatures, 29°C or 25°C, their cytoplasm becomes reconditioned and they express the antigens appropriate to the new temperature.

In this interpretation, the role of the cytoplasm in the control of nuclear activity has been emphasized. Whether in this case temperature operates initially on the nucleus, or initially on cytoplasmic constituents, is unknown, but it certainly involves cytoplasmic changes that control nuclear gene expression and this operates either at transcription or RNA processing, since each antigen type contains only its own specific species of messenger RNA in the cytoplasm.

Most organisms respond to stresses like sudden heat shock by synthesizing a characteristic set of proteins. This is known as the heat shock response. Remarkably there are great similarities in these proteins in all species examined, including *Drosophila*, bacteria and man. Although the heat shock genes are on the nuclear chromosomes in eukaryotes, some of them code for proteins which become bound to the mitochondria. In *Drosophila heidei* mid-third-instar larvae, temperature shock ($37°C$), anaerobiosis and inhibitors of mitochondrial metabolism cause a similar metabolic imbalance, which leads to accumulation of *sn*-glycerol-3-phosphate, since this substance is not oxidized under the abnormal conditions. When the metabolic block is removed by a return to normal temperature, or some other means, oxidation of the accumulated *sn*-glycerol-3-phosphate occurs at a high rate and leads to a sudden drop in its concentration. This in turn causes release of a subunit from the mitochondrial oxidizing enzyme, glycerol-3-phosphate oxidase. This released subunit induces five chromosome puffs on the salivary gland polytene chromosomes (see Chapter 10), indicating transcription of mRNA coding for the heat shock proteins, and about one hour later the five new proteins appear in the cytoplasm. Three of these have been identified: NADH dehydrogenase, NAD^+-dependent isocitrate dehydrogenase and tyrosine aminotransferase. All three become bound to the mitochondria and carry out their activities at that site.

As we will see in Chapter 12, the structural genes coding for the heat shock proteins are each situated adjacent to a characteristic inverted repeat sequence of nucleotides. It is a fair assumption that transcription of these genes occurs following some kind of interaction between the subunit released from glycerol-3-phosphate oxidase and the adjacent common sequences, although this has not yet been directly demonstrated. The heat shock response provides a dramatic illustration of the intimate interrelationship between nuclear and cytoplasmic activities.

A quite remarkable recent discovery is that the sex of alligators, tortoises and probably many other reptiles is determined by the temperature of incubation of the egg, which depends partly on its position in the nest. One of the most recent and plausible explanations for the extinction of the magnificent dinosaurs, that were such a dramatic feature of life in the Mesozoic era around 250 million years ago, is that a slight change in environmental temperature may have produced a dearth of one sex or the other. If these species were also dependent upon this one factor for the establishment of their sexual dimorphism, all could have become extinct within a few generations due merely to a small change in climate.

In those vertebrates with stabilized body temperature, the role of sex determination is taken over by the genome. It is probably a general rule that the functions of physical cues from the environment, which are used by lower species to guide development, become subsumed by the genome in more highly evolved forms (see Chapter 14). For example, in Chapter 2 we saw how in amphibia

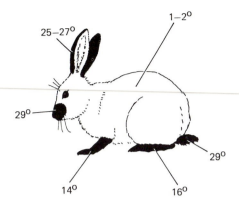

Figure 4.5. Distribution of temperature thresholds for pigment synthesis in the skin of the Himalayan rabbit.
Temperatures indicated are in degrees centigrade. (Redrawn from Schmalhauser, 1949.)

gravity partitions the egg components, creating animal and vegetal poles which are important reference points for morphogenesis. Gravity is unlikely to have such a significant role in mammals since the internally fertilized ova would be constantly subject to re-orientation. However, even the most highly evolved vertebrates receive important developmental instructions from physical features of the environment. In some vertebrates, notably fish, amphibians and mice, the number of vertebrae in the spinal column depends on environmental temperature. In Siamese cats, and rabbits and mice with the same 'Himalayan' pattern of colouring (Figure 4.5), there is a temperature-sensitive gene which actually operates at different temperature thresholds in the different regions of the body to produce the characteristic pattern of dark extremities (see Chapter 7). Reproductive activity in probably the majority of animals is directed in some still mysterious manner by sunlight acting through the eyes and endocrine organs, such as the pineal gland, on the organs of reproduction.

Phenocopies

Genes operate predictably only under standard conditions, which may include factors such as temperature, food supply and many others. If environmental conditions become changed to a marked degree during development, an abnormal phenotype can result which resembles the sort of organism that would have been produced if a gene had undergone mutation. Developmental abnormalities produced in this way, which resemble forms associated with known mutant genes, are called phenocopies.

If larvae of some strains of the two-winged fly *Drosophila melanogaster* are exposed to a temperature of 40 °C in the first four hours of embryogenesis, some develop an abnormal thorax with four wings, resembling the effect of a known

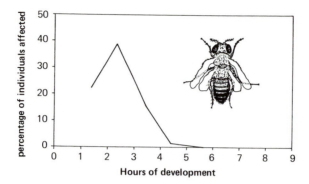

Figure 4.6. Temperature-induced phenocopy of *tetraptera* in *Dropsophila melanogaster*. *Dropsophila melanogaster* flies of a particular strain developed a double mesothorax phenotype when transferred from the normal temperature regime of 25°C to 40°C for a short period during their development. The sensitive period is in the first four hours. This four-winged phenotype resembles that produced by the mutant allele *tetraptera*. The wings are not properly expanded as the fly was dissected out of the pupal case. (Redrawn from Hadorn, 1961.)

mutant allele called *tetraptera* (see Chapter 5). The sensitive period for this effect is shown in Figure 4.6. A more familiar example of a phenocopy is the stunted arms and legs (phocomelia) of human babies born to mothers who had taken the drug thalidomide during pregnancy (Figure 4.7).

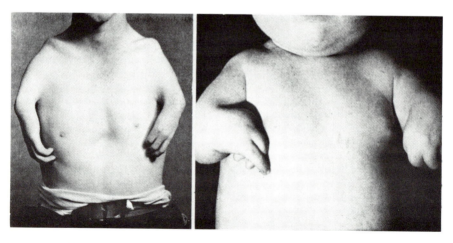

Figure 4.7. Phocomelia in man caused by genetic and environmental factors.
On the left is an inherited defect called phocomelia. A similar defect on the right is caused by exposure to the drug thalidomide at 39–44 days gestation. (Reproduced from Ursprung 1965, with permission.)

4.2. Genetic assimilation

The phenotypic expression of genes is labile not only to environmental factors but may also be influenced by substituting other genes in the background genotype. Thus a specific allele may have no noticeable phenotypic effect when working against most genetic backgrounds, but if the background genotype is selected to favour expression of the allele it can instead exert a major effect. Selection of a background genotype that favours expression of an otherwise hidden allele can come about by formation of a phenocopy promoted by the same genetic background.

Experiments to demonstrate this effect were carried out by Waddington, who exposed *Drosophila* pupae to a 40°C temperature shock 17–23 hours after puparium formation. When the flies emerged, a small proportion were found to have a break in the posterior cross-veins of their wings (Figure 4.8). Artificial selection was then applied by breeding from only those flies that had broken cross-veins. Waddington found that the frequency of individuals who developed the defect following the same treatment increased in the next generation and continued to increase in subsequent generations if the same conditions were applied (Figure 4.8). When selection was applied *against* the appearance of the abnormal phenotype, by breeding only from those flies with complete cross-veins, the defect decreased in frequency. Eventually the strain of flies selected for upward expression of the abnormality gave rise to the defect in well over 90% of treated individuals. At about this stage it was observed that the cross-veinless phenotype was appearing in this stock even when the selected flies were given no temperature shock. By selective breeding from flies which were cross-veinless when raised at normal temperatures, stocks were developed which showed the abnormal phenotype at high frequency without treatment. In one line of flies, close to 100% of

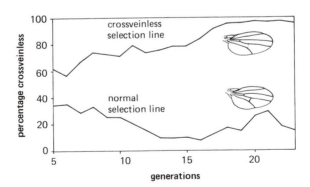

Figure 4.8. The results of selection of *Drosophila melanogaster* for and against the ability to respond to temperature shock by formation of a broken posterior cross-vein.
Pupae were exposed to a 40°C environment for four hours at 21–23 hours after the beginning of pupation. Flies were bred for and against the appearance of the cross-veinless phenotype. (After Waddington, 1975.)

individuals showed the defect, even when raised at normal temperatures. This effect, in which the cause of a phenotypic feature initially due to an environmental factor has been taken over by the genome, is called genetic assimilation.

The explanation for this strange phenomenon is that the original fly population contained hidden genetic variation concerned with the development of the cross-veinless phenotype, which was revealed only under the abnormal conditions of temperature shock. In the line selected for high expression, the character which had originally been an acquired one, exhibited only under abnormal conditions became converted through reorganization and selection within the same population into an inherited character that developed in the normal environment.

In another experiment, *Drosophila* larvae were raised in medium with a high content of sodium chloride. In this case *natural* selection produced a strain of flies whose pupae were more tolerant of high salt concentrations and which developed large anal papillae, even when raised in standard medium. The anal papillae are involved in regulating the osmotic pressure of the body fluids, which becomes disrupted when organisms consume an excess of salt.

4.3. Evolution and development

The concept of genetic assimilation is of profound importance in understanding how pre-adapted individuals can develop in the absence of stimuli for adaptation. In our own species, newborn babies have thickened skin on the soles of their feet, as if they had already had the experience of walking on hard surfaces. Flexion creases are already present on the inner surfaces of the fingers and toes, and this occurs even on digits with fused joints that could never have bent! Embryo wart hogs have callosities on their elbows like their parents who lean on their elbow joints when digging for food. It was the existence of callosities in embryonic ostrich chicks at precisely the correct places to cushion them in adult life that led Waddington to the *Drosophila* experiments, but the need for such a theory had long been recognized.

Estimates of the rates of random substitution of nucleotides in mammalian DNA yield average figures of around 1% every 10 000 000 years, although many of these changes are incompatible with biological function. The average rate of acceptable replacement of amino acids in the haemoglobins is about one in every 7 000 000 years (see Figure 11.6). Despite this very stable situation at the molecular level, animals as highly specialized as the blue whale, the giraffe, the bat and man have evolved from a common ancestral mammal which lived only 70–80 million years ago! A grossly simplified version of the conventional explanation of evolutionary advance is that random genetic mutations produce modified phenotypes, some of which are more suited to survival, so that favourable mutant alleles are passed on in increased numbers to the next generation. Clearly the time-scale required for evolution by this means is very much longer than that actually observed.

If you compare animals of different evolutionary backgrounds that have opted

Shark

Ichthyosaur

Dolphin

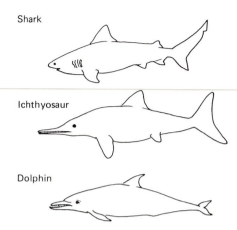

Figure 4.9. Convergent evolution of body forms in large marine animals.
The diagram illustrates how species from widely separate evolutionary backgrounds can evolve similar characteristics when living in a similar environment. The species are representative of cartilagenous fish, reptiles and mammals.

for similar life-styles, it is quite astonishing how their evolutionary paths have converged on similar forms. In some cases it is almost as if the environment itself has moulded them for that particular ecological niche. For example, compare the sharks, the extinct aquatic reptile, *Ichthyosaurus*, and the dolphin (Figure 4.9). They have similar ways of life, similar torpedo-like forms, similar vertebral articulation, and a variety of other common anatomical features, such as their numerous teeth which are an adaptation to their diet of fish. The insect world also contains some truly remarkable examples of convergent evolution, for example, in the mimicry by palatable species of prominent features of distasteful or dangerous neighbours, and in the adoption of cryptic colouration and camouflaging shapes (Figure 4.10). It is difficult to believe that such modifications could have occurred in the time available just by random mutation and random reassortment of favourable alleles, even allowing for natural selection weeding out the conspicuous weaklings.

Waddington's experiments show that adaptation to abnormal stresses can bring into play hidden variant alleles which are not expressed under normal conditions. Natural selection of well-adapted individuals then causes the relative frequency of these adaptive alleles to increase in the breeding population, until individuals arise who carry such a complement of suitable modifying alleles that they express the adaptive phenotype even before those stresses have operated on that individual. Thus, through the processes of natural selection, adaptable ancestors give rise to pre-adapted descendants.

Waddington's theory of genetic assimilation is not well known and still awaits complete acceptance by the scientific community, but it provides one of

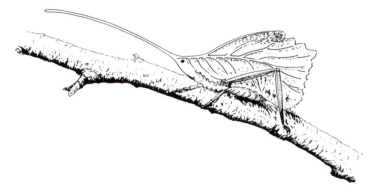

Figure 4.10. An example of camouflage in an insect.
The long-horned grasshoppers have evolved body forms which very closely resemble leaves, enabling them to escape from their predators. This is just one of a multitude of examples of body camouflage. (After Wigglesworth, 1964.)

several necessary complements to Darwin's theory of evolution. If offers a genetic explanation for the remarkably rapid capacity of species to develop forms and physiologies appropriate to the stresses of their particular ways of life, and reveals how organisms can arise which show functional grace and integral harmony between their various parts. Some of the ways in which the genes controlling body form operate at a developmental level are described in Chapters 5 and 6. The implications of this work are discussed more extensively in Chapter 13.

PART II. CELLULAR FACTORS

4.4. *Properties of the cytoplasm*

Unfertilized frog eggs can be activated to undergo parthenogenetic development into essentially normal individuals, merely by pricking with a needle or by exposure to chemical stimuli. Surprisingly, in frogs and probably many other species, neither the nuclear nor the cytoplasmic contributions of the sperm are essential for normal development. The functions of the sperm include activation of the egg and introduction of new genetic information, which can be reassorted within the new individual, so providing scope for novel genetic combinations in the next generation. Another function is to provide a genetic mask for deleterious recessive alleles that may be present in the ovum, but this seems to be a relatively minor feature. In contrast, extra-nuclear cytoplasmic contributions from the maternal system are absolutely essential, together with a haploid set of nuclear material from one or the other parent. Cytoplasmic components are of several types, some merely act as sources of energy, or as raw materials for growth, but others have a much more specific role. The latter includes cytoplasmic polynucleotides (i.e., DNA and RNA), preformed structures and preformed patterns.

Cytoplasmic RNA

As suggested in Chapter 2, maternally derived RNA is probably one of the important constituents of morphogenetic gradients in the egg. In *Xenopus* oocytes 95% of the RNA is ribosomal, 4% is transfer RNA and only 1% is messenger, with the potential for transmitting real genetic information. The mRNA is not translated before fertilization because although it does carry ribosomes, these are blocked by protein. Polyribosomes (or polysomes) which are blocked in this way are known as informosomes, and are converted into active polysomes at fertilization or activation of the egg.

If nuclei from differentiated amphibian cells at the tadpole stage are transferred into enucleated activated eggs, the egg cytoplasm sets processes in motion which cause expression of nuclear genes that were not formerly expressed. This experiment reveals the presence of gene-switching molecules in the egg, which are also presumably essential constituents of the maternally derived cytoplasm (see Chapter 7).

A variety of experiments with cell hybrids produced in the laboratory by fusing dissimilar cells together (Chapter 7) and the observation that the nuclei are synchronized in natural syncytia, like muscle cells, strengthen the general theory that nuclear activity is controlled by extra-nuclear cytoplasmic factors. In turn, the composition of the cytoplasm depends upon the activity of the nucleus, but it appears that in most cases the cytoplasm is the dominant partner.

Cytoplasmic DNA

It will probably be a surprise to most readers to learn that 99% of the DNA in a *Xenopus* oocyte is cytoplasmic, not nuclear. In mammals also some 90% of the DNA in an ovum is extra-chromosomal. Cytoplasmic DNA exists in several forms: as viruses, as small loops like the plasmids found in bacteria and possibly in yolk platelets, but the majority is in the mitochondria.

Mitochondrial DNA resembles that of bacteria in lacking attached proteins and existing as a single closed loop (Chapter 1). It is translated by special mitochondrial ribosomes and mitochondrial transfer RNA, but several enzymes that are functional only in mitochondria are coded in the nucleus. The degree of independence, or inter-dependence of mitochondrial and nuclear genomes is still not fully understood, but there are well-described differences in the genetic codes utilized by the two systems (see Chapter 8).

In Chapter 2 we considered some of the experiments that gave rise to the concepts of 'mosaic' and 'regulative' development. Electron-microscopic examination reveals that the distinction between organisms that develop according to these alternative plans coincides with the relative distribution of mitochondria at cleavage. In regulative embryos, such as amphibia, the mitochondria are distributed fairly evenly between the blastomeres, but in mosaic embryos it is far from uniform. At the first cleavage division of the ascidian, *Ciona intestinalis*, most of the visible mitochondria and 70% of mitochondrial enzyme activity are segregated into the half of the embryo that gives rise to neural structures and muscle somites, while the yolk-enriched cells derived from the other partner develop into endoderm.

4.5. *The role of mitochondria in cytodifferentiation*

Quantitative differences in mitochondrial numbers can also arise later in development. For example, during maturation of lens epithelial cells into fibres (see Chapter 3) mitochondria are ejected and the cells change their metabolism from aerobic to anaerobic. This change is accompanied by a change in the pattern of protein synthesis (see Chapter 7). Some types of tissue differentiation involve an increase in numbers of mitochondria. A 50-fold increase occurs during the formation of adult muscles in locusts.

In addition to their numerical differences, the activity and structure of mitochondria also vary greatly between tissues. Their differentiation is most obvious during organogenesis and other critical stages, such as moulting and hatching of insects and birth of mammals. Maturation and differentiation of activity of the mitochondria occur in response to changes in both intra-cellular and extra-cellular conditions, some of which are described below. Since they play such a vital role in energy metabolism, even a slight modification in their activity can have far-reaching consequences for the cell, controlling a very large proportion of its activities.

Mitochondria begin to undergo oxidative phosphorylation soon after activation of the egg. In sea urchins activation involves degradation of a small molecule that inhibits cytochrome oxidase, the terminal enzyme in the respiratory chain. Within three to six minutes, oxygen uptake increases two- to four-fold and the ATP/ADP ratio increases, heralding the onset of all the many life processes that depend upon these ubiquitous molecules. Thereafter oxidative metabolism is controlled by the relative concentrations of ATP and ADP, by hormones (see below) and by a variety of other conditions.

In view of the enormous importance of mitochondria to practically every biochemical process in most eukaryotes, it is astonishing that so little attention has been paid to their role in development. The heat shock response described above illustrates how cytoplasmic and nuclear factors are integrated with respect to gene expression. In addition, mitochondria have another quite distinct function relating to the metabolism of calcium, which is also of great significance in the regulation of many activities in the cell.

4.6. *Preformed and spontaneously assembling structures*

Some components of the cytoplasm cannot be regenerated if they are lost from the cell. These are acquired initially from the cytoplasm of the maternal gamete and include the mitochondria, centrioles, ciliary basal bodies and, in plants, the chloroplasts and other plastids. It was pointed out in Chapter 1 that these organelles may well have arisen millions of years ago from prokaryote endo-symbionts and they still maintain a considerable degree of autonomy. Although mitochondria contain DNA which codes for mitochondrial functions and many of their structural components and enzymes are coded on nuclear genes, neither they nor the nucleus carry instructions for their assembly. The component parts of mitochondria can become incorporated only into the structure of growing mitochondria, which multiply by fission.

Less elaborate preformed structural elements are also present in the cytoplasm, such as the glucose polymer, glycogen, which acts as an essential primer for synthesis of more glycogen, a piece at least 4 units long being necessary for this to occur.

Experiments with *Paramecium* reveal that this animal's cortex maintains continuity of structure from generation to generation without direction from either nucleus or cytoplasm. This genus reproduces by binary fission, which can occur in any of three situations: (1) following conjugation, a process analogous to mating, when nuclear material and sometimes a small amount cytoplasm is exchanged between two individuals; (2) after a type of self-fertilization known as autogamy; or (3) by a process analogous to mitosis. *Paramecium* carries cilia arranged along its length in rows called kineties, which differ in number between individuals (Figure 4.11). If the number of kineties per individual were defined by nuclear genes, we would expect them to be inherited according to Mendelian rules, and if they were controlled by cytoplasmic factors, numbers should be inherited maternally, although incorporation of a partner's cytoplasm at conjugation might produce detectable effects. What is found is that an individual with a given number of kineties always produces progeny with the same number, regardless of the method of reproduction or incorporation of co-conjugant's cytoplasm. In other words, inheritance of kinetie number is strictly maternal and probably occurs in this way since extension of the cortex occurs by building on to the previous structure, without further instruction, as in mitochondrial growth.

Sometimes conjugation occurs in a head-to-toe fashion, instead of the normal head-to-head, toe-to-toe situation, and occasionally when this occurs small pieces of cortex are transferred from one individual to the other. After the partners separate they can be recognized, because the cilia beat in the wrong direction in

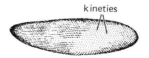

Figure 4.11. Kinetie distribution in *Paramecium*.
The diagram illustrates the dorsal surface of *Paramecium* and reveals the distribution of rows of kineties. The number of rows varies between individuals and is not determined by nuclear or cytoplasmic genes, although faithfully maintained in daughter cells.

the transferred patch, making the recipient swim in a spiral fashion. When these animals undergo fission their daughter cells also carry reversed patches of cilia and the same feature is continued indefinitely in all descendant cells. It seems that the number and arrangement of *Paramecium* kineties depends upon their previous pattern and is not dictated by nuclear or cytoplasmic genes.

The microtubules and microfilaments (see below), the phospholipid bilayer of the plasma membrane, the ribosomes and several other internal and external cell products have the remarkable property of assembling spontaneously from their individual subunits. The best understood class of self-assembling biological structures consists of those composed of the glycoprotein, collagen. This is described in Chapter 8.

4.7. *The extra-cellular matrix*

If you cut a thin section through almost any superficial part of the body of a vertebrate, mount it on a slide, stain the tissues and examine them under a high-powered microscope, you will usually see some well-defined cells, perhaps of several types. They usually have a clearly recognizable external border, a visible nucleus and enough sub-cellular detail to attract attention. But packed around and between them is an amorphous mass of material that looks about as interesting as the shredded paper and polystyrene chips that manufacturers use for cushioning expensive instruments in transit. This material is the extra-cellular matrix and despite appearances it promises to hold the key to many of the intractable problems of development that have plagued biologists for centuries. When clarified, extra-cellular matrix material forms the major component of the cornea of the eye, when calcified it solidifies as bone, when assembled into ropes it forms tendons and when spread out in sheets it makes the molecular filter of the kidney glomerulus and the basal laminae upon which epithelial cells spread and grow.

The matrix consists basically of a network of fibrous proteins embedded in a hydrated polysaccharide gel, but the composition of its constituents are so varied and its many structures so complex that we still know very little about them. What we do know is that the form and composition of the matrix represent an extra-cellular record of the metabolic conditions inside the cells which secreted it. The order which it records can be re-imposed on the cells, so stabilizing their phenotype, or imposed on other cells, so creating similar order in them. The extra-cellular matrix probably plays a critical role in the definition of 'positional information' (see Chapter 6), provides highways for migrating cells, and assembly points for cell aggregates such as those that initiate formation of dermal papillae (see Chapter 3). In the skin basal lamina it acts as a cell-selective filter, obstructing contact between fibroblasts below and epidermis above, while allowing lymphocytes and nerve endings to pass through. It probably also carries information for the differentiative 'programming' of several types of cells that contact it, such as those that migrate out from the neural crest (see Chapter 3).

The gel component of the matrix is a mixture of glycosaminoglycans (formerly called mucopolysaccharides) and proteoglycans (Figure 4.12). There are seven recognized groups of glycosaminoglycans, each composed of serial repeats of amino sugars and uronic acids. This forms a very hydrophilic molecule that takes in large volumes of water and swells very considerably, conferring turgor and filling any available spaces. Proteoglycan is glycosaminoglycan covalently linked to protein, a typical molecule carrying one glycosaminoglycan chain for every 12 amino acids along its length. In principle, these molecules have the potential for almost limitless chemical heterogeneity, and so have the potential to retain detailed records of cell metabolism, which may enable cells to recognize their whereabouts and act accordingly. One unusual type of glycosaminoglycan is hyaluronic acid. This is a very long chain of alternating glucuronic acid and *N*-acetylglucosamine residues. It seems to promote cell migration and is particularly important in wound healing.

Within the gel are fibres of collagen, elastin and in basal laminae, laminin. The collagen monomer is built up as a highly repetitive sequence of amino acids that enables individual monomers to assemble side by side into fibres of various

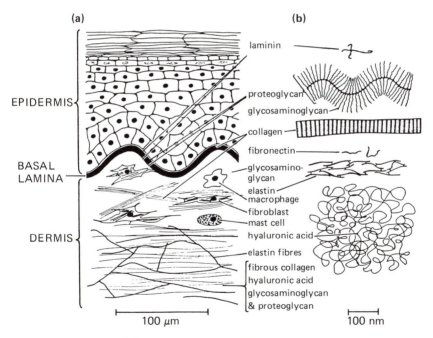

Figure 4.12. Extra-cellular matrix in mammalian skin.
(a) a vertical section through keratinized skin. (b) The molecular components of the extra-cellular matrix. Single molecules of laminin, fibronectin, proteoglycan and hyaluronic acid are illustrated. The collagen and elastin are represented as homopolymers. Proteoglycan consists of many molecules of glycosaminoglycan linked to a protein backbone.

dimensions (see Chapter 8). Elastin is a glycoprotein which adopts no regular secondary structure, but cross-links readily at two sites with neighbouring elastin molecules to form an elastic mesh. Laminin is another large glycoprotein. In epithelial basal laminae collagen forms the filling in a sandwich between laminin and proteoglycan molecules on each surface. These substances seem to be secreted by the epithelia adjacent to them, and all are sometimes present on the surfaces of cells where they probably play a part in cell–cell and cell–matrix recognition.

Another molecule called fibronectin is found on cell surfaces. This is a kind of molecular adaptor, with binding sites for cell membranes and collagen, as well as other extra-cellular molecules, such as heparin and fibrin, that are important in blood clotting at wound surfaces. Fibronectin is particularly interesting since it binds to the same sites on the external surface of the plasma membrane as are bound on the inside by the actin microfilaments of the cytoskeleton. These are like elastic thongs which exert tension across the cell, in opposition to the microtubules of tubulin that push outwards from the centrioles like miniature telescopic tent-poles. Enmeshed within this sub-cellular scaffolding is a relatively vast area of folded plasma membrane carrying enzymes and ribosomes and enclosing tiny pools of metabolites at high concentrations. There seems to be a two-way exchange of instructions between the intra-cellular actin microfilaments and the extra-cellular strands of fibronectin and collagen that line up parallel to them. Intra-cellular order is thus transmitted to the extra-cellular matrix through the orientation of the macromolecules the cells secrete. By the operation of the reverse procedure, the matrix imposes internal order on its adjacent cells. This is one way in which intra-cellular order may be propagated from cell to cell in the establishment of homogeneity of function within tissues. It is also a means by which a state of determination may be imposed on a cell by its external environment.

4.8. Cell junctions

Cells can also control one another's internal structure through the disposition of inter-cellular adhesion sites, which are called desmosomes. Some of these are linked to elements of the cytoskeleton, so that adjacent cells have similar internal structure (Figure 4.13).

A more potent influence is by the direct feeding in of molecules which may be capable of exerting effects on cell metabolism, or even of altering the recipient cell's pattern of RNA transcription. Inter-cellular transfer of molecules is accomplished through the agency of minute tubes within what are known as gap junctions or junctional complexes (Figure 4.14). Molecules smaller than 1500 Da, such as sugars, amino acids and nucleotides, can pass freely between cells through gap junctions, although RNA and proteins are too large to be transported in this way. It is highly probable that the metabolism of cells within tissues is co-ordinated by this means and the gap junctions may also provide a route by which the morphogenetic constituents of pattern-forming gradients pass throughout organs during their development and regeneration (see Chapter 5).

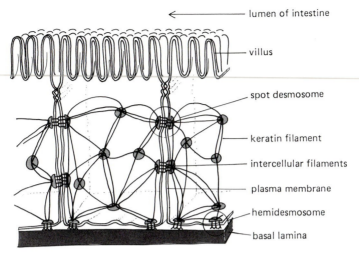

Figure 4.13. Inter-relationship between desmosomes and intra-cellular structure in intestinal epithelium.
The distribution of keratin filaments depends on that of the spot desmosome cytoplasmic plaques, which are linked to those on adjacent cells. The arrangement of filaments is therefore similar in adjacent cells. (Redrawn from Alberts *et al.*, 1983.)

The existence of junctional links can be established by detection of electrical conductivity between cell interiors. Most cells are electrically coupled in the early embryo, but uncouple selectively as development proceeds. This occurs, for example, in amphibians at closure of the neural tube, when the presumptive neural cells become isolated from the surrounding skin ectoderm. At an earlier stage the cells of the neural plate have different membrane potentials along its length. This means that an electric current must presumably be flowing along the neural plate in association with gradients in the concentrations of ions in the cells. Either this current, or the ion gradients, or both, are almost certainly concerned with early events in embryogenesis, but we know next to nothing about how they might operate. It would be interesting to know what would happen if the current were reversed.

The proteinaceous tubes that pass through the gap junctions are called connexons (Figure 4.14). Each of these is composed of six subunits and they link up in register with those on adjacent cells to establish communication of cell contents, while still maintaining a gap of 2–4 nm between the plasma membranes. This gap distinguishes them from other types of inter-cellular linkages such as the desmosomes. Gap junctions can form in minutes, by assembly of preformed connexon subunits, and break communication in seconds.

Intra-cellular p_H is one condition that controls permeability, another is the intra-cellular concentration of free Ca^{2+} ions. As we will see below, the calcium

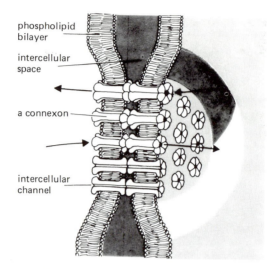

phospholipid
bilayer

intercellular
space

a connexon

intercellular
channel

Figure 4.14. An inter-cellular junctional complex.
A junctional complex is represented in section, revealing the inter-cellular channel which runs through the centre of each connexon. Connexons are each constructed of six subunits. (Based on Rogers, 1983.)

ion is coming to be recognized as having an extremely important role in cell and tissue metabolism.

Vitamin A and its derivatives have particularly interesting effects on animal development, affecting the differentiated states of epidermal cells (see Chapter 3) and morphogenesis of the limb (see Chapter 6). At least one derivative, retinoic acid, causes gap junctions to close and one wonders if this may be one way in which these biochemicals exert their effects (see below).

PART III. ORGANISMIC FACTORS

4.9. *Communication between distant cells*

With evolutionary increase in body size it became necessary for systems to develop for co-ordination of physiology in distant parts of the body. Movement towards or away from stimuli required not only the development of sense organs for their appreciation, but also some kind of system of communication between the sense organs and the organs concerned with locomotion, defence and feeding. Sexual reproduction requires co-ordinated behaviour between individuals, while the acquisition of sexual maturity and certain other events, such as birth and metamorphosis from one state of development to the next, require profound physiological changes to occur in harmony. Two major systems evolved to meet these requirements: the nervous system and the endocrine system. The principles on which they operate are basically similar, both involve secretion by one cell type of a biochemical that exerts an effect on the metabolism of another cell. But whereas endocrine cells secrete their hormones into the blood stream, which carries them to their targets, the nervous system is highly specialized for very localized chemical stimulation by elongated cells that receive their stimulus at the opposite end. The concentration of the effector biochemical at the target cell is thus very different in the two situations. Whereas specificity in response to hormones requires a cell-surface display of hormone-specific receptor proteins, in the case of nerve-stimulated events the target is identified by the pattern of innervation. Some biochemicals, such as epinephrine (or adrenalin, see Chapter 3) act both as neural transmitters and hormones and the two functions are integrated in a complex fashion.

Each endocrine organ secretes a different hormone into the blood and some secrete several (see Table 4.1). For example, the pituitary gland produces somatotropin (growth hormone), adrenocorticotrophic hormone (ACTH), follicle-stimulating hormone (FSH), luteinizing hormone (LH), thyroid-stimulating hormone (TSH) and vasopressin. However, different hormones may produce similar effects in any one cell type. For example, epinephrine, ACTH, glucagon and TSH all cause breakdown of triglycerides to fatty acids in adipose tissue. This is because, having bound to their specific receptors on the surfaces of adipose cells, all four operate within the cells in a similar way.

Amphibian metamorphosis

Conversely, different tissue types may respond differentially to a single hormonal stimulus. For example, metamorphosis of tadpoles into frogs involves rapid and profound changes in both body form and metabolism, which all occur in reponse to a 10-fold increase in the thyroid hormone thyroxine (see Figures 4.15 to 4.19). This acts in opposition to another hormone, prolactin, which, like juvenile hormone in insects sustains larval characteristics and promotes their growth. When thyroxine levels rise the tail, horny teeth and gills are quickly resorbed. The eyes shift from the sides to the top of the head. The larval visual pigment, porphyropsin, in the retina is replaced by rhodopsin and adult haemoglobin appears in the blood, due to regression of one type of erythropoietic tissue and proliferation of another (see Chapter 11). There are also profound changes in the respiratory and digestive systems as the animal changes from an aquatic herbivore to a terrestrial carnivore. Meanwhile new biochemical pathways become mobilized in the liver for conversion of the larval waste product ammonia into urea, which is more suitable for excretion by terrestrial adults.

A remarkable feature of amphibian metamorphosis is the variety of reponses to the same stimulus. In the tail there is wholesale destruction of tissues, destruction of epidermal collagen and regression of growth, so that the whole tail is eventually resorbed into the body. This is an example of what is known as 'programmed cell

Table 4.1. Some hormones and other molecules used for information transfer in animals.

Hormone or other molecule	Function	Source
VERTEBRATES		
Amino acids and their derivatives		
Histamine	Local mediator	Mast cells
Glycine	Neurotransmitter	Nerve terminals
Noreprinephrine	Neutotransmitter	Nerve terminals
Dopamine	Neurotransmitter	Nerve terminals
γ-Aminobutyric acid (GABA)	Neurotransmitter	Nerve terminals
Acetylcholine	Neurotransmitter	Nerve terminals
Thyroxine	Hormone	Thyroid
Epinephrine	Neurotransmitter and hormone	Adrenal medulla
Small peptides		
Eosinophil chemotactic factor	Local mediator	Mast cells
Encephalin	Neurotransmitter	Nerve terminals
TSH-releasing factor	Hormone	Hypothalamus
LH-releasing factor	Hormone	Anterior pituitary
Vasopressin	Hormone	Posterior pituitary
Somatostatin	Hormone	Hypothalamus
Glucagon	Hormone	Alpha cells of pancreas
Proteins and glycoproteins		
Insulin	Hormone	Beta cells of pancreas
Somatotropin, growth hormone	Hormone	Anterior pituitary
Adrenocorticotrophic hormone (ACTH)	Hormone	Anterior pituitary
Follicle stimulating hormone (FSH)	Hormone	Anterior pituitary
Luteinizing hormone (LH)	Hormone	Anterior pituitary
Thyroid stimulating hormone (TSH)	Hormone	Anterior pituitary
Prolactin	Hormone	Anterior pituitary
Somatomedins	Hormone	Liver
Erythropoietin	Hormone	Metanephric kidney
Parathormone	Hormone	Parathyroid
Epidermal growth factor (EGF)	Hormone	Unknown
Nerve growth factor (NGF)	Local mediator	Tissues innervated by sympathetic nerves
Fatty acid derivatives		
Prostaglandin E_2	Local mediator	Many different cell types
Steroids		
Cortisol	Hormone	Adrenal cortex
Estradiol	Hormone	Ovary, placenta
Testosterone	Hormone	Testis
INSECTS		
Fatty acid derivatives		
Juvenile hormone (JH)	Hormone	Corpus allatum
Steroids		
Ecdysone, moulting hormone (MH)	Hormone	Prothoracic gland

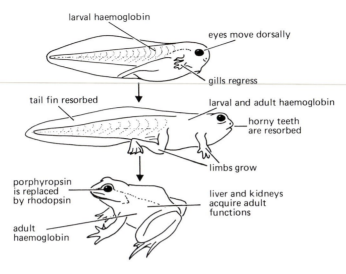

Figure 4.15. Action of thyroid hormone in amphibian metamorphosis.
The diagram illustrates the co-ordinated set of morphological changes that take place as a tadpole metamorphoses into a frog. These changes depend on the increasing ratio of thyroxine to prolactin.

death', a bizarre form of reversed growth that occurs in many very different situations during normal development. In contrast, in the tadpole trunk, collagen synthesis is promoted and the limb buds are stimulated to undergo rapid growth and differentiation. This diversity of responses to the same stimulus is another example of differential competence (see Chapter 3).

In one experiment which demonstrates differential sensitivity to thyroxine, a tadpole eye was grafted into the tail fin of another animal. At metamorphosis, when the tail regressed the eye was unharmed and merely moved backwards, ending up on the animal's rump.

Another interesting feature of amphibian metamorphosis is its dependence on the environment. Iodine is an essential constituent of thyroxine and in geographical regions where there is no iodine in the water, or where low environmental temperatures inhibit the thyroxine receptor, tadpoles can grow larger and larger without entering metamorphosis (see also Figure 4.18). Some species of the Mexican salamander, the axolotl, even reproduce in the larval state, never reaching the adult form unless they are given iodine or thyroxine (see Chapter 13).

Lipophilic hormones

The products of the endocrine glands, the hormones, can be divided into two major classes, lipophilic and hydrophilic. The former include thyroxine and the steroid hormones, oestrogen, progesterone, testosterone and ecdysone. Being

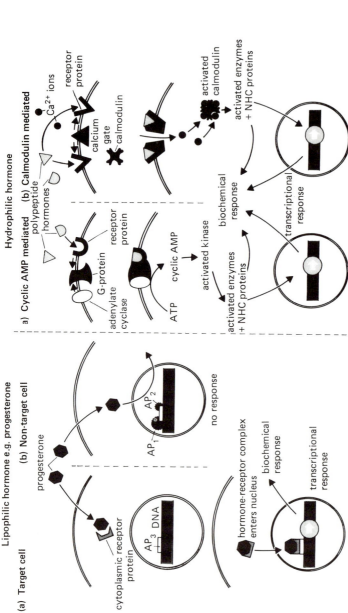

Figure 4.16. Mode of action of hormones.

(a) Lipophilic hormones, such as the steroids, enter the cytoplasm. In target cells they bind to a cytoplasmic receptor protein, enter the nucleus and bind to non-histone chromosomal (NHC) proteins (see Chapter 10), which are themselves bound to specific sites on the DNA. This initiates a transcriptional response. Non-target cells lack cytoplasmic receptors and the DNA-bound protein is masked by other NHC proteins. The diagram illustrates the progesterone system in mammalian oviduct cells. (b) Hydrophilic hormones operate through second messenger systems. In the response mediated by cyclic AMP, the hormone binds to a membrane-bound receptor which activates G-protein, that in turn activates membrane-bound adenylate cyclase. The latter converts ATP to cyclic AMP. This allosterically activates protein kinases which phosphorylate effector enzymes and NHC proteins. These can produce both physiological and transcriptional responses. In calmodulin-mediated responses, binding of hormone to surface receptors operates a 'calcium gate' in the membrane. Ca^{2+} ions then enter the cytoplasm and bind to calmodulin which, in the modified form it then adopts, activates effector enzymes and NHC proteins.

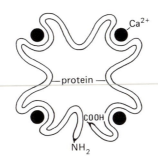

Figure 4.17. The calmodulin molecule.
This protein has four very similar domains, each with a site that binds a calcium (Ca^{2+}) ion. The three-dimensional structure of the molecule depends on the number of Ca^{2+} ions bound. (After Cheung, 1982.)

small and lipid-soluble, these molecules pass through plasma membranes and into the cell cytoplasm.

Most of our knowledge of the action of steroid hormones comes from work with progesterone and oestrogen in the hen. Target cells responsive to progesterone, such as those of the oviduct, are distinguished from non-target cells by the presence of a specific receptor protein in the cytoplasm, a typical target cell containing perhaps 10 000 receptor molecules. Another feature is accessibility of a non-histone chromosomal protein called AP_3, bound to the DNA (see Figure 4.16 and Chapter 10). In non-target cells AP_3 is masked by two other non-histone proteins, AP_1 and AP_2. The hormone–receptor complex will not bind to AP_3 or DNA in isolation, but only to the DNA–AP_3 complex. In this system, therefore, competence to respond to progesterone requires the simultaneous presence of the appropriate receptor protein in the cytoplasm and AP_3 in the nucleus, but absence of AP_1 and AP_2. Tissue specificity in response probably also requires some additional 'priming' of certain gene sequences, such as demethylation of cytosine residues and modification of histones in that region (see Chapters 10 and 12).

Binding of steroid hormone plus its specific receptor to gene sequences is believed to regulate their transcription (see Chapter 12). Evidence for this has been provided in several ways, but perhaps the most convincing is the observation that radioactively labelled insect moulting hormone, ecdysone, binds to sites on polytene chromosomes which puff on exposure to the hormone (see Chapter 9). These sites vary with developmental stage and one site regressed when it bound ecdosone, showing that steroid hormones can switch transcription off as well as on.

A point that is often not appreciated is that the cessation of a hormone-induced response is as critical as its activation. When extra-cellular hormone levels decline, the cytoplasmic hormone–receptor complex dissociates, DNA-bound complexes are released into the cytoplasm and transcription of the hormone specific sequences is switched off.

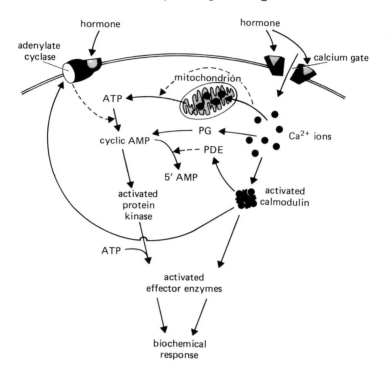

Figure 4.18. Integration of cytoplasmic control systems.
The diagram illustrates interactions that occur between the cyclic AMP and calmodulin systems. A protein or peptide hormone binds to its receptor on the cell surface and activates adenylate cyclase, which converts ATP into cyclic AMP. This intra-cellular messenger activates a protein kinase, which phosphorylates effector proteins either stimulating or inhibiting their biological activity. Another hydrophilic hormone activates a calcium gate in the membrane and Ca^{2+} ions enter the cytoplasm. The calcium of the cytoplasm is normally kept low due to sequestration by mitochondria. Calmodulin is activated by the Ca^{2+} ions and itself activates effector proteins to initiate a biological response. Calcium stimulates the activity of adenylate cyclase and also that of phosphodiesterase (PDE) which breaks down cyclic AMP. Calcium modulates the metabolism of prostaglandin, which also influences adenylate cyclase metabolism. Cyclic AMP may also influence availability of intra-cellular Ca^{2+} ions in other ways. ATP is produced largely by the mitochondria, its synthesis being negatively linked to Ca^{2+} uptake. (Based on Cheung, 1982.)

Hydrophilic hormones

Neural transmitters, the great majority of hormones and the tissue-specific 'growth factors' are hydrophilic and so do not easily pass through the plasma membrane. Instead they trigger responses in target cells by activating receptor

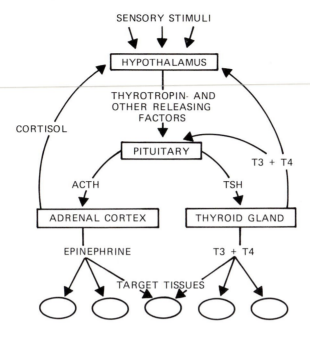

ACTH: Adrenocorticotrophic hormone
TSH: Thyrotropic stimulating hormone
T3: Thyroxine
T4: 3, 3', 5 — triiodothyronine

Figure 4.19. Hierarchical and feedback control of hormone action.
The hypothalamus secretes thyrotropin-releasing factor and other releasing factors in response to sensory stimuli. These act on the pituitary, causing it to produce a variety of polypeptide hormones, including TSH and ACTH. The adrenal cortex produces cortisol in response which inhibits further secretion by the hypothalamus, and epinephrine which acts on other target tissues. The thyroid gland secretes T3 and T4, which act on somatic target tissues and also exert negative feedback inhibition on pituitary and hypothalamus.

proteins embedded in the membrane itself. Apart from the growth factors, whose receptor proteins may themselves phosphorylate cytoplasmic proteins, all are believed to operate through second messenger systems (see Figure 4.16). In some cases the second messenger is the Ca^{2+} ion, in others a cyclic nucleotide called cyclic-AMP (cAMP). These also produce their effects by phosphorylating enzymes, so changing the activities of the enzymes.

Camp-mediated systems

In the cAMP-mediated response the sequence of events is as follows. Hormone binds to the external surfaces of receptor protein molecules embedded in the plasma membrane of the target cell. This alters the conformation of the receptor and enables it to link with and activate another membrane protein called G protein. In its active state, G protein binds GTP (guanosine triphosphate) on its cytoplasmic side and in turn activates a membrane-bound enzyme called adenylate cyclase, which converts ATP to cyclic AMP. The cyclic AMP in turn activates cytoplasmic enzymes called protein kinases. These catalyse the transfer of a phosphate group from other molecules of ATP to specific serine or threonine residues in other cytoplasmic proteins. In some cases enzymes activated by phosphorylation then carry out secondary modifications of additional proteins, so that a cascade of reactions is set up.

An important distinction between this mode of action and that of steroid hormones is that the primary impact is on cell physiology. It is assumed that histones, or genotropic non-histone proteins, become activated as part of this response and that these invade the nucleus to establish new patterns of transcription (see Chapter 10). However, *transcription seems not to be a necessary aspect of the action of hydrophilic hormones, although it is an integral feature of lipophilic hormone action.*

In general, enzymes that are activated by phosphorylation are concerned with degradative metabolism, while those that are inactivated operate in pathways of synthesis. However, the action of hormone is not always to increase cAMP production, in some cases exposure to hormone causes a drop in cAMP levels. There is also a cAMP-regulated system for destroying cAMP after its function has been achieved.

CA^{2+}-mediated systems

Calcium levels in the cytosol are regulated by several means, but one of the most important is by the hormonal regulation of 'gates' in the plasma membrane that control its entry (Figure 4.16). In nerve cells transient opening of a calcium gate causes a wave of depolarization of the membrane potential which is transmitted along the axon. In other cells there is a massive influx of Ca^{2+} ions, many of which become bound by a specialized protein called calmodulin that constitutes as much as 1% of the cytoplasmic protein of a typical cell (Figure 4.16).

The function of calmodulin is to regulate the activities of enzymes in yet another way. The enzymes which fall into this category become functional only when bound to a calmodulin molecule that is in a specific three-dimensional conformation. This depends on the number of Ca^{2+} ions bound, each molecule accommodating up to four ions. In this way calmodulin can convert quantitative differences in intra-cellular free calcium into qualitatively different cellular responses. One of the enzymes it regulates is adenylate cyclase, so forming a link between the two second messenger systems (Figure 4.18).

Those functions that operate through the Ca^{2+} ion second messenger system can do so only when calcium is scarce in the cell. Normally the extra-cellular fluids contain massive quantities, so a means is required for keeping down the levels in the cytosol. A very important intra-cellular calcium 'sink' is the mitochondrial matrix within which, it is thought, Ca^{2+} ions become locked up as insoluble calcium phosphate.

Sequestration of calcium in the mitochondria is interlinked with the metabolic activity required for the production of ATP, there being insufficient energy available to operate both ATP synthetase and the calcium-sequestrating pump at the same time. When calcium is present in profusion, coping with this emergency takes precedence. The mitochondria thus represent an important element in a very complex web of metabolic systems interrelating hormone action, energy metabolism and nuclear activity (Figure 4.18). It will be appreciated that the segregation of mitochondria at early cell divisions, as described above, must have profound implications for the cytodifferentiative control of metabolic activity and gene expression.

4.10. *Control of hormone action*

The endocrine system is largely under direct or indirect control by the nervous system. Changes in external and internal conditions initiate stimuli in the sense organs. Impulses then travel to the brain, where they make connections that activate suitable responses. Regulation of pituitary secretion occurs through the hypothalamus, the portion of the brain that lies immediately above the pituitary gland, to which it is directly linked by both blood vessels and nerves (Figure 4.19).

Another mechanism for control of hormone secretion is hormonal feedback, which can be either negative or positive in its mode of action. One of the products of the pituitary is ACTH, which increases cortisol secretion by the adrenal cortex. In turn cortisol inhibits secretion of ACTH (Figure 4.19). This is therefore an example of negative feedback. One type of positive feedback occurs during the birth of a baby, when the pituitary secretes another hormone, oxytocin, that acts on the uterus to increase its contractions. As the baby's head begins to dilate the neck of the uterus, nerve impulses pass from this region up the spinal cord to the brain. There they stimulate further oxytocin synthesis, which in turn increases the force of the uterine contractions. This causes even greater sensory stimulation and increased secretion of oxytocin. Eventually things reach a climax as the moment of birth approaches, then dies away with the expulsion of the baby from the mother's body.

It might be wondered why organisms have evolved such complex means for relaying extra-cellular signals into the target cell. There are probably two reasons for this. One is that the possibilities afforded for fine control are optimized by increasing the number of links in the chain. The other is that such a system facilitates enormous amplification of the original signal at the site where it is required to act. One of the disadvantages of hormonal communication is that dilution in the blood necessarily results in only a very small number of molecules reaching each target cell. If the hormone operates through the cyclic AMP second messenger system, the original signal can be amplified (a) by the activation of several G protein molecules, (b) by each activated G protein activating several adenylate cyclase molecules, (c) by each activated adenylate cyclase molecule generating many molecules of cyclic AMP, (d) by each cyclic AMP molecule causing the phosphorylation and hence activation (or inactivation) of many enzymes, and (e) by each enzyme acting on many molecules of substrate (Figure 4.20).

Another aspect of the amplification of hormonal signals is the existence of

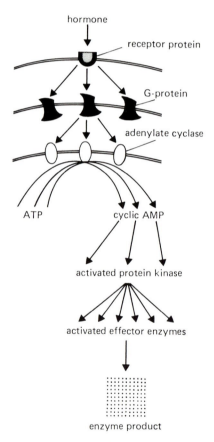

Figure 4.20. Amplification of hormone signals.
The diagram shows how the signal generated by a single molecule of polypeptide hormone is amplified within the cell. Each activated receptor protein can activate several molecules of G-protein, each of which can activate several adenylate cyclase molecules. Each of these can produce many molecules of cyclic AMP, but each cyclic AMP molecule activates only one molecule of protein kinase. Each protein kinase can phosphorylate many effector enzymes and each of these can produce many molecules of product. (After Alberts *et al.*, 1983.)

hierarchical systems of control between the different endocrine organs, as illustrated in Figure 4.19. It will be appreciated that systems like this allow enormous increase in the strength of the stimulus at the level of the target cell.

4.11. *The hormonal control of development*

The similarity between hormones acting at a distance from their source and embryonic inducers acting on proximal tissues (see Chapter 3) is underlined by evidence that many aspects of cytodifferentiation are promoted by hormones. We have already considered the effects of thyroxine on metamorphosis. In the oviduct, oestrogens are required for differentiation of the tubular gland cells, which later secrete ovalbumin under oestrogen stimulus. Another oviduct cell which differentiates in response to oestrogen is the goblet cell, which later secretes the protein avidin when exposed to progesterone.

Non-steroid hormones also control cytodifferentiation, one example being erythropoietin, a soluble glycoprotein produced by the adult kidney in response to erythrocyte deficiency. This stimulates both the proliferation of committed red cell precursors and their differentiation into mature erythrocytes.

An important feature of the action of hormones in development is that frequently two major hormones of contrary effect act in opposition and a gradual change in their relative concentrations causes events to be mobilized in ordered sequence. In amphibians, metamorphosis occurs as a response to secretion of high levels of thyroxine by the thyroid gland. During larval stages this response is kept in check by the larval hormone prolactin. (The reason it has this name is that in mammals it has the completely different function of activating milk secretion in the female after giving birth.) Insect development proceeds under a similar type of control. In this case the two hormones are juvenile hormone (JH) and moulting hormone (MH), or ecdysone. Insect larval epidermal cells exposed to ecdysone in the presence of large concentrations of JH remain in a larval state. When ecdysone is present with only a trace of JH the pupal form appears, but when exposed to ecdysone alone adult structures develop (Figure 4.21). Ecdysone also activates differentiation of the imaginal discs into adult structures (see Chapter 5).

Conditions sometimes prevent or delay the normal maturation process and this is the basis of some of the 'castes' that exist in insect societies (see below). The soldier termites with their Herculean jaws develop in this way in response to excessive juvenile hormone.

Stress due to crowding can sometimes sway the hormone balance and promote maturation. In some insects wing growth is triggered and individuals can escape to less crowded conditions. It is this event which is reponsible for the gigantic and terrifying swarms of migratory locusts of the tropics, but it also causes drastic bodily reorganization in our more homely pests, the aphids. As described earlier in this Chapter, aphids reproduce parthenogenetically when the day length is long. In uncrowded conditions they produce wingless females, but when crowded, their daughters develop wings and fly away to other food plants. As the days

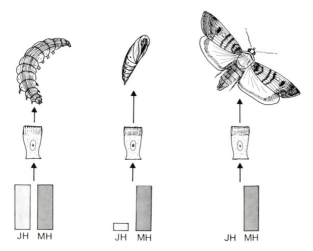

Figure 4.21. Hormonal basis of metamorphosis in insects.
The morphological stage adopted by insects following a moult depends on the relative concentrations of juvenile hormone and moulting hormone. When large concentrations of juvenile hormone are present epidermal cells differentiate into larval structures. A trace of juvenile hormone, together with a lot of moulting hormone causes pupation. Moults induced by moulting hormone in the absence of juvenile hormone result in epidermal cells differentiating into adult cell types. This is associated with differentiation of the imaginal discs (see Chapter 7). (After Wigglesworth, 1964.)

become shorter, sexual reproduction supervenes and the offspring which form are all of the wingless type (Figure 4.3).

4.12. Regulation of organ size

Harmonious integration of metabolism in the body requires that the various organs acquire balance with respect to one another's activities. This involves imposition of limits to growth of the different organs. One regulator of overall rate of body growth is the pituitary hormone, somatotropin, which acts particularly efficiently upon the chondrocytes that lay down the skeleton. Another growth hormone of general effect is thyroxine, while various tissues of the reproductive system respond specifically to the male and female sex hormones. In addition, there are the tissue-specific growth factors, such as epidermal growth factor (EGF) which promotes proliferation and maturation of that tissue specifically.

Inhibition of growth of specific organs is believed by some authorities to be carried out by soluble glycoproteins called chalones, produced by the organs upon which they exert their effects. Chalones were discovered through an experiment involving compensatory hypertrophy of the liver in a pair of rats whose blood systems had been artificially joined, so that the same blood flowed through both animals. When part of the liver of one partner was surgically removed, compensatory growth occurred and the bulk of liver tissue was restored. However, liver growth occurred not only in the hepatectomized rat, but in its partner as well. It was concluded that the balance between stimulatory and inhibitory factors specific to liver growth had become disrupted by hepatectomy, and the conditions in the hepatectomized rat had become communicated through the shared blood stream to its partner, so stimulating its liver to grow also.

It has been suggested that organs normally emit chalone into the blood, which then circulates back to the same organ and specifically inhibits its own growth. Presumably this control is concentration dependent, the blood volume acting as an indicator of total body size. If growth of all organs is controlled by such a system, it could ensure that all parts of the body grow and are maintained in a state of physiological and anatomical equilibrium in relation to total blood volume. This is a particularly interesting aspect of physiological regulation which has received little attention.

4.13. *Control of development by other organisms*

Developmental instructions are passed not only between parts of the body, but also between individuals. Substances which are transmitted between individuals for purposes of communication are called pheromones. Some of these have properties similar to those of hormones. In honey bees, for example, the distinction between a queen and a worker bee is imposed by feeding abundant food plus 'royal jelly' to future fertile queens throughout the larval period. Those larvae which will be raised as sterile female worker bees are given a poorer food supply. They receive royal jelly only in the early stages, then 'brood food', pollen and honey thereafter. Both brood food and royal jelly are a mixture of the secretions of the hypopharyngeal and mandibular glands of young adult female 'nurse' bees, but they probably represent different proportions of the two. Royal jelly contains a highly labile small molecule which seems to stimulate production of juvenile hormone and possibly other hormones that control development.

Queens develop functional ovaries, but workers normally do not. The infertility of the workers is due partly to their early feeding, but also to their consumption as adults of a secretion produced by the queen which suppresses ovarian maturation in her retinue. This queen substance is a mixture of fatty acids, mainly 9-ketodecanoic acid, plus a small amount of 9-hydroxydecanoic acid. If a honey bee colony is deprived of its queen, some of the workers develop functional ovaries and the special wax cells in which they raise queens are constructed. Both features are suppressed by queen substance. Queen substance is also a major component of the perfume used by the virgin queen to attract drones on her mating flight, and later in guiding the swarm when she takes the colony to a new home. Worker bees use pheromones of related biochemistry to provide a signal beacon that advertises the location of the new hive and there are alarm pheromones that communicate distress when the colony is disturbed.

In some cases of parasitism and symbiosis, transmission of biochemicals from one species to the other is required before development can proceed to its normal conclusion. For example, some blood-sucking insects such as the female of the yellow fever mosquito, *Aedes aegypti*, attain sexual maturity only after they have inbibed blood laced with the appropriate biochemical from the unfortunate on which they have fed. In European voles, oestrus in the female is triggered in the spring by oestrogenic substances consumed in growing wheat shoots. Many plants contain juvenile hormone and it is a moot point whether the insects have adopted a freely available food substance for control of their own physiology, whether the plants have responded to the depredations of insect attack by synthesizing a substance that prevents their attackers from breeding, or whether it is a coincidence of similar biochemistry in evolutionarily widely separated species.

4.14. *Control of differentiation by common metabolites*

Some aspects of differentiation seem to be controlled by extra-cellular substances that are too common to be classed as hormones. In Chapter 3 we saw that synthesis of mucus can be promoted in epidermal cells normally specialized for keratin synthesis, if they are exposed to vitamin A. Vitamin A (or retinol) and its derivatives, known as retinoids, exert several unusual effects on development. In Chapter 6 we will deal with the zone of polarizing activity, or ZPA, which emits signals in the limb bud that specify the order and positions of the digits. The action of the ZPA is mimicked by artificial implants that emit retinoic acid. According to some authorities, retinoids become complexed with specific cytoplasmic binding proteins and enter the nucleus in the manner of steroid hormones. Retinoids are also known to block gap junctions (see above), to modify glycoprotein synthesis and to release destructive enzymes from the cytoplasmic 'suicide capsules' called lysosomes, which play a major part in programmed cell death.

Limb mesenchyme contains a common pool of progenitor cells that give rise to both muscle and cartilage (and eventually bone). Commitment to one or other developmental pathway occurs in the chick wing at developmental stage 25, at about four and a half days' incubation. If the pre-myogenic and pre-chondrogenic tissues are exchanged before that stage, the wing will develop without mishap. However, if this exchange takes place after stage 25, cells from the chondrogenic zone differentiate as cartilage and those from the myogenic zone as muscle, despite their new situations. This is another example of the acquisition of determination (see Chapter 2). Work by Caplan and colleagues suggests that adoption of one or the other pathway of development depends on intra-cellular concentrations of nicotinamide adenine dinucleotide (NAD). If limb mesenchyme is incubated in medium containing nicotinamide at high concentration, it differentiates as muscle. If the concentration of nicotinamide in the medium is low, it forms cartilage instead. Definition of the bony and muscular elements of the digits therefore seems to derive from the action of a substance with properties similar to those of retinoic acid, emitted by the ZPA, which produces effects similar to those caused by differing levels of NAD, at positions corresponding to the future digits. In Chapter 6 we will review some of the experiments and theories that have been devised to explain the establishment of the intriguing patterns displayed by the pentadactyl vertebrate limb.

4.15. Summary and conclusions

Some authorities claim that the nuclear DNA contains all the information necessary to code for a complete organism, but many non-nuclear factors are also known to be essential for normal development. These include materials produced by the mother and accumulated in the egg, preformed structures and patterns, sub-cellular organelles, and external physical factors such as light and temperature. The distinction between mosaic and regulative development appears to some extent to be due to the relative partitioning of mitochondria at cleavage. The nature of the substratum and the identity of a cell's neighbours are also important in controlling cell phenotype.

The heat shock response in *Drosophila heidei* illustrates one way in which nuclear and cytoplasmic activities are integrated. As a result of metabolic changes related to mitochondrial activity, an enzyme subunit is released, enters the nucleus and stimulates transcription of specific structural genes coding for mitochondrial enzymes.

Some hormones perform functions similar to those of embryonic inducers, except that they act at a distance. Lipophilic hormones, such as steroids, pass through the plasma membrane into the cytoplasm of all cell types, but competent target cells contain cytoplasmic receptor proteins to which they become bound. The hormone–receptor complex then enters the nucleus, where it becomes associated with specific sites in the DNA. Competence for inducibility by steroid hormones depends on the presence of cytoplasmic receptor protein, together with the presence of some and the absence of other acidic proteins bound to specific sites in the chromatin. Hydrophilic peptide hormones do not enter the target cell, instead they bind to specific protein receptor molecules embedded in the plasma membrane and from that situation activate an intra-cellular second messenger system. So far two such systems have been recognized: one depends on influx of Ca^{2+} ions that bind to the allosteric enzyme-activating protein calmodulin, the other involves activation of adenylate cyclase, which regulates intra-cellular levels of cyclic AMP. The latter molecule modifies cell metabolism by catalysing the phosphorylation of enzymes and other proteins. Where second messenger systems are used, the primary effect would seem to be modification of cell phenotype at physiological levels, transcription is presumably controlled only as a subsequent response. In contrast, the primary response to steroid hormones seems to be the control of certain aspects of transcription, which may involve both the switching on and the switching off of specific genes.

The extra-cellular matrix mediates many processes of embryonic induction and filtering of cells and molecules. It probably represents a three-dimensional network of instructive determinants, providing a reference grid of positional information that reflects the metabolic activities of cells which secreted it. Firbronectin is an important component of the matrix, with a range of affinities for other extra-cellular materials and the capacity for orientation in parallel to intra-cellular microfilaments. It is thought that fibronectin may be important in the two-way communication of order between the cell interior and exterior. Co-ordination of intra-cellular structure and physiological activities within tissues is probably achieved partly by the extra-cellular matrix, partly by the pattern of desmosomal links, and partly by inter-cellular physiological communication through gap junctions.

The mitochondrial population is of vital importance for both development and homeostasis. Mitochondria are inherited maternally, are always produced from extant mitochondria, and show considerable functional differentiation between tissues. One of their major functions is as a calcium sink, maintaining low levels of Ca^{2+} ions in the cytosol and permitting the action of calmodulin as a physiological regulator. Mitochondria are also the major regulators of energy metabolism within the cell, and in that role control practically every other cellular activity. Both functions are interlinked and can be regulated from outside the cell. The much neglected cytoplasmic mitochondrion therefore seems to be very much a major element in the controlled translation of nuclear genotype into phenotype.

In this chapter we begin to gain an inkling of the molecular bases of the traditional embryological concepts of competence and induction, and of how events in different parts of the body, or even in different members of a community, become integrated to produce defined body forms and physiologies. In the next two chapters we will develop these concepts further.

Bibliography

Alberts, B., Bray, D., Lewis, J., Raff, M., Roberts, K. and Watson, I., *Molecular Biology of the Cell*. Garland, New York (1983).

Albone, E. S., *Mammalian Semiochemistry: The Investigation of Chemical Signals Between Mammals*. John Wiley, Chichester, (1984).

Beale, G. H., *The Genetics of* Paramecium aurelia. Cambridge University Press, Cambridge (1954).

Bennett, M. V. L. and Spray, D. C., *Gap Junctions*. Cold Spring Harbour Laboratory, New York (1985).

Berridge, M. J., The interaction of cyclic nucleotides and calcium in the control of cellular activity. *Adv. Cyclic Nucleotide Res.* **6**: 1 (1975).

Bullough, W. S., *The Dynamic Body Tissues*. MTP Press, Lancaster (1983).

Campbell, A. K., *Intracellular Calcium, Its Universal Role as Regulator*. John Wiley, Chichester (1983).

Caplan, A. I. and Ordahl, C. P., Irreversible gene repression model for control of development. *Science*, **201**: 120 (1978).

Cheung, W. Y., Calmodulin. *Sci. Am*, **246**: 48 (1982).

Cold Spring Harbour Symposia on Quantitative Biology, *Organization of the Cytoplasm*. Cold Spring Harbour Laboratory, New York (1982).

Deuchar, E. M., *Cellular Interactions in Animal Development*. Chapman and Hall, London (1975).

Gilbert, L. I. and Frieden, E., *Metamorphosis. A Problem in Developmental Biology*, 2nd edn. Plenum, New York (1981).

Gilham, M. W., *Organelle Heredity*. Raven Press, New York (1978).

Grant, P., *Biology of Developing Systems*. Holt, Rinehart and Winston, New York (1978).

Gronemeyer, H. and Pongs, O., Localization of ecdysterone on polytene chromosomes of *Drosophila melangaster*. *Proc. Natl Acad. Sci. USA*, **77**: 2108 (1980).

Hadorn, E., *Developmental Genetics and Lethal Factors*. John Wiley, New York (1961).

Hay, E. D., Extracellular matrix. *J. Cell Biol.* **91**: 205 (1981).

Hay, E. D., *Cell Biology of the Extracellular Matrix*. Plenum, New York (1981).

Higgins, L. G. and Riley, N. D., *A Field Guide to the Butterflies of Britain and Europe*, 3rd edn. Collins, London (1975).

Kakwichi, S., Hidaka, H. and Means, A. R., *Calmodulin and Intercellular Ca^{++} Receptors*. Plenum, New York (1982).

Lerner, R. A. and Bergsma, D. (eds), The molecular basis of cell–cell interactions. *March of Dines Birth Defects Original Article Series*, **14**: No. 2 (1978).

Maclean, N., *Control of Gene Expression*. Academic Press, London (1976).

McKerns, K. W., *Regulation of Gene Expression by Hormones*. Plenum, New York (1983).

Müller, H. J., Die Saisonformenbildung von *Araschnia levana*: ein photoperiodisch gesteuerter Diapause–Effect. *Naturwissenschaften*, **42**: 134 (1955).

Pitts, J. D., Burk, R. R. and Murphy, J. P., Retinoic acid blocks junctional communication between animal cells. *Cell Biology International Reports*, Supplement A5: 45 (1981).

Pitts, J. D. and Finbow, M. E., *The Functional Integration of Cells in Animal Tissues*. Cambridge University Press, Cambridge (1983).

Pollack, J. K. and Sutton, K., The differentiation of animal mitochondria during development. *Trends Biochem. Sci.*, **5**: 23 (1980).

Poste, G. and Nicholson, G. L., *Cytoskeletal Elements and Plasma Membrane Organization*. Elsevier, Amsterdam (1981).

Preer, J. R., Preer, L. B. and Rudman, B. M., mRNAs for the immobilization antigens of *Paramecium*. *Proc. Natl Acad. Sci. USA*, **78**: 6776 (1981).

Rockstein, M. (ed.), *Biochemistry of Insects*. Academic Press, London (1978).

Rogers, A. W., *Cell and Tissues*. Academic Press, London (1983).

Roy, A. K. and Clark, J. H., *Gene Regulation by Steroid Hormones*. Springer Verlag, New York (1980).

Schlesinger, M., Ashburner, M. and Tissieres, A. (eds), *Heat Shock from Bacteria to Man*. Cold Spring Harbour, New York (1982).

Schmalhausen, I. I., *Factors of Evolution*, translated by I. Dordick, edited by T. Dobzhansky. The Blakiston Company, Philadelphia (1949).

Slavkin, H. C. and Greulich, R. C. (eds), *Extracellular Matrix Influences in Gene Expression*. Academic Press, London (1975).

Slonimski, P. and Attardi, G., *Mitochondrial Genes*, Cold Spring Harbour Laboratories, New York (1982).

Sonneborn, T. M., Gene action in development. *Proc. R. Soc. Lond. (Biol.)*, **176**: 347 (1970).

Stryer, L., *Biochemistry*. Freeman, San Fransisco (1975).

Tata, J. R., Hormonal regulation of metamorphosis. In *Symposia of the Society for Experimental Biology*, Vol. 25, edited by D. D. Davies and M. Balls. Cambridge Univesity Press, Cambridge, pp. 161–181 (1971).

Tickle, C., Alberts, B., Wolpert, L. and Lee, J., Local application of retinoic acid to the limb bond mimics the action of the polarizing region. *Nature*, **296**: 564 (1982).

Tzagoloff, A., *Mitochondria*. Plenum, New York (1982).

Ursprung, H., Developmental genetics. In *The Biologic Basis of Pediatric Practice*, edited by R. E. Cooke, McGraw-Hill, New York (1965).

Waddington, C. H. *The Evolution of an Evolutionist*. University Press, Edinburgh (1975).

Weber, R. (ed.), Biochemistry of amphibian metamorphosis. In *The Biochemistry of Animal Development*, Vol. 2, Chap. 5. Academic Press, London (1967).

Wigglesworth, V. B., *The Life of Insects*. Weidenfeld and Nicolson, London (1964).

Yamada, K., Cell surface interactions with extracellular materials. *Ann. Rev. Biochem.*, **52**: 761 (1983).

Chapter 5 The establishment of invertebrate body patterns

5.1. Aspects of symmetry

In Chapter 1 we considered the distribution of cell types in the bodies of sponges. This group, in common with many other primitive animals, shows radial symmetry: it is not possible to distinguish a left from a right side. Most of the animal species which we consider to be more advanced show instead three other striking features of anatomical organization: bilateral symmetry, segmentation and cephalization. In other words, we can distinguish a left and a right side, which are more or less mirror images of one another, a significant part of the body is built up as repetitions of a basic structural unit and the primary nervous functions are concentrated at the front end. Figure 5.1 (a) and (b) illustrates these elements of body pattern in two organisms which have contributed a great deal to our understanding of developmental genetics, the dipteran fly, *Drosophila* and a urodele amphibian, *Salamandra*.

A symmetrical body is necessary for balance with respect to gravity and for coping with resistance by the medium — air, water or the land surface — through or over which the animal moves. Increase in body size and the acquisition of specialized organs is achieved more efficiently in evolutionary terms and more economically from a developmental point of view, by duplication, or multiplication and modification of an already established structure, than by the construction of radically new body parts. Symmetrical duplication is one aspect of this natural economy.

Many species which move less actively, such as adult echinoderms, have retained radial symmetry. Figure 5.1(c) shows the sea urchin, *Echinus*, which is constructed on a radially symmetrical plan with five elements. It is interesting that the motile larvae of many species, which are radially symmetrical and sessile as adults, show bilateral symmetry. Active movement seems to be most readily achieved by the adoption of a bilaterally symmetrical body form, usually accompanied by elongation of the body. Sensory organs and the brain then become concentrated near the mouth, at the fore-end, in order to assist with feeding and avoidance of obstructions.

Longitudinal segmentation is another aspect of body economy in evolution,

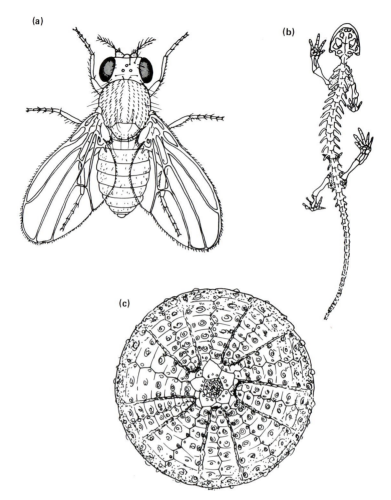

Figure 5.1. Examples of body patterns showing bilateral symmetry, cephalization, segmentation and jointed limbs in (a) *Drosophila melanogaster* and (b) *Salamandra* sp., and radial symmetry (c) in *Echinus esculentus*.
((b) And (c) redrawn from Parker and Haswell.)

being the most economic way to achieve elongation, rather like adding extra carriages to a train as an alternative to redesigning the whole vehicle.

Rather similar appendages, such as jointed legs, are found in the vertebrates and arthropods, but there is very good evidence that these apparently similar structures and their arrangement have evolved quite independently. The vertebrates certainly evolved from invertebrates, possibly from relatives of the ascidians, but the arthropods have quite a different ancestry and almost certainly arose from segmented annelid worms (see Chapter 13). Although the legs and wings of insects have similar functions to the legs of land vertebrates and the wings of birds, these structures are analogous rather than homologous, that is their developmental origins are quite different.

Both vertebrates and arthropods show segmentation. In the insects this is revealed in all structures from the skin down to, but not including, the gut. In the vertebrates segmentation is confined primarily to the muscles, and secondarily to elements of the skeleton and nervous system. The gut is not segmented and the nipples in mammals are perhaps the only indication of segmentation in the skin.

The source of mesoderm is also different in insects and vertebrates. This tissue plays an important part in inductive interactions in both groups (see Chapter 3), but it would seem to be premature, as yet, to consider these interactions to be of a similar nature. In this chapter and the next we will consider the development of some of the major features of body patterning in insects and vertebrates as examples of very different devices by which rather similar phenotypic features are attained. Before that, however, we will consider a much more basic feature of body form.

5.2. Orientation of the mitotic spindle

In some species the entire body plan can be governed by the orientation of the mitotic spindle during the earliest rounds of cell division. This arises as a feature of spiral cleavage, characteristic of the early embryos of several vertebrate groups, the annelids, molluscs, flatworms and nemerteans, which were considered by the early embryologists to show mosaic development (see Chapter 2). In other groups new centrioles arise perpendicularly to pre-existing ones, so that cleavage spindles are orientated perpendicularly to previous spindles. In spirally cleaving embryos cleavage spindles are orientated obliquely, usually at a 60° angle, so that blastomeres become stacked in an alternating pattern. Not all spiral-cleavage embryos develop into spirally shaped adults, but this is the case in the snails, in which some individuals are coiled in a clockwise, and others in an anticlockwise fashion. The best-known example is *Limnea peregra*, in which coiling is usually dextral (i.e., with the opening of the shell to the right), although in rare individuals it is reversed (sinistral).

The direction of coiling is determined by alleles at a single locus, with the wild-type dextral allele (D) acting as dominant. This gene, however, acts not on the individual which carries it, but on the F1 progeny derived from the eggs which it lays. For example, a Dd snail which receives sperm from another Dd heterozygote will produce offspring of genotypes dd, Dd and DD in the proportions $1 : 2 : 1$, but all of these will be coiled dextrally since she has the genotype Dd. Her progeny, when acting as female parent (snails are hermaphrodite), will produce offspring that are respectively sinistral, dextral and dextral, regardless of the genotypes of their mates. Figure 5.2 illustrates this relationship.

The direction of coiling depends on the orientation of the second cleavage spindle (see Figure 5.2). This is determined at oogenesis, apparently by the establishment of cytoplasmic structure within the oocyte. Experiments involving transfer of cytoplasm between eggs of different genotypes reveal that snails carrying the *D* allele incorporate a product into their cytoplasm which is responsible

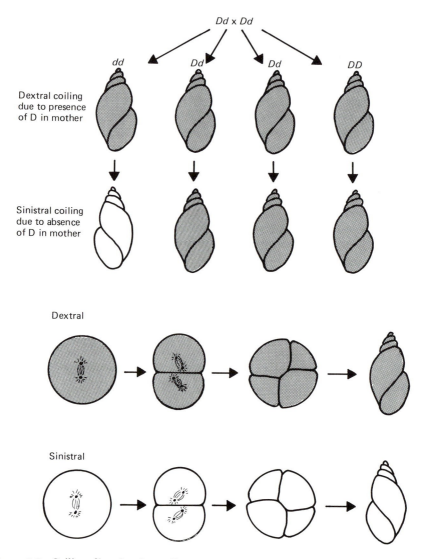

Figure 5.2. Coiling direction in snails. Dextral coiling is due to action of the dominant allele *D* and the presence of its product in the egg cytoplasm contributed by the maternal system. The influence of *D* product is shown by grey shading. Coiling direction is initiated at the second mitosis, due to the orientation of the mitotic spindle.

for dextral coiling. Transfer of this material into an egg produced by a *dd* individual reverses the natural direction of coiling of its progeny. This example illustrates one way in which genetic dominance operates at a biochemical level, by the presence, as opposed to the absence, of a particular morphogenetic determinant.

5.3. The genetic control of body segmentation in insects

Bithorax complex BX-C

There is a fascinating class of mutations in arthropods called homeotic mutations, in which entire organs are misplaced in the body. For example, in *Drosophila melanogaster* the mutation *antennapedia* causes the antennal rudiments to develop as legs (Figure 5.3). *Nasobemia* is named after a fairytale giant, Nasobem, who walked on his nose. This mutant has a leg where its proboscis should be. One mutant called *cockeyed* has its eyes replaced by genitalia! A study of some of the homeotic genes provides remarkable insight into the way in which positional information in the egg is normally interpreted to allow correct development of the different body segments with their appropriate appendages.

As we saw in Chapter 2, the basic body plan of insects is initially laid down in terms of morphogenetic gradients within the egg. These gradients are associated with RNA and probably other substances produced by the maternal system and injected into the egg case through the micropyle at the anterior end. The nuclei that are the progenitors of the somatic cells become determined as, or just after, they become enclosed by the growth of cell walls, during the formation of the blastoderm (Figure 5.5(i)). When cell walls form, each nucleus obtains a sealed micro-environment composed of the cytoplasm taken from a particular part of the gradient. The different nuclei then respond to gene-switching molecules contained in this cytoplasm, so that they each follow a pathway of development appropriate to the specific part of the body.

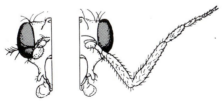

Wild type   *Antennapedia*

Figure 5.3. The *antennapedia* phenotype.
The head of an extreme example of the *antennapedia* homeotic mutant is shown on the right, in comparison with that of a wild-type fly on the left. The mutation causes the antenna to develop as a leg. (After Alberts *et al.*, 1983.)

The head of an adult *Drosophila* larva is a complex structure containing the equivalent of at least five body segments. In the thorax are three segments, the prothorax, mesothorax and metathorax, each of which carries a pair of legs. The mesothorax also carries a pair of wings and the metathorax a pair of drumstick-shaped appendages called balancers or halteres. By contrast, the abdomen is a relatively simple structure, with seven very similar segments lacking legs, wings and halteres, plus a terminal complex of three segments containing the genitalia (see Figure 5.5).

The two-winged flies, or Diptera, almost certainly evolved from insects with four wings, which in turn came from millepede-like arthropods that had many legs instead of just six. These organisms evolved from the annelid worms mentioned above. It is thought that evolutionary reductions in numbers of legs probably required the evolution of leg-suppressing genes active in the abdominal segments while wing-suppressing genes acted in the metathorax. We might therefore expect the two-winged flies to have genes for suppressing leg development, which act in the abdomen, and genes suppressing development of wings, which act in the metathorax. Such genes do seem to exist and it is mutations of these that we know as homeotic mutations.

Mutations of the *bithorax* (*bx*) locus transform the anterior part of the metathorax so that it resembles an anterior mesothorax, while mutations at the *postbithorax* (*pbx*) locus change the posterior metathorax into something resembling the posterior mesothorax (see Figure 5.7). These two loci are closely linked in what is known as the bithorax complex or BX-C, together with several other leg- and wing-suppressing genes. This complex presumably arose by tandem multiplication of one or more ancestral gene sequences, followed by diverse mutation, to produce a set of master genes which control the special properties of the different segments of the thorax and abdomen.

Very careful analysis of a whole series of combinations of mutations in the BX-C enabled E. B. Lewis to put forward a fascinating model for the specification of the thoracic and abdominal segments, which relates the order of the segments to the morphogenetic gradients in the egg and the order of the BX-C genes along the chromosome. The bithorax complex is situated on the right arm of chromosome 3 and has the postulated structure shown in Figure 5.4. It contains a number of regions of particular importance. Working in a proximo-distal direction along the chromosome, they are *Ultrabithorax* (*Ubx*), *bithoraxoid* (*bxd*) and *infra-abdominal 2,3* and *8* (*iab-2*, *iab-3*, *iab-8*). It should be remembered here that loci are traditionally named by reference to the effects produced when occupied by a deficient allele. Whether or not these regions are loci in the strict sense is somewhat contentious, but for simplicity we will consider them as simple loci and call them A, B, C, D and E. It is postulated that each has an 'upstream' (with respect to the direction of transcription) regulator sequence which we will denote by Ra, Rb, Rc, Rd and Re (Figure 5.5). The wild-type alleles of genes A–E code for products which we will call a, b, c, d and e.

According to Lewis's model (illustrated in Figure 5.5), all five products are ex-

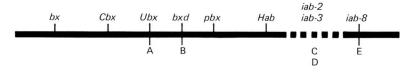

Figure 5.4. The genetic map of the bithorax complex, BX-C.
The symbols correspond to the positions of the following loci: *bx*, *bithorax*; *Cbx*, *Contra-bithorax*; *Ubx* (A in Figure 5.5) *Ultrabithorax*; *bxd* (B), *bithoraxoid*; *pbx*, *postbithorax*; *Hab*, *Hyper-abdominal*; *iab-2* (C), *iab-3* (D), *iab-8* (E), *infra-abdominal 2, 3*, and *8*. This diagram represents a simplification of the known situation. Some of these loci are transcribed from one strand and some from the other strand of the DNA. Different methods of analysis indicate different numbers of functional units defined in terms of RNA transcripts. (From Lewis, 1978.)

pressed in the eighth abdominal segment (AB-8), but none are expressed in the prothorax or head. In the mesothorax, product a alone is produced and segments posterior to the mesothorax are considered to be defined by specific combinations of products a to e, as shown in Figure 5.5. Products expressed at low level are shown in brackets. Expression of progressively longer sections of the gene complex therefore defines progressively more posterior segments of the body.

According to this model, controlled expression of the structural genes is governed by the existence of two gradients: an antero-posterior gradient of repressor within the embryo (Figure 5.5(ii)) and a proximo-distal gradient along the chromosome, in the binding affinity of the upstream regulatory elements for repressor (Figure 5.5(iii)). The high concentration of repressor at the fore-end of the embryo is considered to repress all except the *Ubx* sequence, A. The low repressor concentration at the tail end represses none of the BX-C genes and in the intervening segments there is assumed to be an orderly sequence of expression of the various units, A to E.

The situation is actually a lot more complex than this, since these 'genes' are on both strands of the DNA double helix and are therefore transcribed in opposite directions with respect to the centromere. The length of individual transcripts is also uncertain, some of these overlap and some are processed into smaller units. The model is very strongly supported however by the recent observation, using specific labelled antibodies, that the distribution of the *Ubx* product among the body segments is more or less as the model proposes.

The action of the BX-C is normally under the control of the wild-type alleles of several other genes, including *extra sex combs* (*esc*) and *Polycomb* (*Pc*). Deficiencies at either of these loci cause transformation of larval thoracic and abdominal segments into AB-8. The requirement for the *esc* product is restricted to the early stages of segment determination, but *Polycomb* product is required throughout development. The *esc* substance could be the postulated repressor laid down as an anterio-posterior morphogenetic gradient in the egg, since the progeny of *esc* mutant mothers sometimes have segments with a more posterior character than normal.

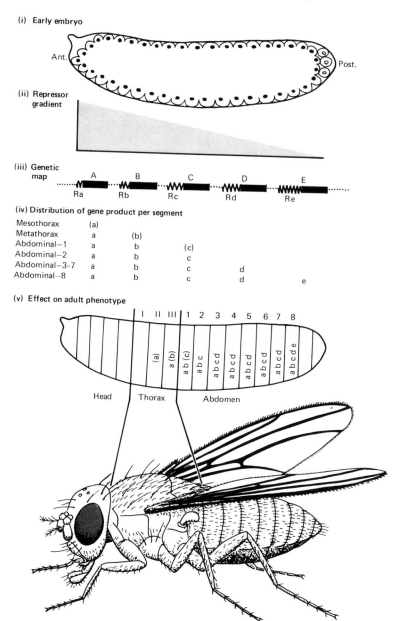

Figure 5.5. The relationship between repressor gradient, the genetic map of the *bithorax* complex and body structure in *Drosophila*, according to the Lewis model.
(i) The distribution of undetermined somatic nuclei just before blastoderm formation. (ii) The postulated gradient of repressor in the embryo. (iii) The genetic map of the *bithorax* complex, illustrating the postulated graded affinity of upstream controlling sequences from Ra (low affinity) to Re (high affinity). (iv) The expression of different combinations of gene products in thoracic and abdominal segments. (v) The relationship between combinations of gene products in the different segments of the larva and phenotype of the same segments in the adult. The number of segments in the head is uncertain.

According to the Lewis model, the bithorax complex is under negative control. All the BX-C genes are considered to be expressed in the absence of repressor, while increasing repressor concentrations inactivate them in the order E, D, C, B, A. Negative control of transcription is unusual in eukaryotes, the usually accepted situation being that genes are non-specifically repressed by histone molecules, specificity being introduced at derepression. Despite this criticism, however, the Lewis model is an inspiring illustration of the way in which molecular and embryological concepts are now becoming integrated to produce explanations for the control of development which were quite unimaginable only a very few years ago.

Sub-segmental compartments

There is genetic evidence that the body segments in *Drosophila* are in fact each built up of an anterior and a posterior subcompartment. The existence of these compartments is indicated by the action of the *bithorax* and *postbithorax* alleles (see above), but this concept is supported by an experimental trick for producing somatic genetic mosaics. In these experiments, larvae which are heterozygous for a recessive trait, such as yellow body colour (y/y^+), are exposed to X-rays (Figure 5.6), and in some cells this irradiation causes chromosome breakage. Portions of the chromosomes sometimes then become exchanged so that

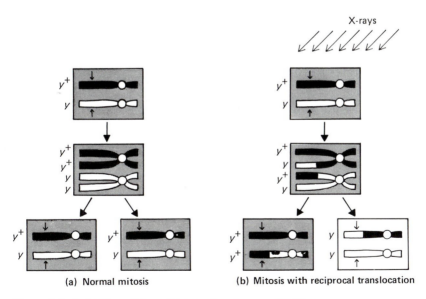

Figure 5.6. Induction of recombinant chromosomes by X-rays.
(a) Normal mitosis of somatic cells of a y/y^+ heterozygote. (b) Mitosis with reciprocal translocation, producing cells homozygous for y and y^+ which give rise to homozygous cell clones. The y locus is indicated.

in effect chromosomal crossing over occurs. Subsequent mitosis then results in some cells which are homozygous (y/y) for the recessive allele. Clones of cells which are homozygous for y and yellow in colour then arise from these among the dark heterozygous cells. Similar clones arise very infrequently due to spontaneous somatic mutations.

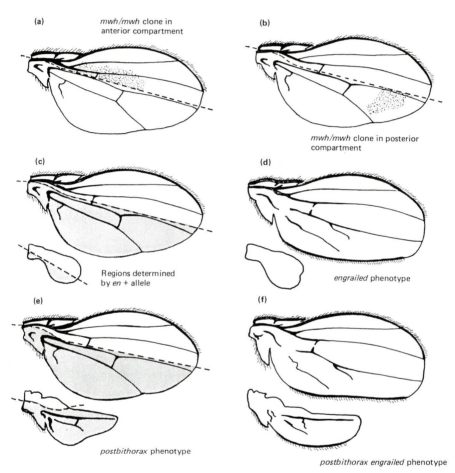

Figure 5.7. Compartments in *Drosophila* wings and halteres. The shaded area represents the region controlled by the engrailed, *en*, locus.
(a), (b) Clones of cells homozygous for the recessive allele for multiple wing hairs induced in heterozygote *mwh/mwh*[+] flies by X-irradiation. The clones are confined to anterior or posterior compartments. (c) Expression of normal allele at the *engrailed* locus is required for correct development of phenotype in the posterior, but not anterior, compartments. (d) Wing and and haltere of *en/en* homozygote. (e) Wing and haltere of *postbithorax*, *pbx/pbx*, homozygote. The wing is normal, but the posterior region of the haltere is transformed into a posterior-wing-like structure. (f) Wing and haltere of *pbx/pbx*, *en/en* double homozygote, showing transformation of the posterior part of the wing and haltere into a form resembling the anterior part of the wing. (After Garcia-Bellido *et al.*, 1979.)

When the locations of such marked clones are examined, they are found to spread up to an invisible line of demarcation down the middle of the segment, in the trunk, legs and wings. Clones apparently never spread over and mingle with those in adjacent compartments, even though there is no obvious physical barrier between the anterior and posterior parts of the same segment. *Multiple wing hairs* (*mwh*) is a recessive mutation which specifically affects individual cells of the wing. Clones of homozygous mutant cells produced by X-irradiating *mwh*/*mwh*$^+$ heterozygotes are illustrated in Figure 5.7((a),(b)). The compartment 'address' of cells is believed to be defined by means of molecules carried on their surface membranes, which preclude association of dissimilar cells.

Sequential determination of nuclei

During the formation of the embryonic blastoderm, the nuclei become determined in a definite sequence. First, a distinction is made as to whether the cells in a particular segment will develop into anterior or posterior structures. The locus which defines this distribution is known as *engrailed* (*en*). Mutants homozygous for *engrailed* lack the gene product which determines posterior features (Figure 5.7(c),(d)). They have legs, wings, eyes and antennae in which posterior regions of segments resemble the anterior parts of the same segment. The basic instruction to a segment therefore seems to be to produce anterior structures, but the product of the *engrailed* locus causes this to be modified in the rear compartment into posterior versions of the same structures.

Flies which are homozygous for both *pbx* and *en* not only have wings in which the posterior regions resemble the anterior, due to deficiency of *en* product, but also halteres modified to resemble the anterior portions of the wings (Figure 5.7(f)). Flies homozygous for *pbx*, but not *en*, have the posterior parts of their halteres modified to resemble the posterior parts of the wings (Figure 5.7(e)).

The second determinative decision is whether blastoderm cells will develop into larval or adult tissues. If determined as adult structures, they become aggregated into small structures known as imaginal discs, which remain in this form until pupation when they become exposed to the insect moulting hormone, ecdysone (see Chapter 4). A third determinative influence specifies the distinction between ventral and dorsal parts of compartments, while a fourth discriminates between the ventral and dorsal regions within that part. A fifth set of instructions determines whether the cell will occupy a position in the trunk of the fly, or an appendage.

The first two of these determinative events occur around the blastoderm stage, the other three during early embryogenesis and the imaginal disc cells normally retain their state of determination throughout development. However, if cultured for long periods they can transdetermine into other states, as described in Chapter 7. As we shall see in Chapter 6, the determination of cells in vertebrates' appendages occurs much later in development and they receive their 'programming' through induction by neighbouring tissues, just before their differentiation into cartilage, bone, muscle, etc. The cells of vertebrates also normally retain their differentiated characteristics, but some tissues, notably those of the eye, have much less rigid controls on phenotype and can transdifferentiate into other tissues (see Chapter 7).

5.4. Summary and conclusions

The most characteristic feature of any organism is its overall body pattern, which in some species derives from features as basic as the orientation of the mitotic spindle during the very earliest developmental stages. Most species can be considered to be built up as multiples of simple structures arranged radially, serially, or side by side in mirror-image form. Such patterns illustrate the exercise of economy in utilization of available genetic information.

The form of the body segmentation in insects is governed by a small set of major genes under the joint control of maternal gene products in the egg cytoplasm and the embryo's own genes. In *Drosophila* the major genes determining segment identity are linked in the bithorax complex. Another locus, *engrailed*, defines the features which are characteristic of the posterior compartments of these segments. Determination of nuclei occurs around the blastoderm stage in insects and follows a sequential pattern. As we shall see in the next chapter, the forces which create the segmented structure of vertebrate bodies are of quite a different type.

Bibliography

Alberts, B., Bray, D., Lewis, J., Raff, M., Roberts, K. and Watson, J., *Molecular Biology of the Cell*. Garland, New York (1983).

Browder, L. W., *Developmental Biology*. Saunders College/Holt, Rinehart & Winston, Philadelphia (1980).

French, V., Development and evolution of the insect segment. In *Development and Evolution*, edited by B. C. Goodwin, N. J. Jolder and C. G. Wylie. Cambridge University Press, Cambridge, pp. 161–193 (1983).

Garcia-Bellido, A., Lawrence, P. A. and Morata, G., Compartments in animal development. *Sci. Am.*, **241**: 90 (1979).

Grant, P., *Biology of Developing Systems*. Holt, Rinehart & Winston, New York (1978).

Lawrence, P. A. and Morata, G., The elements of the bithorax complex. *Cell*, **35**: 595 (1983).

Lewis, E. B., A gene complex controlling segmentation in *Drosophila. Nature, Lond.*, **276**: 565 (1978).

Meinhardt, H., *Models of Biological Pattern Formation*. Academic Press, London (1982).

Chapter 6 The establishment of vertebrate body patterns

6.1. Establishment of the major body axis and somitogenesis

The characteristic feature of the vertebrates is a central segmented axis, the vertebral column, which is derived from an initially unsegmented rod of mesoderm called the notochord that lies just below the dorsal nerve cord. The muscle blocks on either side of the vertebral column are also segmented and are supported by segmentally arranged bones and innervated by segmentally distributed nerves. In those vertebrates considered to be at more complex levels of organization than the fish, there is usually also a pair of fore and hind limbs which have a characteristic bone structure.

We discussed the very early development of amphibians in Chapter 2; later stages of vertebrate development have been most thoroughly studied in the chicken, but it is generally considered that the basic principles of development are probably similar in all vertebrates (see Chapter 14).

In birds and mammals, the body axis is defined by the path of travel of the primary organizer, Hensen's node, through an essentially structureless disc of blastoderm from which derives the body of the embryo (see Chapter 2). As Hensen's node travels slowly in a caudal direction, a ridge arises in the ectoderm on each side. The two ridges reach up and meet in the midline, then fuse to form the neural tube (see Figure 3.11). Before they fuse, a very special group of cells, known as the neural crest cells, move out of the crests of the ridges. The fate of these cells is quite extraordinary and is outlined in Chapter 3. On either side of the neural tube the cells of the mesoderm rotate and aggregate into blocks or somites which give rise to the body segments. Other mesoderm cells move forwards or backwards to surround the nerve cord and join with the notochord. These become the vertebrae, each vertebra being derived from cells of the posterior part of one segment plus the anterior part of the next segment behind.

Hensen's node is considered to be homologous with the dorsal lip of the blastopore in amphibians and to have similar properties in initiating the morphogenetic events of gastrulation (see Chapter 2). If a small piece of blastoderm containing a Hensen's node is transplanted into the blastoderm of another embryo, it will initiate an additional axis in the host tissue. However, somites will arise similarly on either side of a cut made in the chick blastoderm with a needle. So far as segmentation is concerned, the node therefore seems merely to divide the mesoderm into two halves, so releasing an innate tendency for it to form into blocks. As pointed out below and in Chapter 7, development (and regeneration) occurs as if to restore equilibrium to a disturbed system. The movement of the node creates physical discontinuity between left and right sides and initiates a temporal discontinuity between anterior and posterior parts of the blastodisc. This disturbance may be all that is necessary to trigger other changes, which produce new disequilibria and further developmental events. However, our experience with other systems suggests that different species may utilize diverse cues to attain the same ends, so the deductions based on chicken embryos may not represent a universal rule (see Chapter 14).

6.2. Development of the vertebrate limb

The vertebrate limb forms from the limb bud, an outgrowth of loose mesenchyme surrounded by an epithelial sheet of ectoderm. The cartilage, some of the muscle and the connective tissue of the limb are formed out of this mesenchyme, under the inductive influence of the ectoderm.

We can describe the position of any cell in the limb by reference to co-ordinates on three axes: the proximo-distal axis from shoulder to fingertip, the antero-posterior axis from thumb (I) to little finger (V), and the dorso-ventral axis from the back to the palm of the hand (Figure 6.1). The axes become specified during

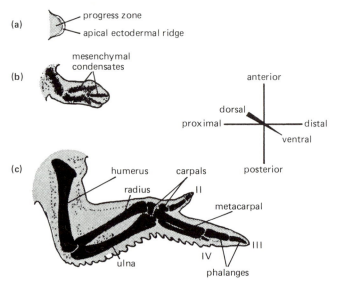

Figure 6.1. Development of a chick wing after (a) 3.5, (b) 6.5, (c) 9.5 days of incubation. (Redrawn from Alberts *et al.*, 1983.)

embryogenesis in this same order. It has been suggested that cell phenotype along the proximo-distal axis is defined in terms of duration of exposure of proliferating mesenchyme to the inductive influence of the ectoderm near the tip, while the antero-posterior axis is defined by a morphogen that has its source in mesenchyme near the posterior boundary of the limb bud. Grafting experiments indicate that by the time the final position of a cell has become established, it has acquired, or then acquires, a unique 'positional value'. The result of this is that tissues which are repaired or regenerated can develop appropriately without reference to the original morphogenetic determinants which formed the limb. This implies that the cells of the limb retain their morphogenetic instructions, possibly through some kind of internal feedback loop response, which reinforces patterns of gene expression originally induced in them. It also implies the existence of systems of inter-cellular communication within the limb. However, these various theories are still controversial.

Definition of the proximo-distal axis

The limb bud grows by proliferation of mesenchyme cells in the so-called progress zone, which is about a third of a millimetre deep, just below the tip. The tip is capped by a ridge of ectoderm called the apical ectodermal ridge, or AER. Removal of the AER prevents morphogenesis of the mesenchyme derivatives and it is deduced that its induction by the AER is necessary for normal development.

If the AER is removed at an early stage, the humerus differentiates normally, but terminates at the elbow; if removed rather later, limb formation terminates at the wrist (Figure 6.2). This shows that *determination of the limb rudiments occurs progressively from humerus to digits.*

If the tip is removed from an old limb bud and transplanted to the stump of a young bud that has had its own tip removed, this causes the limb to form without its middle elements. The digits develop directly after the humerus, without intervening radius, ulna and carpals (Figure 6.3). The reverse experiment, of transplanting a young limb bud to an old stump, causes tandem duplication: humerus, radius and ulna, humerus, radius and ulna, carpals, then digits.

Such experiments as these suggest that, in stark contrast to the situation in insect appendages, *determination of vertebrate limb cells occurs during their development.* You will remember that insect limb cells are determined at the blastoderm stage, but remain undifferentiated until pupation.

Determination of limb structures also seems to occur without reference to the already determined tissue further up the limb. Remarkably, this is very different from the situation in limb regeneration, when morphogenetic information does seem to flow down the limb (see below).

The most acceptable explanation of limb determination during normal development is that put forward in the Progress Zone Model of Summerbell, Lewis and Wolpert (1973). This suggests that determination of mesenchyme cells occurs progressively within the progress zone, during their proliferation in the

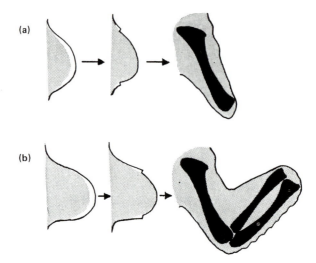

Figure 6.2. The effect of removing the apical ectodermal ridge from an early (a) and a later (b) chick wing bud.
This experiment shows that the proximal parts are determined before the distal elements, under the influence of the AER. (Redrawn from Alberts *et al*, 1983.)

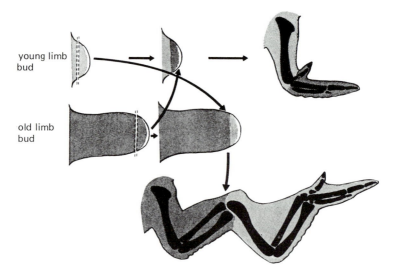

Figure 6.3. The effect of exchanging limb bud tips, including progress zones and apical ectodermal ridges, between young and old limb buds.
The graft and host tissues behave autonomously, showing that the influences which confer positional information along the proximo-distal axis operate only over small distances and are not closely related to previously determined structures.

neighbourhood of the AER. Cells laid down after *brief* exposure to AER develop as upper arm cells, those laid down after one or two further mitoses become forearm, those after three or four mitoses become wrist and those laid down after five, six or seven mitoses give rise to those elements distal to the wrist (see Figure 6.4 (c)).

The various segments of the limb are not very different in size when they emerge from the progress zone, but they then undergo different degrees of growth, so that after about 10 days the chick ulna is about 16 times longer than it was when it emerged, whereas the wrist only doubles in size in that time. If these regions are excised and cultured in isolation, they show similar differences in growth rate, which shows that their inductive experience under the influence of the AER also involves the imposition of a programme for future growth.

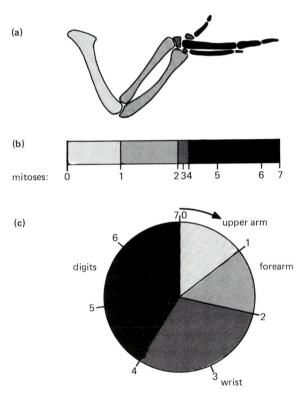

Figure 6.4. The Progress Zone Model for determination of limb parts in the proximo-distal axis of the vertebrate limb.
(a) The skeletal elements of a normal chick wing. (b) The relationship between determination of skeletal elements and the number of mitotic divisions cells have undergone in the progress zone by that stage. (c) Another representation of the information shown in (b), relating determination of upper arm, forearm, wrist and digits to the completion of mitotic cycles by mesenchyme cells in the progress zone. (After Ede, 1978.)

Despite the appeal of the progress zone model, it is not the complete explanation of proximo-distal axis determination, since a limb bud which has been surgically disrupted before cytodifferentiation can undergo internal readjustments that enable its elements to differentiate normally, in the correct positions. The self-organizing capacity of the early limb bud mesenchyme is discussed in the next section.

Definition of the antero-posterior axis

At the posterior margin of the limb bud is a zone of necrotic cells, which roughly coincides with a region named by J. W. Saunders as the zone of polarizing activity, or ZPA. Mesenchyme in this area seems to produce a morphogen which diffuses anteriorly through the limb bud and defines the form, number and location of the digits.

If an additional ZPA is grafted on to the anterior margin of an early chick wing bud, a symmetrical wing with duplicated distal parts develops, often with apparent fusion of the digits at the mid point (Figure 6.5). The most clearly developed digits in a normal chicken wing are numbers II, III and IV, the equivalent of the thumb (I) and little finger (V) being missing. The mirror image limbs have the sequence IV, III, II, II, III, IV, or IV, III, II, III, IV. When a ZPA

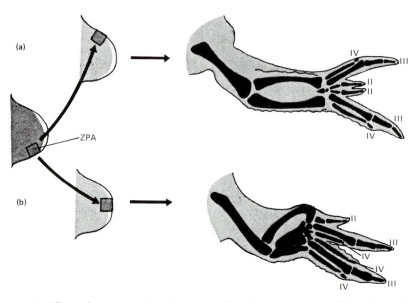

Figure 6.5. Effect of transplanting the zone of polarizing activity (ZPA) into (a) the anterior and (b) the central tip region of a recipient chick limb bud. In both cases, complex patterns of skeletal elements develop, but digit IV forms adjacent to the natural, or implanted ZPA in each situation. (Redrawn from Wolpert, 1978.)

was grafted into an early limb bud tip, this caused formation of digits in the sequence III, IV, IV, III, IV, or II, III, IV, IV, III, IV (Figure 6.5). In these and all other transplantation experiments a digit IV always formed adjacent to the ZPA.

The simplest explanation for these effects is that the ZPA acts as a source of a diffusible substance which defines digit number IV at high concentration and the other digits at progressively lower concentrations (Figure 6.6). In this form, however, the model is unsatisfactory. One objection is that to maintain a simple biochemical gradient over such a long distance (about 1 mm) would require an inordinately high morphogen concentration at source. Another problem is that the formation of the mesenchymal condensates, which give rise to the skeletal elements at intervals across the limb bud, suggests that the mesenchyme cells can somehow interpret the gradient as a series of thresholds. From a biochemical point of view this would be extremely complex. There could be a mechanism based upon regulator genes with graded affinities for repressor morphogen, as was postulated as an explanation of segmentation in insects, but such a model would be unjustified on the available evidence.

An alternative explanation is the Wave-Form Gradient Model of Wilby and Ede (1975). According to this model, cells are sensitive to their internal concentrations of a freely diffusible morphogen, M, which originates in the ZPA. If the intracellular concentration of M exceeds a threshold T1, cells are activated to synthesize more M, but at concentrations in excess of another threshold, T2, they actively destroy M instead. The transformation from inactive to synthetic, at T1, and from synthetic to destructive, at T2, are both irreversible changes, so that once embarked on the destructive course they continue to destroy morphogen even when its concentration has been reduced. As shown in Figure 6.7, the first gradient peak behind the wave front, then splits into two as the upper threshold

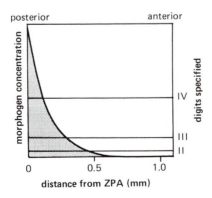

Figure 6.6. Specification of digits in the chick wing by a hypothetical morphogen gradient. The ZPA is assumed to be the morphogen source and the anterior of the wing bud a sink, so that a gradient becomes established across the wing. The different digits are assumed to be specified at particular thresholds. (After Wolpert, 1981.)

is reached and destruction is initiated. The interactions of synthesis, destruction and diffusion then cause the peak closest to the ZPA to move backwards and downwards, to stabilize between the two areas of destruction, while the leading peak moves forwards and upwards to initiate a new area of destruction. The end result is a stable alternating series across the limb in which the mesenchyme cells are either actively accumulating or actively destroying M. According to the model, the morphogen-destructive phase is related to formation of the mesenchyme condensates which give rise to cartilage and then bone, so that bones arise at a defined number of spaced intervals.

A prediction of this model, as opposed to the idea of a simple gradient with thresholds, is that the number of digits formed would be related to the breadth of the limb bud. This is the case in the hen mutant, *talpid*³, which has broad limb buds. A deficiency of the model is that it fails to explain the origin of the individual differences between the digits.

However, the notion that cells can appreciate and mimic the metabolism of their near neighbours is very interesting, in that it suggests a means by which information about cell position in an organ can be continuously monitored and maintained.

Both the simple gradient and the wave-form gradient model predict that no digital elements should form except under the influence of a localized ZPA source, but this does not fit recent observations. The cells of the *talpid*³ mutant apparently do not respond to an implanted normal ZPA, yet they still form a crude limb. If an impermeable barrier is inserted in the midpoint of a normal chick leg bud, separating the anterior and posterior regions, bony elements anterior to the barrier still form. Furthermore, if leg bud mesenchyme cells are dissociated, reaggregated and placed back inside their ectodermal jacket, they will form separate, spaced digits even in the absence of a ZPA, although provision of a ZPA improves their form. In this situation the number of elements is related to the width of the limb bud at that level. It is likely therefore that two antero-posterior pattern-specifying systems co-exist in the limb bud, one becoming established by cell interaction, in a manner perhaps similar to that suggested for the wave-form model but without a localized source of morphogen, plus a 'fine-tuning mechanism' that specifies individual differences between the digits, possibly in response to a simple ZPA morphogen gradient. Examination of the crossopterygian fish ancestors of the limbed vertebrates suggest that the former system may be the more primitive (see Chapter 15). This system probably also assists in specification of the proximo-distal axis, especially during regenerative reconstruction of the early limb bud.

6.3. Positional information

Studies on the determination of the antero-posterior and the proximo-distal axes of the chicken limb bud therefore both indicate that control of pattern definition occurs at more than one level. Some pattern specification takes place under the direction of the ZPA and the AER, but in addition an inter-cellular 'social' phenomenon seems to cause the cells to group and differentiate in a periodic fashion independent of outside influence. As a result of these combined effects the cells acquire what has been called positional information. Cells which have their positional values established with respect to the same co-ordinates or boundaries constitute an embryonic field (see Chapter 14).

If the relative positions of cells within such a field are disrupted before their cytodifferentiative states have become determined they may acquire new positional values appropriate to their new positions by a process of morphallaxis (see below) and the limb develops normally. If the damage occurs later it may be irreparable, or regeneration by epimorphosis may supervene, guided by the positional information which the cells retain (see below).

The nature of the positional information and the means by which it is acquired and stabilized are subjects of some considerable interest. Chondrogenesis in limb bud mesenchyme is promoted at low levels of nicotinamide adenine dinucleotide (NAD), while high levels of NAD promote the alternative differentiation into muscle (Chapter 4). A standing wave of NAD concentrations across the limb bud might therefore produce the observed effects (cf. Figure 6.7). Chondrogenesis is also stimulated by dibutyryl cyclic AMP. Cyclic AMP is detectable transiently during the cartilage precondensation phase and could easily affect intra-cellular NAD concentrations (see Chapter 4). Both NAD and cyclic AMP are small molecules and could pass freely between cells linked by gap junctions, which would be a requirement for the establishment of a biochemical standing wave within a tissue.

One way in which such a distribution could be stabilized would be by the blocking of gap junctional communication. Retinoic acid has this property (plus others), and implants that emit retinoic acid mimic ZPA activity. A function of the ZPA could therefore be to stabilize biochemical differentials that have already arisen within the limb bud.

Another way in which cytodifferentiation could become stabilized is by contact with extra-cellular matrix materials (see Chapter 4). The condensations which give rise to the bones are first characterized by localized chondroitin sulphate synthesis, then accumulation of type I collagen and fibronectin. The concentration of hyaluronic acid then decreases in the condensations, but increases in the intervening regions. Cartilage differentiation is marked temporally by the appearance of type II collagen and proteoglycan, coupled with the disappearance of fibronectin and collagen type I.

Extracts of chicken ZPA which have morphogenetic activity contain a low-molecular-weight fraction and a glycoprotein of high molecular weight produced by association of a widely distributed molecule with the low-molecular-weight moiety.

The observation that ZPA implants function in other species and on both fore and hind limb tissues, coupled with the remarkable finding that Hensen's node will substitute for a ZPA, has given rise to the deduction that the source of

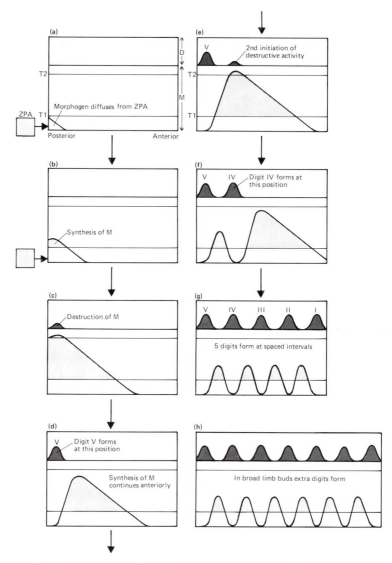

Figure 6.7. The Wave-Form Gradient Model for determination of digits along the antero-posterior axis of the vertebrate limb.

A morphogen (M) synthesized in the zone of proliferating activity (ZPA) at the posterior margin diffuses into the limb bud (a), where other cells respond to it in the following way: at concentrations of M greater than threshold T1, cells actively synthesize M, as shown by the light shading in (b); at concentrations above threshold T2, cells actively destroy M. Destructive activity is shown by the dark shading in the upper part of the diagram beginning in frame (c). This is followed by a localized decrease in M as shown in (d). Synthesis of M continues anterior to this zone of destructive activity until T2 is exceeded again as in (e). This pattern of activity continues along the antero-posterior axis, producing zones in which M is either actively synthesized or destroyed ((f), (g)). The destructive phase is considered to be linked to formation of digital condensations in the mesenchyme. The final stable pattern for a 5-digit limb of normal width is shown in (g) and for a broad-limbed mutant in (h).

specificity of response to ZPA morphogen resides in the responding tissue (see Chapter 3). If a major role of the ZPA morphogen is to block gap junctions this is what one would expect.

6.4. Specification of fore and hind limbs

When a piece of presumptive thigh from a chick leg bud was grafted into the progress zone of a wing bud it developed into a toe! This shows that the specification of fore and hind limbs is distinct from the definition of their parts. Apparently, the information concerning the type of limb that will develop is provided by substances that diffuse from adjacent body somites, since this influence is blocked by impermeable inserts although not by permeable barriers.

The somites also contribute muscle cells to the developing limb. They become organized in relation to the mesenchymal derivatives that are already there and, together with muscle cells and tendons that derive from the limb mesenchyme, stimulate growth of the cartilaginous elements by providing tension. Whether or not these muscle cells introduce further instructions for 'wingness' or 'legness' has not been established.

6.5. Limb regeneration

All species seem capable of some degree of regeneration of damaged tissues, but in general this capacity declines with age and evolutionary advance. The Urodele amphibians are an exception, retaining a remarkable power of regeneration into the adult stage (see also Chapter 7). Most of our knowledge of vertebrate limb regeneration therefore comes from experiments with these animals — the newts and other salamanders.

Regeneration occurs by two main processes, morphallaxis, or reorganization of the tissues remaining, and epimorphosis, or regrowth of missing parts. In the adult urodele limb, some morphallaxis probably occurs during structuring of regenerative tissue, but for the most part regeneration occurs by new growth. In the early embryo morphallaxis is probably of major importance.

In the regenerating vertebrate limb new tissue arises either by dedifferentiation of differentiated tissues, followed by their redifferentiation into the same or other cell types, or alternatively by proliferation and late differentiation of cells, which have until then remained undifferentiated. The supply of cells which can be used for regeneration seems to be unlimited, since repeated amputation of the same limb is followed by unimpaired regeneration.

The initial morphogenesis of complex organs must utilize whatever guiding forces are available at that particular stage of development. Some of the forces which were involved in the initial formation of the organ may persist into later stages, but others do not, so that during regeneration the system must instead utilize the network of positional information which was established as a *response* to the embryonic forces.

As an illustration, areas within a sheet of epithelium could be defined like a map reference, by two co-ordinates, which we could calibrate 1–5 and A–E as in Figure 6.8. These two co-ordinates could represent two non-coincident biochemical gradients which still exist, or which existed previously in the embryo. In the latter case, if the cells within each section of the diagram responded appropriately to the morphogens in those sections, then once these responses had become established the induced pattern could remain, even though the initial gradients had disappeared. If region C3, for example, were then damaged, cells moving into this area could obtain accurate positional information by interaction with the surrounding cells in sections B3, C4, D3, etc.

Some such system does seem to exist in multi-cellular organisms. Transmission of positional information probably involves surface contact with other cells, or surface contact with extra-cellular material secreted by them, as well as exchange of small molecules through gap junctions (see Chapter 4).

Figure 6.8 represents a two-dimensional system. The natural co-ordinates utilized by cells in regeneration of three-dimensional organs can be represented by an all-embracing model, the Polar Co-ordinate Model of French, Bryant and Bryant (1976).

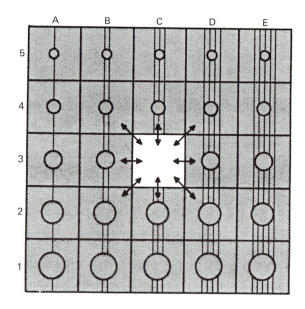

Figure 6.8. The use of positional information in regeneration.
The diagram represents the array of unique combinations of positionally informative molecules which could arise from two gradients orientated at right angles to one another. Cells which come to occupy a damaged area (C3) can obtain information about the normal phenotype of cells in this position by interaction with cells in neighbouring regions.

6.6. *The Polar Co-ordinate Model*

Some of the most puzzling examples of regeneration are shown following grafting, for example when the right foreleg of a newt is amputated and replaced by the left foreleg of another individual. If the grafted left leg is attached with the thumb pointing backwards and palm facing downwards (Figure 6.9(a)) it becomes integrated into the body, but two extra right limbs also develop, one from the anterior and one from the posterior junctions between graft and host. The newt then has three limbs on the right side of its body, two with thumbs pointing forwards and one with its thumb pointing backwards. If the grafted left leg is attached upside down so that the thumb points forwards (Figure 6.9(c)) two supernumerary right limbs again develop at the junction, but this time one develops from the dorsal and the other from the ventral interface.

The reason for these bizarre effects is that the newly formed undifferentiated cells at the graft junction obtain conflicting positional information from their neighbours. In 'attempting' to fill the gap between the locations specified, they create new tissue with intermediate positional values. In two locations around the limb the discrepancy between the positional values of host and graft tissue is so extreme that the discontinuity can be remedied only by creation of a complete new set of intermediate positions, that is by growth of a new limb.

Figure 6.9 shows a modified version of the polar co-ordinate representation of the graft–host junction. The outer circle represents the circumference of the right limb stump at the cut surface and the inner central circle that of the left graft. The two cut surfaces are drawn to different sizes for clarity only. Positional values are marked by numbers 0–12 clockwise for the left limb and anticlockwise for the right. Regeneration occurs so that all intermediate positional values are intercalated by the shortest possible sequence. At those positions where this requires introduction of a complete circle of values, new limbs develop. It can be seen that in both transplant situations there are only two positions around the circumference of the interface where the discrepancy between the two sets of values is that great. At these positions, new limbs develop and their laterality depends on the numbering system derived from the above rules.

This remarkable model does not seem yet to be fully accepted by developmental biologists, perhaps because it is rather difficult to conceive a network of positional information as complex as the clockface. However, as Figure 6.9 shows, the numbers can be replaced by two (or more) simple gradients of positionally informative molecules, without loss of any major feature.

Regeneration in the proximo-distal direction occurs by a related set of rules: positions down the limb, from proximal to distal ends, are defined by concentric circles of diminishing diameter. Most of the work on this aspect has been done with insect legs, and each section of the leg seems to have a closed set of information, as shown in Figure 6.10(a),(b). Regeneration occurs to produce all positional values central to the circle representing the cut surface and can occur only when a complete circle of values is exposed at an amputation site, or emerges because of inter-calary growth during regeneration.

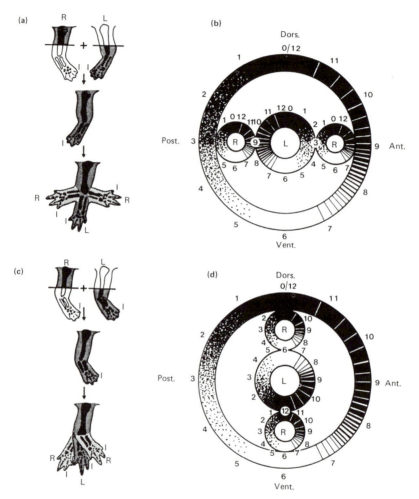

Figure 6.9. The Polar Co-ordinate Model for supernumerary limb development following limb transplantation in newts.
(a) A left limb is grafted on to the stump of a severed right limb, with the palm downwards, so that the thumb (I) points backwards. An extra right limb develops at both the anterior and posterior junctions. (b) The outer ring represents the circumference of the right stump and the central inner ring the circumference of the transplanted left limb. The other two inner rings represent the supernumerary regenerates. Positional information is represented by numbers 0–12, anticlockwise on the right and clockwise on the left limb, and also by two gradients, as shown. New limbs develop anteriorly and posteriorly, where there is maximum incongruity between adjacent tissues. At these two locations a complete set of gradients, or numbers, is produced when intermediate positions are intercalated by the shortest possible route. (c), (d) Regeneration of supernumerary limbs at the dorsal and ventral surfaces, following transplantation of a left limb in reverse orientation. (Modified from French *et al.*, 1976.)

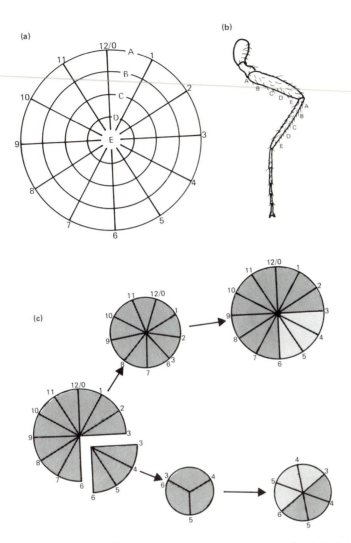

Figure 6.10. The Polar Co-ordinate System for positional specification ((a), (b)) and regeneration in imaginal disc fragments (c).

(a), (b) Each cell acquires features which can be specified, relative to those of other cells in the organ, with respect to its position on a radius (A–E) and on a circle (0–12). Positions 12 and 0 are identical, the others are as shown. Positions on radius A correspond to locations near the proximal end and those on radius E to the distal end of a limb or limb section. (c) A *Drosophila* wing imaginal disc is cut and a 90° sector taken out. When the 270° sector is cultured in the abdomen of a host fly, the cut edges fuse, bringing positions 3 and 6 together. Inter-calary growth then reconstitutes positions 4 and 5. When the 90° sector is similarly cultured, fusion of the cut edges again brings positions 3 and 6 together, but inter-calary growth by the shortest route then produces a mirror image duplication. (From French *et al.*, 1976).

The existence of these rules suggests a one-way flow of positional information along the proximo-distal axis: information flows down, but not up the limb.

Birds and mammals show a much reduced capacity for regeneration compared with urodele amphibians. The evidence is that the same positional rules apply, but the capacity for new growth is restricted. Thus when the distal tip of a chicken left wing bud is grafted to the contralateral limb stump, with the dorsoventral axes opposed, supernumerary muscle and skin structures form as in salamanders, but bone does not.

In amphibians the cellular process of fracture healing is directly related to the electrical phenomena produced by fractured bone. This observation has led to experiments with adult rats in which partial regeneration of bone was stimulated by application of an electrical current to the fracture surface. This raises fascinating possibilities for therapeutic limb regeneration in man which as yet, however, remain merely possibilities.

The Polar Co-ordinate Model has been shown to apply also to the limbs of insects and even to insect imaginal discs, even though these are essentially two-, not three-dimensional. In the case of the imaginal discs, the circular diagram is superimposed directly upon the disc (Figure 6.9(c)). If a sector is cut out of the disc, for example the quarter between 3 and 6 o'clock, and the fragments incubated within the abdomen of a host fly, both fragments will heal their wounds by fusion of the cut surfaces. This now brings positions 3 and 6 adjacent in each, and inter-calary growth replaces the missing positions, 4 and 5, in both cases. The result is that in one case a perfect disc is restored, whereas in the other a mirror-image quarter is formed. If these discs are allowed to differentiate, by implanting in a third instar larva (see Chapter 7), the restored disc develops into a normal organ, whereas the mirror-image partial disc develops into an organ which is deficient, but with mirror-image symmetry.

6.7. *Summary and conclusions*

Most of the studies on pattern formation in vertebrates have been performed on the chicken. In this species the segmentally arranged somites arise due to an apparently innate tendency for somitic mesoderm to form into condensates, following its division by the passage of Hensen's node during definition of the major body axis.

The pattern of organization of the vertebrate limb depends partly on the self-organizing capacity of limb bud mesenchyme, which is refined by exposure to two major influences. Definition with respect to the proximo-distal axis is related to exposure of proliferating mesenchyme to the inductive influence of the apical ectodermal ridge (AER) at its tip. The number, position and form of elements along the antero-posterior axis depends on a morphogen originating in the ZPA at the posterior margin. Distinction between fore and hind limbs is made by response to a factor that diffuses from adjacent body somites. In both insects and vertebrates, when appendages regenerate following injury or surgery, this involves reference to positional information held by neighbouring cells and probably expressed in their metabolism and extra-cellular products. This regeneration occurs in accordance with general rules that apply to both groups.

The experiments described in this chapter illustrate in a most dramatic way the situation of dynamic equilibrium which develops within each living organism and maintains it as an integral whole. An extremely complex web of tissue interactions maintains correct functioning of every system, the contribution of each to the whole being adjusted with respect to those from all other systems. In the next chapter we will consider some of the controls which operate at a cellular level to maintain each differentiated cell type in that state.

Bibliography

Alberts, B., Bray, D., Lewis, J., Raff, M., Roberts, K. and Watson, J., *Molecular Biology of the Cell.* Garland, New York (1983).

Becker, R. O., Stimulation of partial limb regeneration in rats. *Nature*, **235**: 109 (1972).

Brenner, S., Murray, J. D. and Wolpert, L., *Theories of Biological Pattern Information.* Cambridge University Press, Cambridge (1981).

Browder, L. W., *Developmental Biology.* Saunders College/Holt, Rinehart & Winston, Philadelphia (1980).

Bryant, S. V., French, V. and Bryant, P. J., Distal regeneration and symmetry. *Science*, **212**: 993 (1981).

Caplan, A. I. and Ordahl, C. P., Irreversible gene repression model for control of development. *Science*, **901**: 120 (1978).

Christ, B., Jacob, H. J., Jacob, M. and Wachtler, F., On the origin distribution and determination of avian limb mesenchymal cells. In *Limb Development and Regeneration; Part B*, edited by R. O. Kelley, P. F. Goetinck, and J. A. MacCabe. Liss, New York, pp. 281–291 (1983).

Cooke, J., The problem of periodic patterns in embryos. *Philos. Trans. R. Soc. Lond. (Biol. Sci.)* **295**: 504 (1981).

Ede, D. A., *An Introduction to Developmental Biology.* Blackie, Glasgow (1978).

Ede, D. A., Levels of complexity in limb-mesoderm cell culture systems. In *Differentiation In Vitro. British Society for Cell Biology Symposium 4*, Edited by M. M. Yeoman and D. E. S. Truman. Cambridge University Press, Cambridge, pp. 207–229 (1982).

Ede, D. A., Hinchliffe, J. R. and Balls, M., *Vertebrate Limb and Somite Morphogenesis.* Cambridge University Press, Cambridge (1977).

French, V., Pattern regulation and regeneration. *Philos. Trans. R. Soc. [Biol. Sci.]*, **295**: 601 (1981).

French, V., Bryant, P. J., and Bryant, S. V., A model for pattern regulation in epimorphic fields. *Science*, **193**: 969 (1976).

Grant, P., *Biology of Developing Systems.* Holt, Rinehart and Winston, New York (1978).

Hinchliffe, J. R., The chondrogenic pattern of chick limb morphogenesis: a problem of development and evolution. In *Vertebrate Limb and Somite Morphogenesis*, edited by D. A. Ede, J. R. Hinchliffe and M. Balls. Cambridge University Press, Cambridge, pp. 293–309 (1977).

Hinchliffe, J. R. and Gumpel-Pinot, M., Experimental analysis of avian limb morphogenesis. In *Current Ornithology*, Vol. 1, edited by R. F. Johnston. Plenum, New York, pp. 293–327 (1983).

Iten, L. E., Pattern specification and pattern regulation in the embryonic chick limb bud. *Am. Zoolog.*, **22**: 117 (1982).

Meinhardt, H., *Models of Biological Pattern Formation*. Academic Press, London (1982).

Saunders, J. W. Jr, The experimental analysis of chick limb bud development. In *Vertebrate Limb and Somite Morphogenesis*, edited by D. A. Ede, J. R. Hinchliffe and M. Balls. Cambridge University Press, Cambridge (1977).

Saunders, J. W., Cairns, J. M. and Gassling, M. T., The role of the apical ridge of ectoderm in the differentiation of the morphological structure and inductive specificity of limb parts in the chick. *J. Embryol. Exp. Morphol.*, **101**: 57 (1957).

Sporn, M. B., Roberts, A. B. and Goodman, De W. S., *The Retinoids*. Academic Press, London (1984).

Summerbell, D., Lewis, J. H. and Wolpert, L., Positional information in chick limb morphogenesis. *Nature, Lond.*, **244**: 492 (1973).

Tickle, C., Alberts, B., Wolpert, L. and Lee, J., Local application of retinoic acid to the limb bond mimics the action in the polarizing region. *Nature*, **296**: 564 (1982).

Torrey, T. W. and Feduccia, A., *Morphogenesis of the Vertebrates*, 4th edn. Wiley, Chichester (1979).

Wallace, H., *Vertebrate Limb Regeneration*. John Wiley, Chichester, (1981).

Wessels, M. K., *Tissue Interactions and Development*. Benjamin, Menlo Park, California (1977).

Wilby, O. K. and Ede, D. A., A model generating the pattern of cartilage skeletal elements in the embryonic chick limb. *J. Theoret. Biol.*, **52**, 199 (1975).

Wolpert, L., Mechanisms of limb development and malformation. *Br. Med. Bull.*, **32**: 65 (1976).

Wolpert, L., Pattern formation in biological development. *Sci. Am.*, **239** (4): 154 (1978).

Wolpert, L., Positional information and pattern formation. *Philos. Trans. R. Soc. Lond.* [*Biol. Sci.*], **295**: 441 (1981).

Chapter 7 Unstable differentiation

A feature of animals that we take very much for granted is that for the most part their skin always stays as skin, their nerves as nerves and their muscles as muscles. Although all the tissues are derived by differential expression of the genes originally brought together for the first time in the zygote, the tissues once differentiated normally remain in that state throughout the life of the animal. This stability is probably maintained by:

(a) stable maintenance of some of the embryonic influences which initially brought about that differentiation (see Chapters 2 and 3);

(b) dissemination of cytoplasmic constituents through gap junctions that link the cells within each tissue (see Chapter 4);

(c) intra-cellular biochemical feedback loop systems that maintain the activity of individual biochemical pathways within tissues (see Chapter 8);

(d) contact with extra-cellular matrices that stabilize cellular activities (see Chapter 4);

(e) stabilized production of controlling biochemicals of major effect, such as hormones (see Chapter 4);

(f) accumulation of metabolite pools, possibly as complexes with enzymes (see Chapter 8);

(g) chemical modification of the DNA or its associated proteins to ensure maintenance of tissue-specific patterns of gene expression (see Chapters 10 and 11);

(h) adoption by the organism of a stable environment and routine life-style.

The few exceptions to the general rule of tissue stability are very informative in revealing how tissue stability is normally maintained and how differential gene expression comes about. This chapter deals with some of these exceptions.

7.1. The Directed Somatic Mutation Theory

One of the earliest theories of cytodifferentiation was that during development the nuclear DNA may become progressively modified in different parts of the body, effectively establishing populations of cells with different genotypes. The modifications were considered to be irreversible and to be inherited by daughter cells. This theory, known as the Directed Somatic Mutation Theory, has for a long time been considered disproved as a general explanation, although cell genotype is recognized to be irreversibly modified in some situations. In Chapter 11 we will deal with some recent evidence which suggests that minor reversible modifications of the DNA may well be a widespread mechanism for control of gene expression, but in this chapter we will accept the better established view that most of the cells in an organism are genotypically equivalent.

The lens of the eye is an exception. Here maturation of epithelial cells into fibre-like cells involves degeneration and loss of the chromosomes, mitochondria and other organelles, so that the cells ultimately contain no genetic material. This is related to the acquisition of appropriate optical qualities in the lens. The nuclei are also lost during the formation of red blood cells in mammals, although not in birds. In some insects and nematode worms, chromosome material is lost from those cells which contribute to the body tissues, a complete chromosome complement being retained only in the germ cells. In the Salmonidae and several other fish, such as cod, the different body tissues seem to lose (or possibly gain) chromosomes, so that they come to contain different modal numbers. Whether or not this is a mechanism for differentiation of the tissues has not yet been established.

The phenomenon of X-chromosome inactivation by heterochromatization in female mammals (Chapter 10) illustrates that inactivation of genetic material can take place and that its patterning can be retained indefinitely in progeny cells.

There are also examples of increases in chromosome material. The genes coding for ribosomal RNA are amplified in vertebrate oocytes (Chapter 9), and in the two-winged flies certain tissues are characterized by massive increase in all the chromatin into what are called 'polytene' or 'giant chromosomes', containing about a thousand chromatids side by side (see Chapter 9). In mammals also some liver cells are polyploid, that is they contain several complete sets of chromosomes.

These examples, however, are exceptions to the general rule: *that all the cells in the body of a multi-cellular organism normally contain essentially the same genetic information.*

7.2. Retention of totipotency in somatic cell nuclei

Several experiments demonstrate that *nuclei* can remain totipotent, that is capable of coding for the properties of all types of tissue, while *whole cells* become progressively more restricted in their repertoire of expressible genes as development proceeds. Although there is not a great deal of evidence for this claim, it is generally considered to be the case in most cell types. This of course implies that a cell's state of determination (see Chapter 2) depends on extra-nuclear rather than nuclear factors.

The experiments which led to this realization involved transfer of nuclei into eggs in which the original nuclei had been destroyed by ultraviolet irradiation or surgery. In this situation the remainder of the enucleated egg has the capacity to mobilize expression of genetic information in the somatic nuclei transplanted into it. This system has been used with great effect by Gurdon, and Briggs and King, in investigating the restrictions upon developmental potential of somatic cells.

In the first experiments of this type nuclei were taken from epithelial cells of tadpole intestine and transferred into enucleated amphibian eggs; some of these developed into normal tadpoles. This experiment demonstrated the very important point that the nuclei of differentiated cells can retain the potency to form complete organisms, but this interpretation initially was not accepted by all authorities. It was pointed out that in frogs and toads the migratory germ cells pass among those which give rise to the gut, so development could have occurred from totipotent germ cells left behind. On the other hand, since most of the transplant embryos developed into *abnormal* tadpoles, some critics suggested that the nuclei were probably deficient in genetic information due to the prior differentiation of the gut cells from which the nuclei were taken.

The abnormalities of the nuclear transplantation embryos were of a variety of types and this is inconsistent with the notion that cytodifferentiation of intestine epithelium occurs by *directed* somatic mutation. Furthermore, when examined under the microscope, their cells were shown to have suffered chromosome losses, which was clearly the basis of their abnormalities. It was suggested that this could have arisen due to reduction in the chromosome replication rate which normally occurs with advancing age, so that after transplantation some chromosomes failed to be incorporated into mitotic spindles and were lost.

In order to test this idea, nuclei were taken for transplantation from frog embryos at different stages (Figure 7.1). When taken from *Xenopus laevis* blastula cells some 80% of the transplanted nuclei supported normal development. If taken from endoderm cells at the neurula stage, this proportion was reduced to about 50%, while only 15% of intestine nuclei from swimming tadpoles produced clonally derived swimming tadpole offspring.

In a modification of the original experiments, nuclei from tadpole intestine cells were transplanted into enucleated eggs, the embryos were allowed to develop to the blastula stage and the blastula nuclei were again transplanted into enucleated eggs. In this experiment, 70% of the original intestinal nuclei gave rise to normal swimming tadpoles, presumably because this manoeuvre allowed adjustment of the chromosome replication rate so that it fell in phase with cell division. The experiment showed that at least seven out of 10 tadpole intestinal nuclei retain all the genetic information required for development of a whole frog.

Later experiments of this type involved nuclei taken from skin cells of the foot webs of adult frogs and from tadpole tail fins, the cells being indicated as fully differentiated by their capacity to bind antibodies raised against the skin protein, keratin. If the cells were cultured *in vitro* for a few days before transfer they supported development of normal embryos. These experiments confirm that *the nuclei of differentiated cells can contain all the genetic information for a complete animal*.

The point is demonstrable in a much more dramatic fashion by experiments

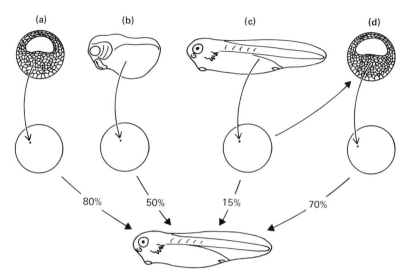

Figure 7.1. Development of *Xenopus laevis* somatic cell nuclei transplanted into enucleated *Xenopus* eggs.
When taken from the blastula, (a) 80% of nuclei produced normal tadpoles, 50% of neurula endoderm cells (b) supported normal development, but when taken from tadpole intestine epithelium (c) only 15% produced normal animals. When transplant embryos derived from intestine epithelial nuclei were dissociated at the blastula stage and their nuclei re-injected into enucleated eggs, (d) 70% of the original nuclei gave rise to normal tadpoles. (Based on experiments by Gurdon.)

with plants. For example, a whole carrot plant can be grown from a single carrot phloem cell. Horticulturists regularly propagate plants from cuttings, establishing numerically enormous genetic clones derived from the leaves, twigs or roots of a single individual, all the clonally derived offspring being genetically identical to the donor parent.

The other particularly important point revealed by nuclear transplantation experiments is that, since somatic cell nuclei are totipotent, the differentiated states of donor cells must be based on properties of the cytoplasm, the membranes, or other extra-nuclear features, not just on the nucleus itself.

7.3. Colour changes

Pigmentation is a prominent aspect of phenotype that has been the subject of much investigation, but some rules derivable from its study are probably of much broader relevance, applying to many other phenotypic features.

Many animal species show colour fluctuations that occur with varying degrees of rapidity. In the Northern and Arctic regions of the globe, some mammals and birds grow white coats in autumn and brown in the spring, moulting out the fur or feathers of the last season. Some lower vertebrates, notably chameleons and flounders, can change in a matter of minutes through a range of dark and light shades, allowing them to merge more effectively with their new surroundings. In these species devoid of fur or feathers, colour is carried in the epidermis in chromatophore cells, which include melanocytes containing the yellow, brown or black pigment, melanin, other cells containing red or yellow carotenoid pigments and guanophore cells carrying crystals of guanine, which alter the effects of the pigmented materials by light reflection.

Within melanocytes melanin is accumulated in cytoplasmic organelles called melanosomes. These develop on fibres of protein to which molecules of the enzyme tyrosinase become bound before they synthesize melanin about themselves. Rapid colour changes are achieved by expanding and contracting the different kinds of chromatophores and by changing the distribution of the melanosomes within them. In some cases control of these changes is a direct response of the skin to sunlight, in others impulses are transmitted via the eyes, possibly with the mediation of hormones, and in some species such changes occur in response to temperature rather than light. Human skin tanning depends initially on photo-oxidation of melanin precursors by visible and u.v. light, but the u.v. light exerts an additional delayed effect in causing the melanocytes to proliferate and increase their extrusion of melanin granules.

Many genes permanently influence the colouring of vertebrates. These control the number and arrangement of the protein fibres in the melanosomes, the distribution of pigment within them and the size, shape and distribution of the melanosomes. Some genes cause feathers or hairs to develop banded patterns. This involves fluctuation in melanin synthesis during growth, and may represent a natural equivalent to the modulation of pigment synthesis that occurs in melanocytes growing in culture (see below). The wild-type mouse hair pattern called agouti involves a switch in synthesis from eumelanin (black) to phaeomelanin (a yellow modification) and back to eumelanin again, during the growth of the hair, producing a shaft that is black at top and bottom, but yellow in the middle. This control seems to be exerted on the epidermal cells by the underlying dermis (see Chapter 3) and can be regulated artificially in skin explants by adjusting the availability of the substrates L-tyrosine and L-DOPA (see Figure 7.2).

Some pigmentation genes are temperature sensitive. For example, Himalayan pattern rabbits and mice are coloured like Siamese cats, with pale fur on the main part of their bodies, but dark fur on the nose, ears, tail and paws. This patterning is related to variation in thresholds of action of a temperative-sensitive gene in different regions of the body, the threshold for expression on the torso being lower than at the extremities (Figure 4.5).

As might be expected, pigmentation also depends on diet. Dramatic examples are the phenocopy of albinism in kwashiorkor patients due to dietary deficiency of tyrosine, and the pink coloration of flamingos derived from carotenoids in their natural food.

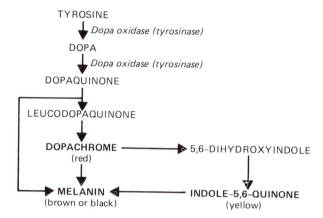

Figure 7.2. Pathways of melanin synthesis.
The production and oxidation of DOPA (dihydroxphenylalanine) are the only steps that require enzyme catalysis. Not the formation of coloured intermediates.

It can be seen that colour variation can occur as the result of a wide variety of differences in cell physiology. Furthermore, there is no reason to assume that pigmentation obeys unique rules. It is to be expected that many other aspects of phenotype may undergo similar long- or short-term fluctuations in the synthesis of specialized cell products, depending on a variety of conditions.

Modulation of pigment cell phenotype

The type of physiology adopted by a cell depends on the conditions to which it is subjected, and when these are such that synthesis of the major tissue-specific products is discontinued or replaced by other activities the cell is said to show modulation. Modulation was defined by Paul Weiss as "a reversible fluctuation within the already established range of determination, together with the covert maintenance of the potentiality to express an original function". In other words, cells which are showing modulation temporarily switch off some of their normal functions, but can switch them on again when the original conditions return.

When cells are dissociated from their fellows and cultured in a dispersed fashion they normally adopt forms and physiologies rather different from those in the intact tissue. For example, monolayer cultures of chondrocytes, the cells that produce cartilage, lose their ability to synthesize the sulphated mucopolysaccharides made *in vivo*; instead, they commence synthesis of DNA and proliferate. However, if reaggregated they recommence mucopolysaccharide synthesis. Chondrocytes retain this capacity even after 30 mitotic generations *in vitro*, and can still differentiate as cartilage if reorganized into aggregates.

In a similar fashion, when retinal pigment epithelial cells are disaggregated and cultured in monolayer, after initially adhering to the dish they spread, proliferate and lose their pigment. Then, when the culture becomes crowded, they cease division and begin again to accumulate melanin. Depigmentation is proportional to culture growth and there seem to be two contributory factors involved: dilution of pigment by increasing cytoplasmic volume and cessation of melanin synthesis.

Melanin is produced from tyrosine by the sequence of reactions shown in Figure 7.2 (see also Figure 3.8). Although this is quite a complex biochemical pathway, tyrosinase (DOPA oxidase) seems to be the only enzyme essential for melanogenesis, the other reactions apparently occurring spontaneously. Depigmentation of proliferating pigment epithelial cultures involves reduction in tyrosinase activity before accumulation of pigment.

A particularly important point is that these modulatory changes occur without fluctuation in synthesis of the tyrosinase RNA transcript. In other words, modulation occurs in response to an influence operating between transcription and translation, the accepted explanation being that tyrosinase mRNA competes for translational facilities with other messengers. It is assumed that when pigment cells are actively proliferating, the urgent messages concerned with growth are translated preferentially, so that messenger RNA coding for tyrosinase is translated only after the cells cease active growth.

In summary then, modulation of pigment-cell phenotype can occur in relation to variation in the supply of enzyme substrates, without fluctuation in tyrosinase activity (see above). Or it can occur through variation in the demands put upon the cells to carry out other activities, such as proliferation, in which case tyrosinase activity may also vary. It should be stressed, however, that neither condition has been shown to bring about a permanent quantitative or qualitative change in the profile of RNA species synthesized, and the phenotype reverts when the unusual conditions are relieved. This modulation is therefore distinctly different from the transdifferentiative changes of a more permanent nature that can also occur in pigment epithelial cultures (see below).

Experiments with hybrid cells

A particularly elegant approach to the control of gene expression is to cause cells of dissimilar types to fuse together, so that nuclei with different histories are together present within a common cytoplasmic pool. The first experiments of this kind involved fusion of avian red blood cells with mouse fibroblasts. Unlike mammalian erythrocytes, those of birds retain their nuclei, but in a condensed and transcriptionally inactive state, together with very little cytoplasm. In contrast, the fibroblast is a highly active, although relatively unspecialized, cell type, with a large cytoplasmic volume. In the heterokaryon formed by fusion, the activity of the fibroblast nucleus remained active and apparently unchanged, but that of the red cell decondensed and began transcribing RNA again. Although the red cell had previously been engaged almost exclusively in haemoglobin synthesis this was not remobilized, only the general 'housekeeping functions' of the cell being reactivated (Figure 7.3(a)). It was concluded that the fibroblast cytoplasm contains

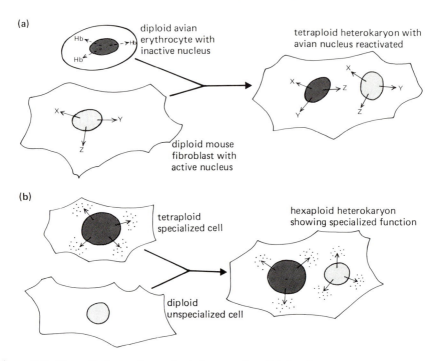

Figure 7.3. Control of protein synthesis in somatic cell hybrids.
(a) A diploid avian red cell with an inactive nucleus which had formerly been synthesizing haemoglobin was fused with a diploid mouse fibroblast carrying out a variety of unspecialized functions. The avian nucleus became reactivated and produced unspecialized products similar to those produced by the fibroblast, but did not resynthesize haemoglobin. (b) A tetraploid cell synthesizing specialized products was fused with a diploid unspecialized cell. In the heterokaryon, specialized syntheses continued in the tetraploid nucleus and in some cases the same syntheses were mobilized in the formerly unspecialized diploid nucleus.

diffusible substances with the capacity to reactivate genes in the condensed red cell nucleus, but only with respect to features already expressed in the fibroblast itself.

In another set of experiments, cells of a pigmented melanoma line (i.e., a tumour of skin melanocytes) containing either a diploid or a tetraploid set of pigment-cell chromosomes were fused with unpigmented diploid fibroblasts. When the two genomes were numerically balanced in the hybrid, pigment synthesis was extinguished, but in some hybrids containing two complete sets of pigment-cell chromosomes, together with only one set from the fibroblast, pigment synthesis continued. In experiments of this type, specialized syntheses have even been switched on *de novo* in the genome represented at inferior ploidy (Figure 7.3(b)). These experiments indicate that specialized syntheses are controlled in a quantitative fashion. The controls have generally been considered to operate either at transcription or translation.

The general deduction from a range of experiments of this type is that genes coding for 'housekeeping functions', that is those common to most cell types, are mobilized by diffusible molecules in the cytoplasm of cells which express those functions. These molecules switch on and maintain those functions apparently without strong competition from inhibitors or suppressors. In contrast, specialized functions, such as the synthesis of haemoglobin and tyrosinase, seem to depend on the ratio between suppressive and expressive factors and suppressor predominates in the cytoplasm of cells not specialized for that function. The expressive factors seem to be specific for each gene, or possibly for a small group of genes contributing to the same function, and are present in sufficient concentration to overcome the diffusible suppressor.

7.4. Definitions relating to unstable differentiation

Apart from the experiments on somatic cell hybrids, which represent an artificial situation, up to this point we have considered modulations of phenotype that fall within Weiss's definition in that they represent reversible physiological variation without permanent change in the state of determination of the cell or tissue. Other changes of a more radical nature can occur that involve alterations in both differentiation and determination, some of which call into question the general understanding of those terms (see Chapters 2 and 3).

The most recent relevant definitions are those given by Yamada in 1982. 'Differentiation' denotes the "appearance and establishment of the structural and functional properties associated with specificities of somatic tissues and their cell types". He points out that differentiation occurs in a step-wise fashion and that the stability of the differentiated state increases in parallel. This implies restriction in the range of realizable pathways of development. The fixation of developmental fate that results from this restriction is called determination (see Chapter 3).

Radical 'reprogramming' of embryonic cells after determination is completed, but before the differentiated state is obtained is called transdetermination. The prototype of this event is the alteration of organ type in *Drosophila* imaginal discs, following their partial disaggregation and culture for extended periods (see below).

Re-routing of the differentiation pathway after a group of cells has begun to express its specific differentiated state is called transdifferentiation. Metaplasia is a related term originally used to denote the pathological alteration of tissue specificity *in vivo*. An example is the change between keratin- and mucus-secreting activities by skin epithelia in relation to exposure to vitamin A (see Chapter 3). At first sight this change may appear similar to the modulations observed in cultured cells. In fact it is quite different, as the effect is produced by calling into play a new group of precursor cells which were not originally committed to keratin synthesis.

It should be noted that none of these definitions includes a molecular explanation of events. Indeed, we really have little idea whether such states, or changes of state, occur by means which are common to several cell types or unique to each one. In the sections which follow we will attempt to make some headway in this direction.

7.5. *Transdetermination of insect imaginal discs*

The life-cycles of insects are divided into phases which are phenotypically distinctly different from one another. In the higher insects (Holometabola or Endopterygota) the distinction between the last larval phase, or instar, and the adult, or imago, is very marked and the transition takes place during a pupal stage when the larval organs break down to be replaced by adult structures (see Chapter 4). This involves the growth and development of packages of cells called imaginal discs, which have remained in a determined but overtly undifferentiated state throughout the larval stages. The imaginal discs are pouches of epithelium shaped like crumpled deflated balloons that evaginate and differentiate at metamorphosis, in response to ecdysone stimulation (see Chapters 4 and 10). In *Drosophila* there is one median disc that forms the genital structures and 10 pairs of major discs corresponding to wings, legs, halteres, mouth parts, eye plus antenna, and dorsal prothorax (Figure 7.4).

Drosophila imaginal disc cells can be cultured if the disc is cut into pieces and the pieces injected into the abdomen of an adult fly. When the host fly grows old the disc tissue can be removed and transplanted into young adults. Under these

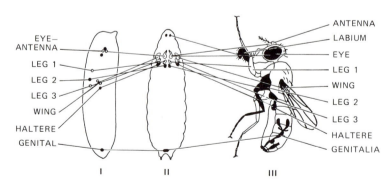

Figure 7.4. The embryonic origin of adult body structures in *Drosophila melanogaster*. (a) The location of imaginal disc progenitor cells at the blastoderm stage. (b) The positions of imaginal discs in the third-instar larva. (c) Organs of the adult fly (Based on Wildermuth, 1970.)

conditions the cells will proliferate, but remain undifferentiated since they are not exposed to ecdysone, which is not synthesized in the body of the adult. If disc cells are injected into a third-instar larva that is about to pupate, they come under the influence of ecdysone and within the pupa will differentiate into cells characteristic of adult organs. Structures derived from donor discs taken from larvae that would develop into dark male adults and raised in larvae that would develop into yellow female hosts can readily be distinguished from host tissues by both their colour and chromosomal constitution.

If fragments of imaginal discs are transplanted only a few times between adults and then placed in a third-instar larva, most will develop into rudiments of the type of organ for which they were originally destined. For example, cells derived from wing discs will usually develop bristles of a type characteristic of the wing and those from a leg disc will grow bristles of leg type. Although the cells of the different discs are by most criteria indistinguishable at the disc stage, they are in fact determined with respect to their ultimate fate. In some cultures of this type determined states have been retained faithfully for eight years or more, through over 1500 mitoses. This is a very potent demonstration that at this early stage imaginal disc cells are really determined with respect to their future development.

Some disc cells seem less firmly determined, however, especially if the cells are isolated by disaggregation of the discs into cell suspensions. When later exposed to ecdysone, instead of following their initially determined pathway such cells develop instead into the rudiments of organs of other types! Thus wing disc cells may develop features characteristic of the eye, antenna or legs. Such cells are said to have undergone transdetermination. Ernst Hadorn, the discoverer of this intriguing property, defined transdetermination as "a change in the determined state of cells to a different state, which will initiate a pathway of differentiation leading to structures that no longer correspond with the initial specificity of determination".

The mechanism of transdetermination

The organ type resulting from transdetermination is strictly limited, depending on the original disc, and follows the pattern shown in Figure 7.5. Inspection of this diagram shows that the resultant organ type tends to be one that is anatomically adjacent to that originally defined, or to be in an adjacent segment, although that is not a strict rule. However, the frequency of transdetermination in one direction is usually very different from that in the reverse. All changes towards mesothoracic features are more frequent than those away from mesothorax and, surprisingly very few if any discs have ever transdetermined to genital specification. There is thus a general tendency for transdetermination to occur away from genital specification and towards mesothorax.

Since only a proportion of the cells in each imaginal disc seem to have the capacity to transdetermine, the relative frequencies which have been reported for the different discs may be unreliable. Nevertheless, transdetermination frequency is directly related to degree of proliferation, and if a mutation is induced in proliferating disc cells it can be shown that transdetermination occurs in groups of adjacent cells that are not necessarily clonally related. This suggests that the effect is a property shared by communication between neighbouring cells (see Chapter 4), rather than being due to a nuclear modification that is distributed to daughter cells.

A very interesting interpretation of disc determination and transdetermination was proposed by Kauffman, who described the determined states of the discs in terms of a combination of four binary alternatives. In this model 0 represents the less stable, and 1 the more stable of alternative determined states. States defined by 0 symbols thus tend to transdetermine into those defined by 1, for the equivalent property. In Kauffman's code, genital discs are represented by 0000, eye–antenna by 1100, leg by 1010, haltere by 0111, and both wing and mesothorax by 1111 (Figure 7.5). This representation may seem little more than another version of the information indicated by the arrows, but this coding was in fact derived independently of transdetermination, by classification of a large number of mutant genes affecting complementary sets of adult organs. These symbols may therefore represent master genes that together define the phenotypes of the adult organs.

Kauffman suggests that the distinction between the 0 and 1 determined states could originate in the position at which the disc cells arise in the blastoderm, with respect to thresholds associated with four gradients in the egg cytoplasm (see Figure 7.4 and Chapter 2). On this theory, two gradients from anterior to posterior, and two running from the ends to the middle of the egg, would be sufficient to define the differences. Further, it has been suggested that the gradients may relate to waves of mitotic activity. The components of the gradients are still unknown, but from what we have learnt from other systems (e.g., see Chapters 2, 4, 5, 9 and 10), we might expect protein or RNA to play a part.

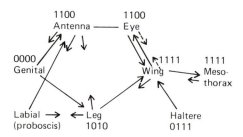

Figure 7.5. Transdeterminations among the imaginal discs of *Drosophila melanogaster*. Lengths of arrows represent approximate reported frequencies of transdetermination. The numbers indicate the binary code model of determined states derived on the basis of pairs of complementary mutant classes affecting combinations of organs (Based on Hardorn, 1978, and Kauffman, 1973.)

A molecular model for transdetermination

All the changes that occur in the discs at metamorphosis can be initiated by ecdysone, provided it is acting within an insect body. At a molecular level then, the state of determination of an imaginal disc seems to be represented primarily by the pattern in which the DNA is 'primed' to respond to ecdysone. There is direct evidence that ecdysone binds to genes that are transcribed in response, and since ecdysone is a steroid hormone we might as a first assumption expect this operation to be basically similar to that of progesterone (see Chapter 4). If that is the case, we would expect the cytoplasm of ecdysone target cells, including disc cells, to contain receptor proteins with which the ecdysone first becomes complexed. Since imaginal discs respond to ecdysone only in the absence of juvenile hormone, the latter may possibly compete for the same cytoplasmic receptor (see Chapter 4).

At the DNA level, avian cells responsive to progesterone are characterized by accessibility of a chromosomal protein called AP3 bound at, or close to, responsive genes, which in non-responsive cells is masked by other chromosomal proteins (AP1 and AP2). When the cytoplasmic receptor–progesterone complex passes into the nucleus of the target cell, it binds to the DNA–AP3 complex where it initiates transcription (see Chapter 4). Whether or not ecdysone acts in a comparable fashion is uncertain, but if we provisionally accept this concept, Kauffman's binary code allows us to postulate a molecular model for imaginal disc determination and transdetermination.

Assume there are four sets of genes — D1, D2, D3 and D4 — corresponding to the four symbols in Kauffman's code, each of which is primed by a separate 'primer protein', equivalent to AP3 in the progesterone system. We could call these P1, P2, P3 and P4 respectively, imaginal disc cells being characterized by the presence of these proteins bound beside each of the D genes. Assume also that disc cells contain ecdysone receptor protein in their cytoplasm. Some of the DNA-bound P proteins are accessible to receptor–hormone complex, but others are masked by proteins equivalent to AP1 and AP2. We can call these masking proteins M1, M2, M3 and M4, depending on which P protein they mask. Presence of masking protein at any D–P site in an imaginal disc cell is indicated by 0 in Kauffman's code and its absence by 1 (Figure 7.6). Thus a leg, defined as 1010, is deemed to have the D2–P2 and the D4–P4 sites masked by M2 and M4 respectively, while the D1 and D3 sites carry their appropriate P proteins (P1 and P3) unmasked and accessible to ecdysone–receptor complex. On exposure to ecdysone, this disc therefore expresses genes D1 and D3.

It is suggested that specific patterns of gene masking are laid down initially with reference to morphogenetic gradients in the egg cytoplasm (see Chapter 2) and are maintained in proliferating cells by control of the synthesis of the individual masking proteins. On this model transdetermination of one type may occur as a consequence of loss of functional masking protein (Figure 7.6). If a group of leg imaginal disc cells (1010) should lose the capacity to synthesize M2 and M4, or

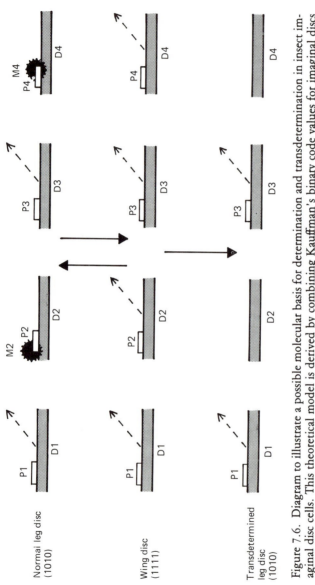

Figure 7.6. Diagram to illustrate a possible molecular basis for determination and transdetermination in insect imaginal disc cells. This theoretical model is derived by combining Kauffman's binary code values for imaginal discs of *Drosophila melanogaster* with current models for transcriptional control by steroid hormones (see Chapter 4). Four master genes, D1, D2, D3, D4, are considered to be primed by primer proteins P1, P2, P3, P4 respectively, which act as attachment sites for complexes of ecdysone with cytoplasmic receptor. Binding of this complex (which is not represented in the diagram) initiates transcription of the master genes. The primer proteins are variously masked by masking proteins M1–M4, depending on the organ type for which that disc is determined. In wing discs (code 1111) all P proteins are unmasked and all D genes expressed on exposure to ecdysone, whereas in normal leg discs (1010), masking proteins M2 and M4 block transcription of D2 and D4 respectively. Transdetermination from leg to wing could occur by loss of functional M2 and M4. Transdetermination from wing to leg could occur either through regaining functional M2 and M4, or by losing P2 and P4.

should lose a co-factor necessary for M2 and M4 to maintain active configurations (see Chapter 8), all the P–D complexes would be unmasked. On ecdysone exposure, all four D genes would then by expressed as in wings, instead of only D1 and D3 as in legs, and the disc would be deemed to have suffered transdetermination.

Hadorn originally suggested that transdetermination may be precipitated by discrepancies in the rate of DNA replication relative to other aspects of cell metabolism. It is here suggested that this could involve cessation of synthesis of specific M proteins, or a failure in their activation, as a result of depletion of cytoplasmic factors through growth (see Chapter 8).

This model allows scope for several means by which transdetermination could occur in the reverse direction — in the change from expression to non-expression of an organ-specific master gene:

1. Conditions within a group of disc cells may change, so that synthesis of a previously unexpressed masking protein becomes mobilized.

2. A specific masking protein may have been present all along, but in an inappropriate allosteric conformation (see Chapter 8). Under new conditions it may change its form and become functional.

3. The conditions which normally maintain synthesis of a specific primer protein may become modified, so that its particular D gene eventually becomes inaccessible to receptor–hormone complex. A wing disc (1111) expressing all four master genes, could thus transdetermine into leg (1010) by loss of P2 and P4, by new synthesis or activation of M2 and M4, or by any combination of these events (Figure 7.6).

Transdetermination occurs co-ordinately in groups of proximal cells, not in isolated cells or cell clones, so whatever the molecular mechanism, it probably involves the agency of cytoplasmic metabolites small enough to pass through gap junctions (see Chapter 4). It is suggested that these small metabolites may be essential for the co-ordinated synthesis or properties of the M and P proteins.

Although this model has been presented in considerable detail, the reader should be aware that it is highly speculative. As yet we are largely ignorant of the way ecdysone acts at the chromatin level, and none of the putative controlling proteins has been isolated or identified. Nevertheless, it is hoped that the model may provide a focus for constructive thought and experimentation.

7.6. Regeneration of the eye lens

The regeneration of damaged tissues and organs provides dramatic indications of the pluripotency of differentiated cells. The best-studied example is that of lens regeneration in the group of Urodele amphibians known as salamanders, which includes the newts. The salamanders are unusual in that the developmental

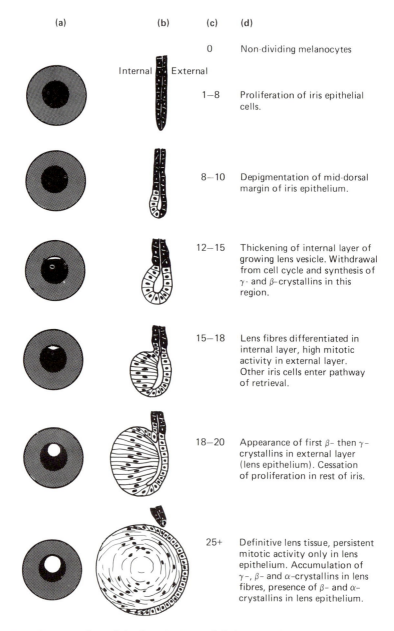

Figure 7.7. Regeneration of the lens in *Notophthalmos viridescens*.
The external appearance of the lentectomized eye (a) is represented alongside a median sagittal section through the dorsal iris (b), the time in days after lentectomy (c) and events occurring in the iris epithelial cells and their derivatives (d). (After Yamada, 1977.)

plasticity characteristic of embryonic states continues into adulthood. Nevertheless, the principles they illustrate are probably applicable to vertebrate embryos in general.

If the lens is removed from the eye of an adult newt, a new one derived from the pigment epithelium of the dorsal iris will regenerate in precisely the correct position (Figure 7.7). This phenomenon was first described by Colucci in 1891, but was rediscovered independently by Wolff four years later and has become known as Wolffian lens regeneration.

Following lentectomy, DNA synthesis recommences in several parts of the iris pigment epithelium, mainly in the dorsal sector. These areas become depigmented and the cells proliferate into rounded masses, most of which later regress. However, the one closest to the dorsal midline swells and begins to synthesize the lens-specific crystallin proteins. As the new lens grows, it detaches from the iris and eventually, within six weeks or so, a perfect functional replacement lens has formed.

Initial changes are most profound in the dorsal pupillary margin. There chromosomes replicate and cell division commences at about day 4. The iris epithelial cells then detach from one another, while their melanosomes and much of the rest of the cytoplasm is lost by several processes together known as cytoplasmic shedding. When all pigment has been lost, crystallin proteins start to appear in the cells and their sequence and timing of appearance follow essentially the same course as seen during embryonic development of the normal lens (see Figure 7.7).

Control of lens regeneration

Following their initial proliferation, iris epithelial cells are faced with two alternatives. They can either embark upon the 'pathway of conversion', lose all their melanosomes and become lens cells, or they can enter the so-called 'pathway of retrieval' and revert back to iris cells. The two pathways are distinguished by the length of their cell-cycle times, which are about 46 and 79 hours respectively, the difference being in the S and G1 phases. Cells in the retrieval pathway complete less then four cycles before they settle down again as pigment cells, whereas those on the conversion pathway go through at least six cycles as they convert to lens phenotype. This difference may be related to mitochondrial density, which is greatest at the dorsal pupillary margin.

It is thought that regeneration may be initiated by the absence of physical contact between lens and iris, together with the absence of a diffusible product of the lens. Lens regeneration occurs only if the original one is removed or displaced, but is suppressed by lens proteins injected into lentectomized eyes. Another important feature is that lens regeneration occurs only in the presence of a diffusible product of the neural retina, as can be demonstrated by inserting barriers between the retina and the iris.

Some important clues to the role of the neural retina were provided by experiments in which iris epithelium was taken out of the eye entirely and cultured in tissue-culture medium. Intact dorsal iris epithelium maintained in isolation shows a limited capacity to change into lens tissues, but this is strongly enhanced by the presence of neural retina. In contrast, intact ventral iris shows no such tendency. On the other hand, if iris epithelium is dissociated into single-cell suspensions and seeded into culture vessels, these cultures will develop lentoid bodies composed of lens-like cells containing crystallin proteins. This occurs readily in the absence of all other tissues and is not promoted by products of the neural retina. Furthermore, under these conditions dissociated pigment cells from ventral iris will produce lentoid bodies equally as well as those from the dorsal sector.

These experiments suggest that the function of neural retina in Wolffian lens regeneration is to assist cytoplasmic shedding and disaggregation of the epithelial cells. Dorsal iris has the greater capacity for this response *in vivo*, but when the tissues are artificially disaggregated the distinction is eliminated and the cells in both sectors express an innate tendency to transdifferentiate into lens. Iris cells in the dorsal sector have the most mitochondria. In the next section we will see that the activity of the TCA cycle enzymes which they carry is critical for another kind of transdifferentiation, from neural retina into pigment epithelium.

Observations on the lens regeneration system therefore suggest that the differentiative stability of iris epithelium depends partly on intra-cellular cytoplasmic factors, partly on factors that operate at the tissue level, and partly on influences associated with other tissues. These include the presence of melanosomes (and possibly mitochondria) and other cytoplasmic material in the iris cells; stable adhesion of these cells to one another and to the basal lamina; appropriate concentrations of diffusible products of the lens and neural retina; and possibly physical contact with the lens capsule. One theory is that the diffusible lens product inhibits transdifferentiation through direct action on the iris. Another idea is that the lens product acts indirectly, by neutralizing a product of the retina that would otherwise cause the iris to form more lenses. As yet, none of the mechanisms involved is understood.

Neural retina seems to have two roles in lens regeneration: first, by promoting cytoplasmic shedding during dedifferentiation, second, by enhancing the differentiation of lens fibres once the initial changes have occurred. In the latter role it probably acts in a similar way to the optic vesicle during normal lens induction, promoting a natural propensity of pigment cells to transdifferentiate into lens (see Chapter 3).

Before they lose their melanosomes the iris cells are really undergoing a form of modulation. When the melanosomes have gone, the cells seem to be in a differentiative limbo before they begin the first tentative moves along the pathway of conversion to lens. It has not yet been established whether melanosomes *per se* contribute cytodifferentiative stability, or whether they are merely markers of other more important cytoplasmic factors that are lost at the same time.

Principles illustrated by lens regeneration

This example illustrates what is probably a very important general feature of

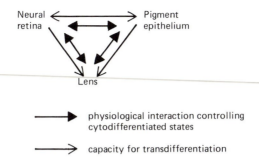

Figure 7.8. The hypothetical tripartite tissue interaction system in the vertebrate eye. A variety of experiments indicate that these three tissues may maintain their differentiative integrity by mutual physiological interaction. Transdifferentiation can occur from both of the retinal components to either of the other tissues, but lens cells do not transdifferentiate.

development: that organs and organisms comprise sets of balanced *systems*, regeneration occurring as a response to disruption of these systems. Regeneration takes place in the eye as if to restore equilibrium where this has been lost by removal of the lens. The 'eye system' appears to comprise three main elements: neural retina, pigment epithelium and lens, which normally exert some influence upon one another that maintains their relative quantitative and spatial representation (Figure 7.8). In salamanders, if the neural retina is removed a new one regenerates from the pigment epithelium behind. A second retina will form even if the first is merely displaced away from the pigment epithelium. The presence of one lens normally inhibits another from forming, by causing the iris pigment epithelium to remain in its original state. If the lens is removed a new one will regenerate from the pigment epithelium of the iris, but only with the aid of the neural retina. During normal development of the vertebrate eye, the lens and the two parts of the retina develop only after inductive interactions between them have occurred, and the integrity and health of lens and retina remain for ever intimately interrelated (see Chapter 3). Cytodifferentiation, growth and spatial position therefore seem to be intricately related in the eye, and the eye as a whole controls the phenotype of its constituent parts.

This interpretation, that cytodifferentiative states are dependent upon tissue interaction, emphasizes another basic principle of embryology; that development proceeds from states of imbalance to one of balance. Organisms adopt standard forms and proportions of tissues of different types because it is only in those states that they achieve physiological and morphological equilibrium in the context of the genetic information available for expression and the predominant environmental influences.

7.7. *Transdifferentiation of vertebrate eye tissues*

Regeneration of lens and neural retina, as described above, have actually been demonstrated only in certain species, and in these regenerative capacity declines with developmental stage. Transdifferentiation is a related phenomenon that occurs in cultures grown from dispersed cells. In one or another species, or stage of development, pigment epithelium taken from the retina will change phenotype in culture into both lens and neural tissue, while neural retina will similarly transdifferentiate into pigment epithelium and lens. Lens fibres represent a terminally differentiated state, having lost their nuclei and many of their cytoplasmic components. They do not change into any other type of cell. The transdifferentiation which takes place in retinal tissues isolated in culture can be viewed as their 'attempt' to replace the other two elements of the tripartite balanced system that normally occurs in the intact eye. From this viewpoint, transdifferentiation in culture and regeneration *in vivo* can therefore be seen as two sides of essentially the same coin.

Neural retinal cultures established from chicken embryos provide a particularly fruitful system for investigating the molecular events that occur during trans-differentiation. This system has been developed with great effect in the laboratories of Okada and Clayton. In these experiments neural retina is dissected out, dissociated by trypsin digestion and inoculated into dishes of tissue-culture medium. The cells stick down singly or as small aggregates, from which some move out to form sheets of cells which in places pile up as multilayers. At about 30 days, lentoid bodies arise from the multilayers or are formed apparently by rounding out of the remaining aggregates. The lentoid bodies have the protein composition and internal structure characteristic of lens fibre cells. Cell division within the monolayer results in an epithelium of 'potential pigment cells'. Among these, melanin synthesis begins in pigmentation initiation centres consisting of up to half a dozen or so cells, from which it then spreads outwards to produce a sheet of pigment epithelium (see Figure 7.9).

The ratio of pigment to lens cells formed can be changed by apparently trivial manipulations, such as increasing or decreasing the inoculum density, slightly modifying the medium composition, or folding the cell sheet. The observation that the area finally occupied by potential pigment and pigment cells is directly

---

Figure 7.9. Transdifferentiation in neural retina cultures.
Long-term mass cultures of chicken embryo neural retina cells grown in minimal medium supplemented with foetal calf serum will transdifferentiate into lentoid bodies containing lens fibre cells (V) and into pigment epithelium (IX). Freshly inoculated cells settle largely as aggregates (I), from which flattened cells (II) spread out. These continue to spread (III), mount up into multi-layers (IV) and form lentoid bodies, or undergo mitosis to pavement epithelium (VI) and potential pigment cells (VII, VIII). Pigmentation begins in pigment leader cells (VIII) from which it spreads outwards. Occasionally lentoid bodies arise within the pigment epithelium. Addition of succinate to the medium and increased bicarbonate concentration are thought to stimulate TCA-cycle activity in the cells, while addition of malonate should inhibit it. From experiments in which culture medium was supplemented with these biochemicals, it was deduced that step VI–VII is promoted by increasing TCA-cycle activity, while steps I–VI are promoted by TCA-cycle depression. Step VII–VIII is promoted by slight depression of the TCA cycle, but this has no effect on VIII–IX. (From Pritchard, 1981.)

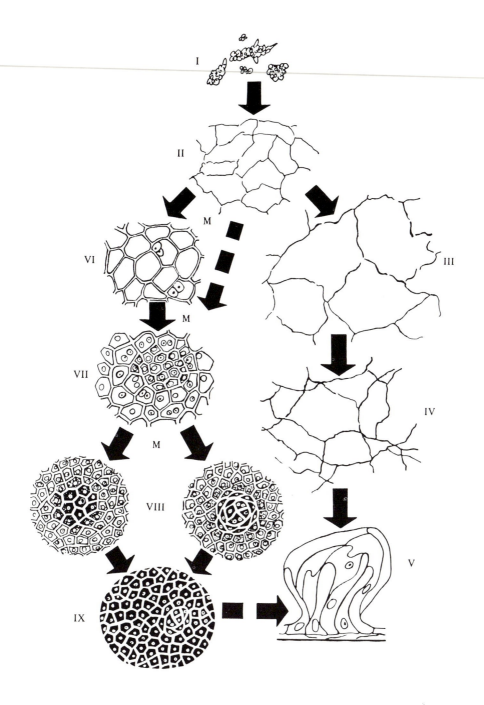

proportional to the concentration of bicarbonate ions in the medium led to a particularly significant discovery. This was that each step in the transition between nerve and pigment cells occurs optimally under conditions which would sustain different levels of activity of the biochemical pathway known as the tricarboxylic acid (TCA) or Krebs cycle (Figure 7.9). The TCA-cycle enzymes are carried on the mitochondria and the cycle itself is vitally important in the production of NAD and ATP, the universal 'currencies' of energy in all forms of life. This observation is one of the first to implicate a specific biochemical pathway in the differentiation of a particular cell type (see also Chapters 4 and 8).

Another very important clue to the molecular events that take place is that the capacity to transdifferentiate into lens fibres from neural retina (and a variety of other starting points, cf. Chapter 3) is related to the presence of small quantities of crystallin mRNA in the cytoplasm. The changes in gene expression that occur during transdifferentiation seem to involve the promotion of selected mRNA species from the 'scarce' or 'intermediate' to the 'abundant' cytoplasmic categories (see Chapter 9), and it is probably this step that is responsive to cell physiology. As yet we have no idea how this operates, but differential polyadenylation of heterogeneous nuclear RNA (hnRNA) molecules is one interesting possibility (see Chapter 9). As the eye gets older, the neural retina nuclei lose their capacity to synthesize crystallin messenger, export it to the cytoplasm, or maintain it there, and the cells' ability to transdifferentiate falls in parallel.

Transdifferentiation and transdetermination

A point of considerable theoretical interest concerns the stage during the trans-differentiative process at which cells acquire, lose or re-acquire states of deter-mination (see Chapters 2 and 3). It is tempting to draw parallels between transdifferentiation of vertebrate eye cells and transdetermination of insect imaginal discs, but this could be a mistake. As we have seen, determination of the imaginal discs seems to depend primarily on the pattern in which a few master genes are programmed to respond to moulting hormone. There is no reason to assume a similar situation in the vertebrate eye.

Experiments with insects also suggest that the states of determination of their discs are derived from 'mosaic' aspects of early development (see Chapter 2), while the differentiation of vertebrate eye tissues clearly depends upon regulative phenomena (Chapter 3).

It is well accepted that imaginal disc cells become determined *before* they differentiate. However, several experiments suggest that vertebrate cells capable of transdifferentiation are not determined so rigidly at stages when, by structural and biochemical criteria, they are partially differentiated. If these observations are valid, final determination of some vertebrate eye cells, such as those of neural retina and pigment epithelium, may occur *after* overt partial differentiation, not before, and this may involve restriction in the range of RNA species transcribed, processed or exported to the cytoplasm (see Chapter 9). However, it should be pointed out that this interpretation represents a radical and major departure from the accepted dogma that determined states precede overt cytodifferentiation (see Chapter 3). Something of a compromise between these views is represented by the Leader Cell Hypothesis below.

In summary then, the transdifferentiation of partially differentiated chicken embryo neural retina depends on physiological changes that occur in the cytoplasm of the dissociated cells. Changes in energy metabolism are important and seem to lead to changes in the recruitment of nuclear RNA into the cytoplasmic mRNA high abundance class. This picture contrasts markedly with the situation in insect imaginal discs, in which transdetermination probably involves a change in the 'programming' for transcription of a few master genes. There are no such master genes known to operate in the vertebrate eye.

The leader cell hypothesis

The final stages in the formation of pigment epithelium by transdifferentiation in neural retina cultures reveals another intriguing principle. Pigment first accumulates in the pigmentation initiation centres scattered within the potential pigment cell sheet (Figure 7.9). The number and density of these foci increases if TCA-cycle activity is slightly inhibited (although strong inhibition blocks all pigment synthesis). Pigmentation then spreads to proximal cells and outwards, always to adjacent cells. The originally pigmented cells then die, but this has no effect on the outward spread of pigmentation. Development of pigment in potential pigment cells thus appears to be stimulated initially by TCA-cycle depression and secondarily by proximity to already pigmented cells. What is particularly interesting is that, although initiation of pigment synthesis is enhanced by TCA-cycle depression, secondary stimulation is not. This observation suggests that the cells at the pigmentation initiation centres are uniquely responsive to TCA-cycle depression and that these 'leader cells' influence their immediate neighbours to become pigmented also, but do this by a variant of the events by which they themselves reached that state. Thus, although the pigment epithelium that develops in neural retina cultures may look homogeneous, its cells have actually reached that state by diverse means.

It is important to appreciate the possible relevance of these observations to normal eye development. Pigment epithelium and neural retina could plausibly differentiate from one another during normal development due to physiological stimulation of leaders with different propensities, within the same initial mixed population of optic vesicle cells (see Chapter 3). In the outer layer of the two-walled optic cup, 'pigment cell leaders' may be stimulated to establish pigment epithelium, whereas in the inner layer the different conditions there fail to stimulate this response. In the inner layer, 'neural retina leaders' may respond by establishing the neural retina phenotype. In both tissues the bulk of the cells could then act as 'followers' which acquire phenotyping instructions in a second-

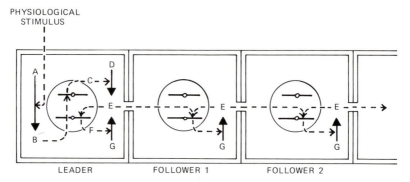

Figure 7.10. The Leader Cell Hypothesis for tissue differentiation. This theory is based on the differentiation of pigment cells in transdifferentiating neural retina cultures (see Figure 7.9).

An inductive stimulus impinges on many cells, but leader cells only are competent to respond, being distinguished by their capacity to carry out the reaction which converts substance A into B. Execution of this reaction modifies the cytoplasmic physiology and leads to transcription of the C gene and translation of enzyme C. The opening up of the genome for transcription of C is considered to represent a determinative step. Enzyme C catalyses formation of the small metabolite E, which activates expression of gene F, coding for enzyme F. Enzyme F also catalyses production of E, so that a positive feedback loop becomes established. Metabolite E is thus produced in quantity and spills over into adjacent cells, where the same feedback loop involving enzyme F becomes established. This occurs without the determinative step during which the C gene is primed for transcription. Substance E is passed into the next adjacent cell and the process is repeated throughout the colony. The differentiated phenotype (e.g., melanogenesis) depends on the operation of the feedback loop involving E, F, and G. At a later stage, the reaction in which substance A is converted into B may also take place in the follower' cells, enabling them also to undergo the determinative step related to transcription of C.

ary fashion from their nearest leaders, possibly by transfer of small molecules through gap junctions (Chapter 4).

According to this model then, 'follower' cells, which constitute the bulk of the tissue, acquire their determined states (if at all) *following* overt manifestation of the differentiative physiology of that tissue. It is suggested that transdifferentiation occurs in culture because embryonic, as yet undetermined, follower cells undergo dedifferentiation when isolated from their leaders and then come under the sway of leaders of other phenotypes stimulated under the dissimilar environmental conditions.

To what extent experiments carried out with biological material cultured in isolation in man-made vessels can provide valid insight into the events that occur during normal development is a debatable point. But if the changes that occur in transdifferentiating cultures bear any resemblance to the events of cytodifferentiation *in vivo*, these cultures may hold the clue to some of today's greatest outstanding problems in developmental biology. The leader cell concept offers an explanation of why dissociation of vertebrate eye cells permits their subsequent transdifferentiation, since, by dissociation, uncommitted cells are separated from their leaders. It also indicates the source of the pluripotent cells involved in the many aspects of tissue regeneration (see above and Chapter 6) and provides a new insight into the early stages of normal cytodifferentiation. This novel hypothesis therefore provides an important new approach to considering how animal development may come about.

7.8. *Summary and conclusions. A new theory of cytodifferentiation*

In this chapter we have considered three kinds of cytodifferentiative instability:

1. Modulation involves fluctuating control at translation, or at physiological levels, without necessarily any changes in the cytoplasmic or nuclear RNA content of those cells.

2. Transdifferentiation involves changes in RNA profiles in the cell cytoplasm. The means by which these changes come about are not yet known, but they could operate initially at the RNA-processing or nuclear-export stages. Transdifferentiative changes are triggered by cytoplasmic physiology, which in turn depends on external influences.

3. The third kind of instability, resulting in transdetermination, probably involves transcriptional control, with primary effects exerted on the RNA constitution of the nucleus and then the cytoplasm.

These three kinds of cytodifferentiative change are therefore really quite distinct at molecular levels, although a culture of chicken embryo pigment epithelial cells, for example, could show all three in turn.

During the early development of vertebrates, all the body cells are initially only partially determined and are capable of following any of a wide range of developmental pathways. As development proceeds this pluripotency becomes restricted, usually in association with the acquisition by the cytoplasm of a specialized physiology. This, however, does not normally include radical and irreversible changes in the nuclear DNA, as proposed by the Somatic Mutation Theory.

Evidence from experiments with vertebrate eye tissues suggests that instructions specifying cell phenotype are probably exchanged between the cells of each tissue and between the different tissues, until a situation of equilibrium becomes established in which the representation of each tissue is balanced with respect to the others. If this balance is disturbed by removal of one or more types of tissue, regenerative adjustments may occur to replace the missing components and these may involve transdifferentiation of already differentiated cell types. The capacity for such changes varies between species and declines with age, as the tissues acquire more rigid states of determination.

The situation in the developing eye may be different from those which obtain in other parts of the body where precise positioning of the component tissues may be less critical. Whether or not that is the case, the evidence from the vertebrate eye argues against the generally accepted theory that determined states always precede overt cytodifferentiation. It is more as if patterns of gene expression become established by physiological interaction with other tissues, while final determination occurs later and involves fixation of already expressed patterns. Gene expression seems to be controlled to some extent by selective accumulation of nuclear transcripts in the cytoplasmic high abundance class, and it is probably this aspect which is responsive to fluctuations in cell physiology. In Chapter 9 the suggestion is made that differential polyadenylation of messengers may affect recruitment of mRNA into the different cytoplasmic abundance classes. Polyadenylation has also been considered to control messenger longevity (Chapter 8). As yet we know little about these considerations with respect to the retina.

In contrast to the situation in vertebrate eyes, insect imaginal discs go through an early determinative step before their differentiation and growth in the pupa. A great deal of attention has been paid to the stability of their determined states, which seem to be based essentially on the pattern of programming of a few master genes to respond appropriately to the steroid moulting hormone, ecdysone. A new molecular model is proposed for determination and transdetermination of imaginal discs in insects.

The picture emerging from studies on unstable differentiation in vertebrates suggests that the following course of events may take place generally during development, or at least in some organs of those species in which regulative, as compared to mosaic, aspects of control predominate (see Chapter 2). In the undifferentiated cells of early embryos, a multitude of genes corresponding to many diverse specializations are transcribed, but only at very low levels. Their RNA products fall in the 'low' or 'intermediate' cytoplasmic abundance classes described in Chapter 9. The cytoplasm of the early embryonic cell thus contains a variety of mRNA species in a metastable state which can be resolved in a variety of ways.

These cells are undifferentiated and undetermined. The acquisition of special-ized patterns of gene expression may well follow utilization of a particular set of mRNA molecules selected from the whole range available, this selection depend-ing on the physiological influences impinging on the cell or tissue. The operation of the translation products of these messengers as enzymes catalysing chemical reactions certainly modifies the cytoplasmic composition of the cells and may set in motion chains of metabolic events that are self-reinforcing. At this point the cells could be described as 'differentiated'. At some stage recruitment of the appropriate nuclear RNA species from the 'low' or 'intermediate' cytoplasmic categories into the 'abundant' class becomes established and consolidated, partly by restriction in the range of mRNA species transcribed. At this stage the cell becomes determined.

It is suggested that some differentiative events may be pioneered by 'leader' cells whose response to a particular set of conditions occurs at a low threshold, these leaders may then influence their less responsive neighbours to follow suit. Extension of cell phenotype from one cell to its neighbours could be done by discharging small effector molecules through gap junctions into neighbouring cells, so short-cutting the metabolic events that initiated differentiation in the leaders. The population of cells in physiological communication may then differentiate with homogeneous phenotype, although its members have acquired that state by more than one means. If 'follower' cells are separated from their leaders at this stage they may be free to dedifferentiate, or adopt distinctly different phenotypes, depending on their own states of determination and the physiological influences to which they are subjected.

If this theory is broadly correct, some tissues may persist in a somewhat embryonic state throughout life, with undetermined follower cells held in subjugation by determined leaders. Other tissues may move to more rigidly determined states, depending on the organ and species to which they belong. States of determination are maintained by forces operating at several levels, which probably include biochemical feedback loops in the cytoplasm (Chapter 8), contact with basement membranes and other extra-cellular matrices (Chapter 4), as well as patterns of association of proteins with nuclear DNA (Chapter 10). Dissociation of the cells from one another, digestion of the extra-cellular matrix and mitotic proliferation all tend to remove these influences and promote dedifferentiation (see also Chapter 13).

This general theory is based largely on observation of transdifferentiating cultures, but it is also to some extent speculative. It amounts to an essentially new concept in cytodifferentiation that in several respects disagrees with standard theory. Its unconventional features include the idea that major changes in gene expression are initiated by physiological responses to external cues, and the suggestion that cells can be overtly differentiated before they are determined (cf. Chapters 2 and 3). The idea that differentiation can be pioneered by leader cells, which then instruct their neighbours to adopt a similar phenotype, is also new, but not without precedent. The pulsation of embryonic heart muscle seems to be lead by 'pacemaker' cells which co-ordinate their neighbours, and in the slime mould, *Dictyostelium discoideum*, the aggregation of isolated amoebae into slug-like multi-cellular organisms is initiated by leader cells that emit cyclic AMP. This stimulates other amoebae to move towards them and to secrete the same biochemical, so that the whole population eventually come to adopt the characteristics of the leaders. The theory outlined here represents a new approach to cytodifferentiation which could furnish explanations for several hitherto intractable problems of development.

Armed with the suspicion that gene expression is controlled at a variety of levels, from the physiological down to the transcriptional and beyond, we are now in a position to examine developmental events in terms of molecules. In this inquiry we will proceed in that same order, beginning with an investigation into the physiological events that occur among the proteins of the cytoplasm.

Bibliography

Burnett, A. L., The acquisition, maintenance and lability of the differentiated state of *Hydra. Res. Probl. Cell Different.*, **1**: 109 (1968).

Clayton, R. M., The molecular basis for competence, determination and transdifferentiation: a hypothesis. In *Stability and Switching in Cellular Differentiation*, edited by R. M. Clayton and D. E. S. Truman. Plenum, New York, pp. 23–28 (1982).

DeHaan, R. L. and Sachs, H. G., Cell coupling in developing systems: the heart cell paradigm. In *Current Topics in Developmental Biology*, Vol. 7, edited by A. A. Moscona and A. Monroy. Academic Press, London, pp. 193–228 (1972).

Eguchi, G., "Transdifferentiation" in pigmented epithelial cells of vertebrate eyes *in vitro*. In *Mechanisms of Cell Change*, edited by J. D. Ebert and T. S. Okada. John Wiley, Chichester, pp. 273–291 (1979).

Gerisch, G., Chemotaxis in *Dictyostelium. Ann. Rev. Physiol.*, **44**: 535 (1982).

Grant, P., *Biology of Developing Systems*. Holt, Rinehart and Winston, New York (1978).

Gurdon, J. B., *The Control of Gene Expression in Animal Development*. Harvard University Press, Cambridge/Oxford University Press, Oxford (1974).

Hadorn, E., Transdetermination. In *The Genetics and Biology of Drosophila*, edited by M. Ashburner and T. R. F. Wright. Academic Press, London, pp. 555–617 (1978).

Hauschka, S. D., Clonal aspects of muscle development and the stability of the differentiated state, In *The Stability of the Differentiated State*, edited by H. Ursprung. Springer-Verlag, Berlin, pp. 37–57 (1968).

Kauffman, S. A., Control circuits for determination and transdetermination: *Science*, **181**: 310 (1973).

Nöthiger, R., The larval development of imaginal disks. In *The Biology of the Imaginal Disks*, edited by H. Ursprung and R. Nöthiger. Springer-Verlag, Berlin (1972).

Okada, T. S., Cellular metaplasia or transdifferentiation as a model for retinal cell differentiation. *Curr. Top. Devel. Biol.*, **16**: 349 (1980).

Pritchard, D. J., Transdifferentiation of chicken embryo neural retina into pigment epithelium: indications of its biochemical basis. *J. Embryol. Exp. Morphol.*, **62**: 47 (1981).

Reyer, R. W., The amphibian eye: development and regeneration. In *Handbook of Sensory Physiology*, Vol. VII/5, *The Visual System in Vertebrates*, edited by F. Crescitelli. Springer-Verlag, Berlin, pp. 309–390 (1977).

Schmalhausen, I. I., *Factors of Evolution*, translated by I. Dordick, edited by T. Dobzhansky. The Blakiston Company, Philadelphia (1949).

Schmid, V., Wydler, M., Alder, H. and Bally, A., Regeneration occurring *in vitro* by cellular transdifferentiation. *Prog. Clin. Biol. Res.*, **85** A: 257 (1982).

Weiss, M. C., The use of somatic cell hybridization to probe the mechanisms which maintain cell differentiation. In *Human Genetics 1977*, edited by S. Armendares and R. Lisker. Excerpta Medica, Amsterdam, pp. 286–292 (1977).

Whittaker, J. R., Aspects of differentiation and determination in pigment cells. In *Concepts of Development*, edited by J. Lash and J. R. Whittaker. Sinauer Associates Inc., Stamford Connecticut, pp. 163–178 (1974).

Wildermuth, H. R., Determination and transdetermination in cells of the fruitfly. *Sci. Prog. (Oxford)*, **58**: 329 (1970).

Yamada, T., *Control Mechanisms in Cell-Type Conversion in Newt Lens. Monographs in Developmental Biology*, Vol. 13: edited by A. Wolsky. S. Karger, Basel (1977).

Yamada, T., Transdifferentiation of lens cells and its regulation. In *Cell Biology of the Eye*, edited by D. S. McDevitt. London, Academic Press, pp. 193–241 (1982).

Chapter 8 Proteins and translation

Scientists of various disciplines disagree when it comes to defining the nature of life, but most would hold that the essence of life, as we know it on this planet, resides in the co-ordinated catalytic action of proteins. The transmission of coded information between the generations is by comparison a higher order function. This implies that living organisms have a certain elemental and molecular composition, that a restricted range of chemical reactions occur within or around them and that these events extend over a restricted time-scale. Indeed, if living organisms exist which have a very different chemical composition and which operate on a very different time-scale, we might well fail to recognize them as alive. The creation of enormously complex networks of interlinked chemical processes, operating harmoniously at speeds very much higher then would be expected on simple chemical grounds, is perhaps the greatest of the accomplishments of living systems.

Spontaneous molecular dissociation, or reaction between potentially reactive molecules is normally inhibited by energy barriers that must be overcome before the reaction can take place. The great secret that living systems discovered is that these barriers can be broken down into series of lesser barriers by the formation of unstable complexes with proteins. At the protein level, evolution has the appearance of a never-ending quest for improvement in biological catalysis, within the context of all the other biochemical processes occurring in the same molecular neighbourhood. The word enzyme (meaning 'in yeast') refers to these proteins with these catalytic properties.

The average single enzyme molecule can process several hundred molecules of substrate per minute. Catalase, which occurs in the cytoplasm of vertebrate cells, holds the record, being able in that time to cause the decomposition of several millions of molecules of hydrogen peroxide, exceeding the performance of the best platinum catalysts.

In order to operate in this way the enzyme must actually combine with its substrate, although after the reaction has occurred the products are released and the enzyme itself remains unchanged, ready for the next substrate molecule. Although their association is transient, enzyme and substrate temporarily become very intimately bound together, the specificity of the catalysis depending on a close fit between the surfaces of the two molecules. This matching includes complementarity in shape and distribution of positive and negative charges, which form electrostatic bonds between the two, and in the distribution of hydrophobic (or lipophilic) groups, which tend to coalesce like oil droplets in water, further bringing the two molecules together. Each enzyme is highly specific in its properties, although like inorganic catalysts they can operate their reaction in either direction and do not influence its position of equilibrium.

All the properties of the protein component of the enzyme derive directly or indirectly from its primary structure, that is the sequence of amino acids it contains. Proteins adopt two main types of secondary structure, the α-helix and the β-sheet. In the α-helix, stabilizing bonds occur between an NH group of one amino acid residue and the CO of another four places away. This produces a spiral of pitch 3.6 residues, with the chemically active groups of amino acids sticking out sideways. The β-sheet is a pleated structure formed by cross-linking between adjacent parallel or anti-parallel polypeptide strands. There are usually between two and 10 strands in each sheet. Most globular proteins contain one or both structures.

Although all enzymes are proteinaceous, not all proteins are enzymes. The most abundant protein in vertebrates is collagen, which has no known enzymic properties, while other 'structural proteins', such as keratin in vertebrates and sclerotin in arthropods, are intrinsic components of the outer integument. One strange protein acts as an antifreeze in the blood of Antarctic fishes. This contains multiple repeats of a group of three alanine residues, followed by aminodeoxygalactosyl threonine or its *N*-acetyl derivative, which binds to the surface of small ice crystals and inhibits growth along the crystal's preferred axis.

The amino acid sequences of all proteins, be they enzymic or structural or have other special properties, are dictated by the nucleotide sequences of the genes and, so far as we know, practically every gene initially exerts its effects by becoming expressed as protein. Exceptions to the latter rule are the DNA sequences coding for transfer, ribosomal and 'processing' RNA (see Chapter 9), which are expressed only as untranslated RNA (see below), and those genomic sequences which control the expression of other genes without themselves being transcribed into RNA. We will deal with RNA and transcription in Chapters 9 and 12, here we will concentrate more on the physiological control of gene expression at the protein level and the way in which proteins are produced, by translation of the genetic message that is relayed from nucleus to cytoplasm in the form of a messenger RNA transcript.

8.1. *Translation*

Translation of the coded information carried in the structure of a messenger RNA molecule into a polypeptide of specific sequence is carried out in small cytoplasmic bodies called ribosomes. Those of eukaryotes differ from prokaryote ribosomes in

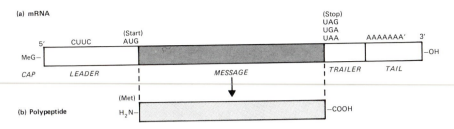

Figure 8.1. Diagram illustrating the main features of cytoplasmic messenger RNA (a) and its polypeptide product of translation (b).

several respects, notably their sensitivity to specific toxins: for example, eukaryote ribosomes are essentially unaffected by streptomycin which is highly toxic to prokaryotes (see Chapter 1). A messenger RNA molecule carrying several ribosomes is called a polyribosome or polysome.

In RNA, the ribose molecules are linked on one side through the 3′ and on the other side through the 5′ carbon atoms. The 5′ free end of the strand is finished off by a 'cap' of methylguanosine, attached in reverse polarity which, together with adjacent sequences in the messenger 'leader', forms an attachment site for the ribosomes (see Figure 8.1, Chapters 9 and 12). Ribosomes thus contact the messenger at the 5′ end and translate it in the 5′ to 3′ direction.

The decoding and translation of the messenger is achieved through the mediation of a class of small RNA molecules called transfer RNA or tRNA. Each tRNA molecule is shaped like a clover leaf (see Figure 12.6), and is charged with an amino acid molecule at the base of the stalk. At one specific region on the centre unit of the clover leaf a specific 'anticodon' is displayed. This consists of a sequence of three bases, which becomes matched with its complementary sequence, or 'triplet codon', on the mRNA. This matching of codon and anticodon occurs within the ribosomes in accordance with the normal rules of base pairing (i.e., A pairs with U and C pairs with G) for the first two bases on the messenger, while the rules of wobble pairing (see below) apply to the third. A ribosomal enzyme, peptidyl transferase, then transfers the amino acid from the tRNA to the growing polypeptide. The spent tRNA is then released from the ribosome, which moves on to the adjacent codon.

For detailed analysis, translation can be subdivided into three phases: initiation, elongation and termination.

Initiation

Between the cap and the translation 'start' signal is a short sequence which is complementary to a characteristic string of nucleotides at the 3′ and of the 18S ribosomal RNA sequence, in the smaller subunit of each ribosome (see Chapter 9).

Initiation commences with the formation of an initiation complex consisting of the mRNA, the smaller ribosomal subunit and methionyl tRNA charged with a methionine molecule (Figure 8.2(a)). Initiation is assisted by as many as nine 'initiation factors' integrated within the ribosomal subunit, plus others in solution. The ribosomal subunit binds to its complementary sequence on the messenger leader (see Table 9.1), then scans along until it finds the sequence AUG, which signifies where translation should start. In other positions in the messenger the AUG instruction is read as 'insert methionine', but close to the cap it means 'start translation by inserting methionine'. The larger ribosomal subunit then joins the small one and translation proceeds.

Messenger leaders vary in their degree of complementarity with the ribosomal binding sequence, and it has been suggested that such variation would impose differential requirements for soluble initiation factors. The importance of this feature in translational control is discussed below.

Translation in prokaryotes differs from that in eukaryotes in that a modified amino acid, *N*-formyl methionine, is inserted at the start. Bacterial mRNA also lacks the methylguanosine cap, but virtually all bacterial mRNA species have an additional characteristic GGAGG sequence at the 5′ end, complementary to CCUCC at the 3′ end of the smaller rRNA moiety.

In both prokaryotes and eukaryotes the large ribosomal subunit contains two sites, known as the A (for aminoacyl) and P (for peptidyl) sites (see Figure 8.2). At the end of initiation the P site of an attached ribosome contains a methionyl tRNA molecule charged with methionine, with its anticodon engaged in the AUG start codon, while the A site is empty. After the start codon has moved out of the ribosome, a new initiation complex forms around it and a second ribosome follows the first along the messenger. Another ribosome then follows this, until a series becomes strung out along the messenger like beads on a string.

Figure 8.2. Stages in the translation of messenger RNA into polypeptide.
(a) Initiation: (i), (ii) Formation of the initiation complex. Charged methionyl tRNA and a small ribosomal subunit containing the sequence GAAG in its 18S RNA component bind to the sequence CUUC in the 5′ messenger leader. The ribosomal subunit slides along the mRNA in the direction 5′ to 3′. (iii), (iv) The anticodon in the methionyl tRNA engages with the AUG translation 'start' signal and a large ribosomal subunit, together with translation initiator factors, bind to the small subunit, with the methionine residue in the P site of the large subunit. (b) Elongation: (v) A molecule of leucyl tRNA charged with leucine enters the A site of the ribosome, its anticodon engaging with the codon 3′ to AUG, which is CUG. Elongation factor EF1 is also incorporated. (vi) The peptidyl transferase reaction: The methionine residue is released from its tRNA and linked to the amino sequence of the leucine in the adjacent A site, with formation of a peptide bond. (vii) The translocase reaction: Elongation factor EF2 enters the ribosome and the leucyl tRNA is transferred from the A to the P site, the methionyl tRNA, now without its amino acid, is released. (c) Termination: When a stop codon, UAA in this case, enters the ribosome, a protein-release factor (RF) also enters and causes disruption of the translation complex and release of the messenger.
In this representation the triplet codons have been spaced out for ease of identification.

8.2a Initiation

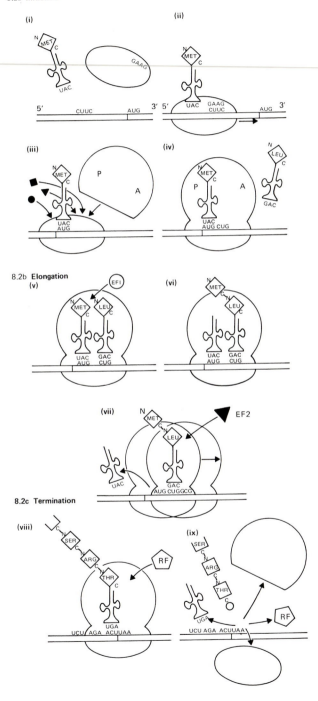

(i)

(ii)

(iii)

(iv)

8.2b Elongation

(v)

(vi)

(vii)

8.2c Termination

(viii)

(ix)

Elongation

Elongation proceeds in three stages (see Figure 8.2(b)). In the first, the appropriate aminoacyl tRNA is located in the ribosomal A site, as dictated by the codon in the mRNA. This requires the agency of a soluble elongation factor called EF1. The next step is complex and still not properly understood. This is the peptidyl transferase reaction, which involves formation of a peptide bond between the amino ($-NH_2$) group of the aminoacyl tRNA at the A site with the carboxyl ($-COOH$) group of the amino acid currently occupying the P site. The reaction therefore attaches the next amino acid to the growing polypeptide and at the same time releases the tRNA previously held in the P site.

The third step is called the translocase reaction and requires another soluble elongation factor calle EF2. This promotes expulsion of the uncharged tRNA from the P site and moves the growing peptide from the A to the P site, leaving the A site vacant. During this stage the ribosome moves three bases down the mRNA locating the next codon in the A site. The translocase reaction is easily disrupted and particularly susceptible to toxins such as diptheria toxin. One of the most poisonous substances known, an extract of castor beans called ricin, interferes with the operation of EF2. Ricin is so potent a poison it is used as an assassination weapon.

Speed of elongation depends on temperature. At $16.4°C$ in sea urchins, each ribosome takes about 20 minutes to traverse an mRNA molecule of average length.

Termination

Elongation of the polypeptide chain continues until a 'stop' codon, or termination codon enters the ribosome (Figure 8.2(c)). There are three different stop codons, UGG, UAA and UGA, and they are recognized by a single multivalent protein-release factor in eukaryotes, although two or three are used by bacteria. When one of these codons enters the ribosome, the release factor modifies the specificity of the peptidyl transferase, so that a water molecule is added to the peptide instead of another amino acid. The ribosome is then released from the messenger and dissociates into its two subunits, so freeing the completed polypeptide chain.

8.2. *The genetic code*

The cracking of the genetic code must rank with the construction of Mendeleyev's Periodic Table and the development of the Theory of Relativity by Einstein as one of the greatest scientific achievements of all time. In a series of elegant experiments carried out in several laboratories, it was shown first that the sequence of bases in the mRNA is colinear with (i.e., in the same order as) the sequence of amino acids in the resultant polypeptide. The 5' end of the mRNA corresponds to the N terminus of the protein (the end with an unbound $-NH_2$ group) and the 3' end to the C terminus (the end with a free $-COOH$ group). Messenger RNA is therefore translated in the direction 5' to 3' and proteins are produced in the direction N to C. Secondly, each group of three consecutive bases (codons) were shown to code for specific amino acids. Thirdly, the specific coding of each theoretically possible codon, for its specific amino acid was deduced (see Figure 8.3).

Tryptophan is represented by only one codon, while serine and leucine are each coded by as many as six different codons. AUG codes for methionine, but also acts

THE GENETIC CODE

First Base	Second Base								Third Base
	U		C		A		G		
U	UUU	phe	UCU	ser	UAU	tyr	UGU	cys	U
	UUC	phe	UCC	ser	UAC	tyr	UGC	cys	C
	UUA	leu	UCA	ser	UAA[1]	*stop*	UGA[2]	*stop*	A
	UUG	leu	UCG	ser	UAG[1]	*stop*	UGG	try	G
C	CUU	leu	CCU	pro	CAU	his	CGU	arg	U
	CUC	leu	CCC	pro	CAC	his	CGC	arg	C
	CUA[3]	leu	CCA	pro	CAA	gln	CGA	arg	A
	CUG	leu	CCG	pro	CAG	gln	CGG	arg	G
A	AUU	ile	ACU	thr	AAU	asn	AGU	ser	U
	AUC	ile	ACC	thr	AAC	asn	AGC	ser	C
	AUA[4]	ile	ACA	thr	AAA	lys	AGA[5]	arg	A
	AUG	met	ACG	thr	AAG	lys	AGG[5]	arg	G
G	GUU	val	GCU	ala	GAU	asp	GGU	gly	U
	GUC	val	GCC	ala	GAC	asp	GGC	gly	C
	GUA	val	GCA	ala	GAA	glu	GGA	gly	A
	GUG	val	GCG	ala	GAG	glu	GGG	gly	G

Figure 8.3. The genetic code in nucleus and mitochondria.
ala, Alanine; arg, arginine; asn, asparagine; asp, aspartic acid; cys, cysteine; gln, glutamine; glu, glutamic acid; gly, glycine; his, histidine; ile, isoleucine; leu, leucine; lys, lysine; met, methionine; phe, phenylalanine; pro, proline; ser, serine; thr, threonine; try, tryptophan; tyr, tyrosine; val, valine. *stop*, termination of a coding sequence. 1, UAA and UAG code for glutamine in *Paramecium* and some other cilates; 2, UGA codes for tryptophan in mammalian and yeast mitochondrial RNA; 3, CUA codes for threonine in yeast mitochondrial RNA; 4, AUA codes for methionine in mammalian and yeast mitochondrial RNA; 5, AGA and AGG both represent 'stop' signals in mammalian mitochondrial RNA.

as the start signal to indicate the beginning of a new polypeptide, ambiguity being avoided by the presence or absence of ribosome-binding sequences just 'upstream' of this codon (see Chapter 9). UAA, UAG and UGA all signal the end of the polypeptide. The complete code dictionary is shown in Figure 8.3.

There are two reasons why some amino acids can be coded by several codons. First, some amino acids are carried by several different species of tRNA displaying different anticodons. Secondly, some tRNA molecules recognize several codons through a less specific type of matching called wobble pairing. The basis of wobble pairing lies in the capacity of G in the anticodon to pair with both C and U in the third base of the codon, and of U in the anticodon to pair with both A and G in the third base position. There is also a modified base, inosine (I), at the 5' end of some anticodons, which can pair with either U, C or A in the third position.

The genetic code is generally believed to be the same for all organisms, and spectacular support for this assertion was provided by an experiment in which rabbit haemoglobin mRNA was translated correctly by rabbit ribosomes using bacterial tRNA. This, and related evidence, has led to the belief that all organisms are descended from a single common ancestor, as it is thought unlikely that precisely the same associations of anticodon sequences with specific amino acids would have arisen independently more than once.

As we saw in Chapter 1, mitochondria and chloroplasts have their own DNA. They also have their own tRNA and recent evidence shows that the genetic code utilized by mitochondria is not quite identical to that used in the translation of nuclear messages (see Figure 8.3). Whether the separate codes have diverged as a measure which ensures segregation of nuclear and organelle functions, or whether they indicate somewhat different ancestral origins, remains a subject for speculation. It is interesting that the AGG sequence which frequently marks the end of an exon in nuclear RNA (see Chapter 9), acts as a stop signal in mammalian mitochondrial RNA.

8.3. *Control of translation*

It is important to distinguish between factors which control the synthesis of specific proteins and those that regulate protein synthesis overall, but the two are in fact interrelated in a complex fashion. Translation of a specific mRNA depends on its supply from the nucleus. Additional controls include its specific and non-specific masking, exposure to or protection from ribonucleases that would degrade it, and regulation of its incorporation into initiation complexes. Different messengers require quite different ionic conditions for their optimal translation and the ratio of available tRNAs is also important. For example, in reticulocytes, which synthesize haemoglobin predominantly, the relative concentrations of tRNA species correlate closely with the frequencies of the corresponding amino acids in globin. In proliferating lymphocytes, protein synthesis in general seems to be regulated up and down in relation to the cell cycle by the supply of the initiator tRNA, methionyl tRNA.

In a typical cell 10–80% of the cytoplasmic mRNA is present as free ribonucleoprotein, the remainder being bound to ribosomes. The distribution of individual mRNA species between the free and the bound fractions is non-random, which implies that either some messengers are preferentially selected for translation, or that some are preferentially barred. In fact both forms of selection seem to take place.

Negative control

Some cytoplasmic mRNA species are not available for translation due to their combination with masking proteins. This is a particularly important feature in the egg and early embryo, in which maternal transcripts are preserved as 'informosomes' pending transcription of the embryo's own genes (see Chapter 9). As we saw in chapter 2, morphogenetic gradients are probably laid down in the form of RNA in insect eggs. When enucleated *Xenopus laevis* eggs were fertilized by *X. borealis* sperm, haploid embryos were produced in which the nuclear and cytoplasmic components were of different species. In these embryos, histone H1 was shown to be produced from the maternal cytoplasmic fraction (and therefore from stored mRNA) up to the mid-gastrula stage. Other maternal mRNA persists even into the late neurula. It has been suggested that if different kinds of messengers are unmasked due to conditions prevalent in different parts of the body of the embryo, this could lay the foundations of regionally distinct patterns of gene expression (see Chapter 9). Skeletal muscle shows evidence of messenger storage, as myosin mRNA is synthesized in the separate myoblasts, but is translated rapidly only after their fusion into myotubes (see Chapter 13).

Incorporation of charged tRNAs into ribosomes is blocked by a 20-base chain of RNA in the brine shrimp, *Artemia salina*, which survives desiccation as an arrested gastrula. However, this control is not known to operate in other species.

To what extent blocking of translation can occur in a messenger-specific fashion in eukaryotes is uncertain. There is evidence that collagen synthesis can be specifically inhibited at elongation or termination by a fragment of pro-collagen. In bacteria certain key ribosomal proteins act as specific blocks to their own translation. Since these proteins will bind to both the mRNA which codes for them and to the rRNA in the ribosome, competition between the two types of RNA may ensure co-ordinate regulation of synthesis of the ribonucleic and proteinaceous components of the ribosome. It is as yet uncertain whether similar systems exist in higher organisms.

In Chapters 4 and 12, *transcriptional* responses to heat shock are described. Heat shock (35 °C) produces equally marked effects at the *translational* level in *Drosophila* cells, in that it apparently causes destruction of a factor necessary for translation of 'normal' mRNA. The mRNA species transcribed in response to heat shock are translated however, since their translation will apparently occur in the absence of the heat-sensitive factor.

There have been reports of translational control by RNA molecules called tcRNA (translational control RNA). These operate non-specifically, most inhibit translation, but some are stimulatory.

Positive control

Haemoglobin provides an interesting example of the way in which protein synthesis can be regulated in a positive and specific fashion at initiation (see Figure 8.4). Adult haemoglobin, or HbA, consists of two α-globin polypeptide chains, two β-globin chains, four molecules of a porphyrin called haem and four iron atoms. For efficient haemoglobin production the red-cell precursor, or reticulocyte, must equalize the synthesis of the two globin chains and co-ordinate their concentrations with that of haem and iron. The cytoplasmic population of α-globin messengers is normally about 40% higher than that of β-globin, but when the cell is starved of iron synthesis of α-globin is preferentially inhibited.

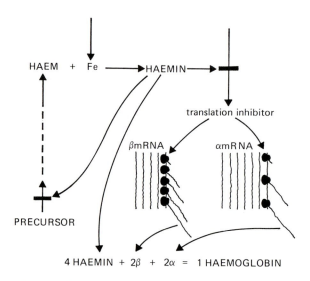

Figure 8.4. Control of production of haemoglobin.
Haem is formed within the cell from a precursor. Iron (Fe) enters the cell and combines with haem to produce haemin. Haemin inhibits synthesis of more haem. Four haem molecules combine with two α-globin and two β-globin polypeptide chains to form one molecule of haemoglobin. The cytoplasmic population of α-globin mRNA molecules is greater than that of β-globin, but the β-globin polysomes carry more ribosomes, so that formation of the two globin chains is about equal. Translation of both globin chains is regulated by interaction of haemin with a translation inhibitor.

Under normal conditions the average β-globin messenger carries five ribosomes, whereas those coding for α-globin carry only three, so that production of α- and β-globin ends up about equal. The basis of the difference in translational efficiency seems to lie in the affinity of the two messengers for soluble initiation factors, which exert a discriminatory control. The β-globin messenger apparently has a higher affinity for certain initiation factors, so that it is always translated more efficiently. When initiation is restricted, α-globin synthesis suffers first. Thus the quantitative distribution of the α-globin message between free and polysome-bound fractions depends on the overall rate of translation initiation.

Some initiation factors are cell-type specific. This can be demonstrated using the cell-free medium produced by lysis of reticulocytes. In this system globin mRNA is translated in preference to mRNA coding for myosin, but if proteins extracted from initiation complexes of muscles are added the system changes its translational preference to myosin.

Uptake and metabolism of iron by reticulocytes are closely regulated by hormones and the iron combines with haem to produce haemin. Synthesis of haem is controlled by feedback inhibition (see below) by haemin acting on the first unique enzyme of porphyrin biosynthesis (δ-amino laevulinate synthetase).

Haemin is also one of the substances that controls the overall rate of protein synthesis by reticulocytes and hence the ratio of production of the two globin chains. This it does by preventing formation of an inhibitor of initiation complexes. Once the α-globin chain is synthesized it interacts immediately with the β subunit, even before the latter is released from its polysome, otherwise it is rapidly degraded. The $\alpha\beta$ pair then combines with haemin and the resultant complex links up with another like itself.

The reticulocyte therefore provides a beautiful illustration of the intricacies of molecular control of gene expression at the translational level. This includes control of the overall rate of protein synthesis, preferential synthesis of cell-type-specific proteins, and discriminatory regulation of the different cell-type-specific proteins.

Messenger degradation

The mRNA of eukaryotes differs from that of prokaryotes in having a cap at one end and a poly-A tail at the other (Figure 8.1). It is thought that these refinements may protect it from degradation by ribonucleases, since eukaryote mRNA has a relatively long half-life, while that of bacteria is only a few minutes, in a bacterial or eukaryotic cell. When injected into *Xenopus* oocytes, foreign eukaryotic mRNA yields translation products for several days at a rate of up to 200 polypeptide chains per messenger per hour. The enormous silk productivity of the silk moth caterpillar (*Bombyx mori*) can be satisfactorily explained only if each silk fibroin message is translated thousands of times in the silk glands.

Eukaryote messengers vary in the speed at which they are degraded and those which are found most abundantly in the cytoplasm have the longest poly-A tails (see Chapter 9). It has been suggested, therefore, that differential polyadenylation is a means by which protein production is controlled, through its effect on messenger survival. Histone messenger carries no poly-A and it is probably this deficiency that favours its rapid degradation in the cytoplasm. Histone messenger degradation occurs at all times during the cell cycle, except the DNA synthetic (S) phase, when it is translated to provide histone that binds with newly formed DNA (see Chapter 10). Most species of mRNA molecules can survive mitosis to be translated again in the subsequent G1 phase.

The mRNAs known to have short half-lives include those for many enzymes inducible by hormones. For example, in the chick oviduct the half-life of the mRNA coding for ovalbumin is increased from 3–24 hours by inducing steroid hormones. Similarly in mammary gland organ cultures the half-life of casein mRNA is increased about 20-fold in the presence of the hormone prolactin. How these controls operate is at present unknown.

Translational controls associated with viral infection

At least two kinds of unusual translation control come into operation, particularly under conditions of viral infection. One of these involves synthesis by the cells of glycoprotein translation inhibitors called interferons. These are produced in response to the presence of double-stranded RNA formed during the replication of certain viruses. The second control involves suppression of termination by the intervention of suppressor tRNA molecules. These cause the insertion of amino acids at positions corresponding to 'stop' codons, resulting in the synthesis of very long 'read-through' proteins in which the stop has been ignored. This facility prevents accurate translation of viral proteins, so presumably reducing their adverse effects. In Chapter 9 we will see that membrane-bound hydrophobic immunoglobulins are regularly produced in vertebrates by reading through the stop signal that operates in the synthesis of the equivalent water-soluble protein (see Figure 9.13), but this operates at transcription of mRNA not at its translation.

Many of the growth factors called somatomedins contain amino acid sequences identical to insulin, but with extra C-terminal portions. These may represent the products of evolutionarily related genes, but it has been suggested that they are actually produced from the insulin gene by read-through of the stop codon. Stop codon read-through, controlled by suppressor tRNA, has the potential to be an important aspect of cytodifferentiation, but as yet we know virtually nothing about whether this possibility has been exploited by living systems.

8.4. Post-translational modification

Some enzymes and peptide hormones acquire their full activity as they spontaneously fold into their characteristic three-dimensional forms. In contrast many others are synthesized as inactive precursors, which in the case of enzymes are called proenzymes or zymogens. This prevents the molecule acting directly back on the cell that produced it, the properties of the final protein being to some extent derived from structural reorganizations, or other groupings, molecules or ions with which the simple polypeptide later becomes complexed. Some types of modification are standard for most polypeptides. The N-terminal methionine is usually cleaved off and sometimes the remaining N-terminal amino acids are acetylated. Disulphide bridges are formed by reactive sulphydryl (SH) groups between appropriately situated cysteine residues and some chemical groups become modified. The amino acids hydroxylysine and hydroxyproline are produced by modification of lysine and proline, there being no codons for these molecules.

The simplest type of post-translational modification is polypeptide cleavage and one of the best-known examples occurs in the formation of the hormone insulin (Figure 8.5). In this case the requirement for processing ensures that the hormone is inactive during storage in the β-cells of the pancreatic islets, where it is transcribed and translated from the insulin gene. The initial product of translation is a polypeptide 84 amino acids long called proinsulin. This simple chain forms a ring that becomes stabilized by three disulphide bonds. A portion 33 amino acids long is then cut out to leave two shorter polypeptides, one of 30 and the other of 21 amino acids, linked by two disulphide bonds. It is only after this processing that the molecule acquires the hormonal properties of insulin.

Chymotrypsin is a digestive enzyme that hydrolyses proteins in the small intestine. The inactive precursor, chymotrypsinogen, is produced in the pancreas as a single polypeptide chain cross-linked by five disulphide bonds. It is activated by another pancreatic enzyme, trypsin, which hydrolyses the peptide bond between amino acids 15 and 16. The newly formed N-terminal group, isoleucine 16, then turns inwards and forms an electrostatic bond with aspartate 194 in the interior of the molecule. This in turn triggers a series of localized conformational changes, creating a pocket for binding an aromatic side-chain of the substrate protein. The

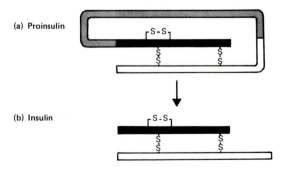

Figure 8.5. Production of insulin by simple processing of a precursor.
The initially formed polypeptide, proinsulin (a), forms a loop stabilized by disulphide bonds between cysteine residues. A portion of the chain is cut out, leaving a two-chain protein, the active molecule of insulin (b).

first formed active enzyme, π-chymotrypsin, acts on more chymotrypsinogen molecules, producing yet more active enzyme. The stable form, α-chymotrypsin, is created by removal of two further small peptides, to produce a three-peptide protein.

Trypsin is the common activator of all the pancreatic proenzymes and co-ordinates their production. In some systems a cascade of activations takes place, the activated form of one activating the next in the series. This occurs, for example, during blood clotting.

Altogether at least 125 different kinds of post-translational covalent modifications are know to occur, including adenylation/deadenylation, methylation/demethylation, and phosphorylation/dephosphorylation. The last is particularly important and can be controlled by extra-cellular hydrophilic hormones acting (through the cyclic-AMP second messenger system and calmodulin, see Chapter 4) on enzymes called protein kinases, which catalyse the attachment of phosphate groups. Dephosphorylation is performed by phosphoprotein phosphatases and controlled by a variety of metabolites and ions that combine with potential substrates to make them fit the enzyme attachment sites. Reversible phosphorylation and acetylation are probably of great importance in modification of chromosomal proteins, such as histones, in the control of transcription (see Chapter 10).

The proteinaceous component of an enzyme is called an apoenzyme. To form a functional 'holoenzyme', most apoenzymes must combine with co-factors. These are of several types: metal ions such as copper, iron and magnesium, moderately sized organic molecules called prosthetic groups, or a special type of substrate called a co-enzyme. Haem is an example of a prosthetic group, which enables the globin chain to adopt a biologically active form with an extremely high content of α-helix. Whereas enzymes are theoretically required only in catalytic quantities (but see below!) and are unaffected by the reaction, co-enzymes such as NAD and ATP participate in reactions in exact (stoichiometric) proportions and do become modified.

Secreted proteins characteristically carry glycosyl (sugar) moieties and are synthesized exclusively by ribosomes that are membrane-bound in 'rough' endoplasmic reticulum. The newly translated polypeptide is extruded into the canals, or cisternae, of the endoplasmic reticulum, where it is directed towards the Golgi apparatus for glycosylation. According to the signal hypothesis, translation of messengers coding for secretory protein begins on unattached ribosomes. The N-terminal, or 'signal sequence' of a secretory protein, it is suggested, is very hydrophobic (i.e., lipophilic) and penetrates the phospholipid membranes of the endoplasmic reticulum, so gaining entry to the cisternae. When this occurs, the membrane contacts binding sites on the lower surface of the large ribosomal subunit and holds on to it. As translation is completed the signal sequence is presumed to be cleaved off and the new polypeptide is released into the cisternal canals.

Most 'genes' code for only a single protein product, although, as we have seen,

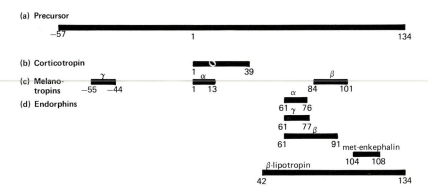

Figure 8.6. Production of pituitary hormones by differential processing of a common precursor polypeptide. Amino acids are numbered with the N-terminal amino acid of corticotrophin as 1.
(After Mainwaring, Parish, Pickering and Mann, 1982.)

some stop signals may be read through to form molecules that are essentially similar, but with an extension at the C-terminus. However, some polypeptides give rise to a whole array of smaller derivatives. This occurs, for example, in the pituitary gland, where a large family of related hormones are formed by cleavage of the primary polypeptide product of one RNA transcript. The DNA sequence concerned seems to have evolved by gene duplication, as it contains four homologous sequences corresponding to four active sites in the proteins produced (see Figure 8.6).

8.5. Association of protein subunits

Many species of protein normally comprise the polypeptide products of more than one gene. Haemoglobin is one such molecule and it will be appreciated that the rate of production of adult haemoglobin could in theory be regulated by limiting the synthesis of either the α- or the β-subunit (see above). Lactate dehydrogenase (LDH) is an example of an enzyme which in vertebrates is present in most tissue types although its composition and properties vary between them, because the subunits are assembled in different proportions. This arises as a feature of differential rates of synthesis of the subunits. A study of LDH illustrates some of the important principles that govern differences in enzyme activities that exist both between different tissues and in single tissue at different stages of development, due to differential assembly of subunits.

Lactate dehydrogenase

LDH is a key enzyme in the metabolic pathways associated with energy metabolism, catalysing the interconversion of pyruvate and lactate. Pyruvate is

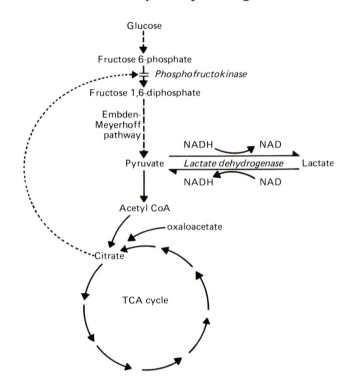

Figure 8.7. The role of lactate dehydrogenase in energy metabolism.
The position of LDH in relation to the Embden–Meyerhoff pathway and TCA cycle is indicated. Citrate allosterically inhibits the rate-controlling enzyme, phosphofructokinase.

produced from glucose through the Embden–Meyerhoff pathway of glycolysis and is linked to the Krebs, or tricarboxylic acid (TCA) cycle, through acetyl co-enzyme A. LDH requires the co-enzyme NADH/NAD for its action, depending on the direction in which the reaction is operating (see Figure 8.7).

If a tissue extract containing LDH is fractionated by electrophoresis in starch or agar gel, its location can be rendered visible by a specific stain that utilizes its enzymic properties. The gel is sliced into two sheets to expose the proteins and either half is incubated in a mixture of lactic acid, NAD and a tetrazolium salt. The latter turns blue in the presence of NADH, which is formed in these conditions only where LDH is present. A blue stain thus appears in those parts of the gel which contain LDH.

With tissue extracts, five bands appear, as shown in Figure 8.8. These represent different forms (isoenzymes or isozymes) of the enzyme which have moved at different speeds in the electric field because they differ in the net charge they carry. The different isoenzymes are numbered LDH 1–LDH 5, depending on their mobility. Isoenzymes 1 and 5 differ in their reaction with analogues of NAD, in

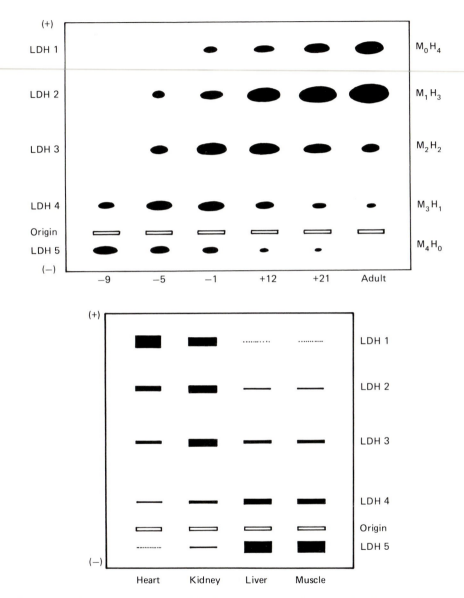

Figure 8.8. Isoenzymes of lactate dehydrogenase (a) in relation to developmental stage in mouse hearts and (b) in heart and body muscle and other tissues of adult rats.
Extracts of the tissues were placed in the wells at the origins of the starch gels, separated by electrophoresis and stained to reveal the location of the enzyme.

their capacity for inhibition by pyruvate, in K_m values for various substrates, and in their change of K_m with p_H, the K_m being the substrate concentration at which the enzyme reaches half its maximum activity. These two extreme forms of LDH therefore clearly represent proteins of quite distinct properties although they catalyse the same reaction. The other enzymes (LDH 2–4) have properties intermediate between those of LDH 1 and LDH 5.

Exposure to urea, or alternate freezing and thawing, causes multimeric proteins like LDH, that are bound by disulphide bonds, to break up into their subunits. If tissue extracts are given this treatment just before electrophoresis, on staining only two bands appear, one at a position corresponding to LDH 1 and the other at LDH 5; the remaining three bands are not seen. This indicates that the five isozymes are derived from only two subunits. The pattern of five bands is interpreted as indicating that each functional enzyme molecule is composed of four subunits of either M (for muscle) or H (for heart) type. LDH 1 can be shown to consist of four H subunits (M_0H_4) and LDH 5 four M (M_4H_0). LDH 2 has the composition M_1H_3, LDH 3 has M_2H_2 and LDH 4 has M_3H_1 (see Figure 8.9).

If isolated M and H subunits are mixed in equal amounts and they were to associate in groups of four entirely at random, they would be expected, on probability grounds, to form the five different isoenzymes in the binomial proportions $1:4:6:4:1$. This in fact is what does occur, showing that association of M and H subunits is random (Figure 8.9). The isozyme patterns produced by most tissue

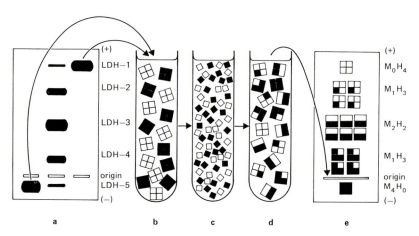

Figure 8.9. The structure of lactate dehydrogenase revealed by dissociation and reassociation of subunits.
(a) The isoenzymes of LDH were separated by electrophoresis. (b) LDH 1 and LDH 5 were eluted from the gel and combined in equal proportions. (c) The subunits were dissociated by alternate freezing and thawing. (d) The enzyme subunits were allowed to reassociate at random. (e) The isoenzymes formed by reassociation were identified and quantified by electrophoresis. There were five isoenzymes in binomial proportions, as expected if two different subunits were to reassociate at random in groups of four.
(Redrawn from Markert and Ursprung, 1971.)

extracts are also normally in binomial proportions, although they are only in the $1:4:6:4:1$ ratio when M and H have been produced in equal quantities. This suggests that association of subunits also occurs at random within the living cell.

A comparison of the relative amounts of isoenzyme in different tissues reveals some very interesting characteristic differences. For example, heart muscle of human and other mammalian adults has a high concentration of the H subunit and the isoenzymes are predominantly of the composition M_0H_4 and M_1H_3. In adult skeletal muscle the M gene is more active, so there is a predominance of M_4H_0 and M_3H_1. The difference in isoenzyme patterns has a functional significance, reflecting differences in the availability of oxygen and the relative importance of aerobic and anaerobic metabolism in the different tissues. The M_4H_0 and M_3H_1 isozymes which predominate in tissues highly dependent on glycolysis for energy have a very high affinity for pyruvate as electron acceptor, while M_0H_4 and M_1H_3 (which predominate in tissues with purely aerobic or respiratory metabolism) have a relatively low affinity. The body muscle type isozymes are thus very effective in re-oxidizing NADH, with pyruvate, to produce NAD and lactate, whereas those characteristic of heart muscle are less reactive with pyruvate and allow NADH to be more readily re-oxidized aerobically by the mitochondria.

LDH isoenzyme patterns also change with age. In heart muscle the M gene is particularly active in early development. That of the H gene increases as development proceeds, there being equal amounts of M and H only around birth (Figure 8.8). A similar change occurs in the lens of the eye. The lens grows by proliferation of epithelial cells in the germinative region, which forms a ring

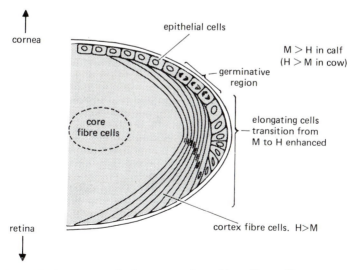

Figure 8.10. LDH isoenzymes during maturation of lens fibre cells.
During maturation of lens epithelial cells into lens fibres in the calf, there is a change from predominantly LDH 5 (M) to predominantly LDH 1 (H).

anterior to its rim. As new cells form they are forced backwards and elongate into fibre cells. As this occurs there is a progressive decrease in the ratio of M to H. In adult lenses there is an excess of H over M in all these cells (see Figure 8.10). The controls upon the relative rates of synthesis of the two subunits have not yet been elucidated, but it will be appreciated that the relative concentrations of the five isoenzymes, each with subtly distinct properties, depends on the conrolled synthesis of only two gene products.

Collagen

Some multimeric proteins are very much more complex than LDH and of these the best studied example is collagen (see Chapter 4). The name 'collagen' means 'to produce glue', this protein being the basis of the old-fashioned furniture glue made by boiling animal bones. There are several different types of collagen constituting extra-cellular fibres and epithelial basement membranes and contributing to teeth, cartilage, bones and tendons. Seven genetically distinct collagen chains each about 1000 amino acids long, have been well described. In theory, these could associate in more than 100 different permutations, but certain combinations seem to be favoured, producing what are known as collagen types I, II, III, IV and V. Altogether collagen amounts to about 25% of the total body protein in mammals.

Each type of collagen is coded by a very long stretch of DNA containing some 50 introns (see Chapter 9). Apart from a short stretch at each end, the primary polypeptide, or pro-α chain, consists of about 1000 amino acids, with glycine at every third position (Figure 8.11(a)). A very large proportion of the remainder are proline, some of which become hydroxylated. Proline and hydroxyproline confer a twist on the polypeptide, to form a helix with just under 3 amino acids per turn, compared with 3.6 in the more tightly coiled α-helix. The glycine residues therefore line up one above the other along one side of the pro-α chain. Three such chains combine together, with the glycine residues adjacent, to form a three-stranded thread known as procollagen (Figure 8.11(b)). Hydroxylysine residues on the procollagen then become glycosylated with glucose–galactose disaccharides before the procollagen is secreted out of the cell. At this stage, the ends of the thread are trimmed, the resultant product being called tropocollagen. Outside the cell the tropocollagen threads link up beside similar molecules to form fibrils, which when examined under the electron microscope show a pattern of dark and light bands with a 67 nm periodicity (Figure 8.11(c)).

When the amino acid sequence of the pro-α chain is examined in detail, it is found that, in addition to the glycine residues at every third position, there is a more subtle repetition of charged and uncharged residues every 234 residues, or 67 nm. This is responsible for the tropocollagen monomers linking side by side with a 67 nm stagger. On negative staining for electron microscopy, the gaps between the ends of the tropocollagen monomers produce the characteristic

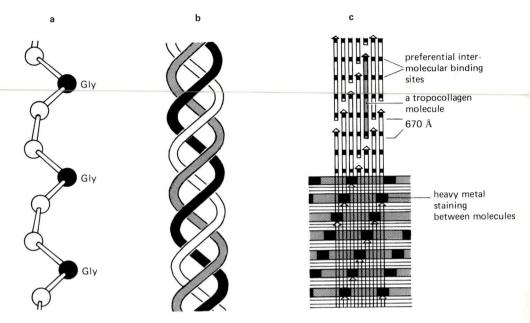

Figure 8.11. The molecule basis of collagen structure and assembly.
(a) The pro-α chain, with glycine at every third position. It is coiled into an α-helix due to the twist confered by proline and hydroxyproline. (b) A strand of procollagen or tropocollagen formed from three pro-α chains. (c) A portion of collagen fibril formed by lateral association of tropocollagen monomers.They line up 67 nm out of phase due to repetition of regions of affinity within each chain. When negatively stained for electron microscopy this produces a banded appearance.

67 nm banding pattern seen in collagen fibrils. The fibrils are further strengthened by cross-linking between modified lysine residues, to produce very tough multi-stranded ropes of glycoprotein. A load of at least 10 kg is required to break a collagen fibre 1 mm in diameter.

The precise form adopted by collagen depends on the solvent conditions in the extra-cellular space where it forms, the density of its carbohydrate groups and the orientation of the cytoskeleton of the cells which secreted it (see Chapter 4). These conditions can vary considerably, so that a wide variety of different types of collagen are formed in the different parts of any individual, some unbanded or with variant banding patterns, in sheets instead of ropes, in small strands, or in amorphous masses. Despite these variations, the gross structure of collagen is in all cases a direct consequence of its primary amino acid sequence and hence of the sequence of nucleotides in the collagen genes. This direct relationship between primary amino acid sequence and secondary, tertiary and higher-order structures, is a general rule of protein chemistry.

8.6.   Control of enzyme degradation

As we saw above, the effective cytoplasmic concentration of a particular species of mRNA depends not only on the rate at which it is synthesized, but also on the rate at which it is broken down and lost from the cell. The effective concentration of its protein product is also a function of the difference between its rate of translation and the rate at which the protein is degraded.

Tryptophan pyrrolase is an enzyme of mammalian liver which converts tryptophan into formylkynurenine. Its activity is 'inducible' by both hydrocortisone and tryptophan, which acts synergistically with respect to enzyme function. In other words a mixture of the two 'inducers' will produce more functional enzyme than either acting alone. Using an antibody specific to tryptophan pyrrolase, prepared by injecting the purified enzyme into rabbits, it was shown that only one type of protein is involved and enzyme activity is proportional to the amount of this protein. This means that the two different inducers regulate the concentration of a single enzyme. The way in which they work was revealed by injecting rats with one or both inducers, together with radioactive amino acids which became incorporated into the newly synthesized protein. By this means hydrocortisone was found to increase the rate of synthesis of the enzyme, whereas in contrast tryptophan delayed its degradation. This experiment therefore demonstrates that *the effective concentration, or 'activity', of an enzyme can be regulated by metabolic controls acting both at its synthesis and at its destruction.* Enzyme degradation is selective and tissue specific. For example, in rat liver, heart muscle and skeletal muscle the average protein has a half-life of 2.2, 1.0 and 22.0 days respectively, whereas that of LDH 5 is 16, 1.6 and 31 days.

8.7.   The mechanism of enzyme action

One of the best-studied examples of enzyme action is that of chymotrypsin, one of the proteolytic enzymes produced by the pancreas. Chymotrypsin is a compact ellipsoid molecule carrying all its charged groups on the surface, except for three that play a critical role in catalysis. These are serine 195, histidine 57 and aspartate 102. These three form a charge relay network, aspartate drawing a proton away from the serine and converting the $-CH_2OH$ side-chain of the serine into the highly reactive $-CH_2O^-$. The operation of the charge relay network enhances the rate of catalysis about 1000-fold.

Chymotrypsin selectively cleaves peptide bonds on the carboxyl side of the aromatic side-chains of tyrosine, tryptophan and phenylalanine and beside large hydrophobic residues such as methionine. These residues interlock with a pocket in the enzyme, positioning the adjacent peptide bond beside serine 195. Peptide-bond hydrolysis starts with an attack of the serine $-CH_2OH$ group on the $-CO$ group of the peptide bond. The charge relay network draws a proton away from the serine charging its side-chains, then histidine 57 donates a proton to the nitrogen atom of the adjacent amino acid, breaking the peptide bond. A reverse process recharges the enzyme and prepares it for another round of catalysis.

This is just one of an enormous number and variety of biological catalytic processes that occur in all living systems.

8.8. *Control of metabolic pathways*

The hundreds of enzyme-catalysed chemical reactions that take place in any cell do not occur independently of one another. Most are linked in some kind of sequence, so that the product of one reaction becomes the substrate of the next, the product of that forming the substrate of a third, and so on. Any atom introduced into the system is thus passed from enzyme to enzyme along a metabolic pathway and on the way may contribute to many different molecular structures before eventually being returned to the environment. Some pathways have as many as 20 reaction steps and the different recognized pathways are linked by common substrates or end-products, or by the end-product of one serving as a regulator for another. A few pathways lead into metabolic blind alleys, where vast accumulations of end-product may build up. This occurs for example with vitamin A in the livers of polar bears. In most instances, however, regulatory mechanisms come into play which inhibit such excesses.

Allostery

In the simplest form of chemical regulation products inhibit the reactions by which they are formed in accordance with the Law of Mass Action. In some cases, however, accumulation of end-product beyond a critical concentration can act as a signal that impinges back on an enzyme earlier in the sequence, so slowing down the whole pathway and preventing accumulation of intermediates also. This is called feedback inhibition or end-product inhibition. The metabolites which carry out this kind of regulation are called effectors, and the enzymes upon which they act are regulators.

Most regulators occur at the beginnings of pathways or at branch points. One example is the pathway which, by five enzyme-catalysed steps, converts L-threonine into L-isoleucine (Figure 8.12). The first enzyme, L-threonine deaminase, is strongly inhibited by the end-product, isoleucine, but not by any of the intermediates. Regulatory enzymes of this type are believed to become modified temporarily by a change in shape, a so called 'allosteric' change, caused by binding the effector, and thereby altering the conformation of its active site.

Some allosteric enzymes are *activated* by an effector (Figure 8.13). The central pathway of energy metabolism—the Embden–Meyerhoff pathway—is regulated by a key rate-limiting enzyme that responds to several controls (Figure 8.7). This is phosphofructokinase, which converts fructose 6-phosphate into fructose 1,6-diphosphate. Phosphofructokinase is inhibited by ATP and citrate, but stimulated

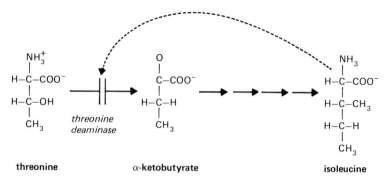

threonine   α-ketobutyrate   isoleucine

Figure 8.12. Feedback inhibition in a biochemical pathway.
Isoleucine inhibits its own synthesis from threonine, by inhibiting the action of threonine deaminase. (After Stryer, 1975.)

by ADP and phosphate. Because it responds to a range of effectors produced by other pathways it can integrate the rates of several enzyme systems. Regulatory enzymes in other systems also respond allosterically to ATP or ADP, and this extends the control over several more pathways. Anabolic and catabolic processes are usually controlled by different regulators, so that rates of biosynthesis and degradation of given cell components are regulated independently.

As we saw above, some proteins are composed of a number of subunits of similar or dissimilar type. Some of these contain several active sites perhaps derived, in evolutionary terms, by duplication of an ancestral gene sequence. An enzyme may therefore be multivalent with respect to a single catalytic property, have more than one propensity integrated within the one molecule, or even have

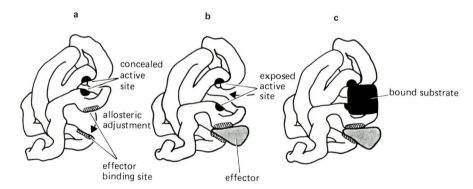

Figure 8.13. Allosteric activation of an enzyme.
(a) A hypothetical allosteric enzyme is represented with its active site concealed. (b) An effector molecule binds to the effector binding site, so exposing the active site at another part of the molecule. (c) A molecule of substrate becomes bound to the active site of the allosterically activated protein.

properties unique to the multimer that are not indicated in the monomers. When an effector binds to one part of an allosteric molecule it causes an active site at another part to become exposed or, alternatively, masked. In some molecules allosteric changes caused by binding substrate are exploited for opening up further active sites. For example, haemoglobin has four oxygen binding sites. Its binding affinity for the first oxygen atom is low, but binding the first oxygen causes an allosteric change such that its binding affinity for a second oxygen is greater. This causes further allosteric modification, increasing the binding affinity for the third atom and then the fourth. The result is that the last oxygen binds with 280 times the affinity of the first. As a tetramer, haemoglobin is far, far more efficient than four monomers of similar sequence would be.

Compartmentation

Different enzymes and enzyme systems are characteristically located in one or another organelle or sub-cellular 'compartment'. For example, replication and transcription of the chromosomes occurs necessarily in the nucleus, translation takes place in the ribosomes and protein glycosylation in the Golgi apparatus. In mouse tissue β-glucuronidase is present in both the lysosomes and the endoplasmic reticulum. It is coded by the same gene in the two locations, but a separate gene determines its integration into the endoplasmic reticulum. The entire glycolytic enzyme system, including the Embden–Meyerhoff and associated pathways, is in the soluble portion of the cytoplasm, whereas the enzymes concerned with the products of glycolysis, those involving electron transport and phosphorylation of ADP to ATP, are in the mitochondria. Within the mitochondria themselves each enzyme of the electron-transport chain is actually located physically adjacent to the next in the pathway, allowing the entire sequence to function in supreme co-ordination. Sub-cellular compartmentation of this sort not only facilitates performance of each activity but also segregates reactions which might otherwise be chemically incompatible.

The functioning of enzymes depends very largely on their molecular environment. In addition to the number and nature of charged amino acids they contain, the charge they carry depends on the local hydrogen ion concentration or pH, which in turn depends on a host of internal and external variables. Enzyme activity also increases with temperature. The expression of a particular gene may therefore be controlled at the protein level by a multitude of immensely complex interacting systems and influences, which operate potentially in the context of all other gene products, and all other molecules of external provenance present in the organism.

8.9. Do enzymes have other functions?

One of the problems in understanding the real importance of enzymes is that they are frequently present in quantities 10–100 times as large as would seem to be required for catalytic function. Often there are almost as many molecules of enzyme as of substrate or product. It has therefore been suggested that enzymes may have another function: that of binding substrates or metabolites to hold them safely in store. The presence of such a store would buffer the cell against fluctuations in substrate supply and would contribute to the stability of its metabolic activities. The real importance of this possibility remains to be evaluated.

8.10. Summary and conclusions

Apart from those DNA sequences that are concerned with the functioning of the genome itself, so far as we know all genes exert their effects through the proteins which they define. The three-dimensional structure of a protein depends ultimately on the sequence of amino acids it contains, as dictated by the sequence of nucleotides in the gene which codes for it.

Protein synthesis in the cell can be controlled in a general fashion through the supply of limiting factors required for translation of mRNA transcripts. In addition, there are specific controls that operate with respect to individual messenger species. When messengers compete for translation-initiation factor, their relative rates of translation can be governed by their affinity for the factor, as well as their numerical representation in the cytoplasm. Other controls on translation include protein masking of mRNA and messenger longevity, which correlates with poly-A tail length. Histone messengers, which are not polyadenylated, are usually degraded rapidly, except during the DNA synthetic phase of the cell cycle. The abundance of amino acid-charged tRNAs is also important in translational

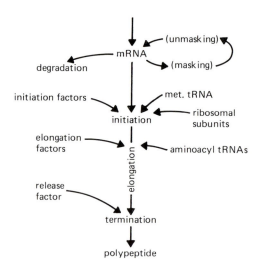

Figure 8.14. A diagrammatic summary of translation, showing potential control points.

control, especially that of the initiator, methionyl tRNA, the supply of which varies during the cell cycle. Some soluble initiation factors exert discrimination between messengers and in some cases other RNA molecules may play key roles in controlling translation in ways which are not clearly understood.

So-called 'induction' of synthesis of specific proteins can involve stimulation of translation of specific messenger RNA species, protection of that RNA from degradation, or protection of the protein itself, apart from controls at transcription (see Chapters 4 and 12) or RNA processing (Chapter 9; see Figure 8.14). Many messenger RNAs survive mitosis to be translated again in the daughter cells.

The raw polypeptides initially produced by translation of mRNA are usually modified in a variety of ways before becoming functional proteins (see Figure 8.15). Post-translational modification includes cleavage, homomeric or

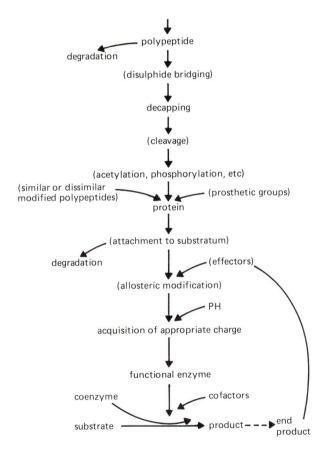

Figure 8.15. Post-translational expression of a typical gene showing potential control points.
Aspects of processing that apply only to certain proteins are enclosed in brackets.

heteromeric association, chemical modification, acquisition of co-factors and allosteric changes. Enzyme activity also depends on a host of other factors, including ionic concentration (and in particular, pH), intra-cellular compartmentation, and supply of co-enzymes. The form, composition and properties of some complex proteins are governed by a variety of environmental factors operative at their creation, although in all cases the three-dimensional structure of proteins has its basis in their amino acid sequence. Conditions that affect post-translational events vary considerably both spatially and temporally, so that one or a few genes can give rise to a whole spectrum of protein products. Collagen is one such protein that contributes to many different extra-cellular materials, contact with which exerts profound effects on intra-cellular metabolism (see Chapter 4).

Some enzymes are present in very much greater quantities than required for their catalytic functions. Why this should be is unknown, but one suggestion is that they may sequester substrates and metabolites, so smoothing out fluctuations in their supply and contributing to the stability of cell metabolism.

It can be seen that the scope for post-translational control of gene expression is enormous, but none of these controls can operate unless RNA transcripts are produced from specific genes, processed within the nucleus and exported to the cytoplasm. In the next chapter we will consider the properties and fate of RNA.

Bibliography

Alberts, B., Bray, D., Lewis, J., Raff, M., Roberts, K. and Watson, J., *Molecular Biology of the Cell*. Garland, New York (1983).

Ashworth, J. M., *Cell Differentiation*. Chapman and Hall, London (1975).

Austen, S. A. and Kay, J. E., Translational regulation of protein synthesis in eukaryotes. In *Essays in Biochemistry*, Vol. 18, edited by R. N. Campbell and K. D. Marshall, Academic Press, London, pp. 77–120 (1982).

Clark, B. F. C., *The Genetic Code and Protein Biosynthesis*, 2nd edn. Edward Arnold, London (1984).

Chothia, C., Principles that determine the structure of proteins. *Ann. Rev. Biochem.*, **53**: 537 (1984).

Cohen, P. (ed.), *Enzyme Regulation by Reversible Phosphorylation — Further Advances*, Vol. 3, *Molecular Aspects of Cellular Regulation*. Elsevier, Amsterdam (1984).

Fersht, A., *Enzyme Structure and Mechanism*, 2nd edn. Freeman, Oxford (1984).

Godrey-Colburn, T. and Thach, R. E., The role of mRNA competition in regulating translation. IV. Kinetic model. *J. Biol. Chem.*, **256**: 11762 (1981).

Goldberger, R. F., *Biological Regulation and Development*. Plenum, New York (1980).

Graham, C. F., Limits of molecular biology. In *Developmental Control in Animals and Plants*, 2nd edn, edited by C. F. Graham and P. F. Wareing. Blackwell Scientific Publications, Oxford, pp. 371–395 (1984).

Hagenbüchle, O., Santer, M. and Steitz, J. A., Conservation of the primary structure at the 3′ end of 18s mRNA from eukaryotic cells. *Cell*, **13**: 551 (1978).

Herman, R. H., Cohen, R. M. and McNamara, P. D., *Principles of Metabolic Control in Mammalian Systems*. Plenum, New York (1980).

Hershko, A. J. and Ciechanover, A., Mechanisms of intracellular protein breakdown. *Ann. Rev. Biochem.*, **51**: 335 (1982).

Krebs, E. G. and Bearo, J. A., Phosphorylation-dephosphorylation of enzymes. *Ann. Rev. Biochem.*, **48**: 923 (1979).

Lehninger, A. L., *Biochemistry*. Worth, New York (1970).

Mainwaring, W. I. P., Parish, J. H., Pickering, J. D. and Mann, H., *Nucleic Acid Biochemistry and Molecular Biology*. Blackwell, Oxford (1982).

Maitra, U., Stringer, E. A. and Chandhuri, A., Initiation factors in protein biosynthesis. *Ann. Rev. Biochem.*, **51**: 869 (1982).

Markert, C. L. and Ursprung, H., *Developmental Genetics*. Prentice-Hall, Englewood Cliffs, New Jersey (1971).

Nakanishi, S., Inoue, A., Kita, T., Nakamura, M., Chang, A. C., Cohen, S. N. and Numa, S., Nucleotide sequence of cloned cDNA for bovine corticotropin-β-lipotropin precursor. *Nature*, **278**: 423 (1979).

Papaconstantinou, J., Molecular aspects of lens cell differentiation. *Science*, **156**: 338 (1967).

Pérez-Bercoff, R., *Protein Biosynthesis in Eukaryotes*, NATO Advanced Study Institutes Series, Series A. Vol. 41. Plenum, New York (1982).

Prescott, D. M. and Goldstein, L., *Gene Expression: Translation and the Behaviour of Proteins, Vol. 4, Cell Biology. A Comprehensive Treatise*. Academic Press, New York (1980).

Rees, A. R. and Sternberg, M. J. E., *From Cells to Atoms*. Blackwell, Oxford (1984).

Rattazzi, M. C., Scandalios, J. G. and Whitt, G. S., *Isozymes, Current Topics in Biological and Medical Research*, Vol. 5. Alan R. Liss, New York (1981).

Schimke, R. T., On the roles of synthesis and degradation in regulation of enzyme levels in mammalian tissues. *Curr. Top. Cellular Reg.*, **1**: 77 (1969).

Schimke, R. T. and Doyle, D., Control of enzyme levels in animal tissues. *Ann. Rev. Biochem.*, **39**: 929 (1970).

Storti, R. V., Scott, M. P., Rich, A. and Pardue, M. L., Translational control of protein synthesis in relation to heat shock in *D. melanogaster* cells. *Cell*, **22**: 825 (1980).

Strickberger, M. W., *Genetics*. Macmillan, New York/Collier Macmillan, London (1976).

Stryer, L., *Biochemistry*. W. H. Freeman, San Francisco (1975).

Suzuki, D. T., Griffiths, A. J. F. and Lewontin, R. C., *An Introduction to Genetic Analysis*. W. H. Freeman, San Francisco (1981).

Taylor-Papadimitriou, J., *Interferons. Their Impact in Biology and Medicine*. Oxford University Press, Oxford (1985).

de Vries, A. L., Role of glycopeptides and peptides in inhibition of crystallization of water in polar fishes. *Philos. Trans. R. Soc. [Biol. Sci.]*, **304**: xxx (1984).

Woodward, D. O. and Woodward, V. W., *Concepts of Molecular Genetics*. Tata McGraw-Hill, New Delhi (1978).

Chapter 9 RNA

The production and fate of ribonucleic acid (RNA) is central to the interpretation of genotype into phenotype. Indeed, it is cogently argued that the original ancestral genetic material may have been ribonucleic acid, rather than deoxyribonucleic acid (DNA). In eukaryotes, some of the RNA occurs as the transcripts of nuclear genes that specify the amino acid sequences of proteins. The translation of these genetic messages requires the agency of two other well-described RNA classes — ribosomal (rRNA) and transfer RNA (tRNA) — which are not translated into protein (see Chapter 8). There are strong indications that cell nuclei contain at least one other type of RNA essential for processing the primary RNA transcript. This falls into the category of small nuclear RNA (snRNA), and will be referred to here as 'processing RNA' to distinguish it from the other snRNA species which may have different functions.

The mitochondria (and in green plants the chloroplasts) produce RNA transcripts of their own genes (see Chapter 1), which are translated by a separate set of ribosomes and transfer RNA molecules.

Many intriguing questions can be asked about the functioning of the different types of RNA, and Chapter 12 will return to the crucial issues related to their production. At this stage, however, we will limit ourselves to some of the controls that are exerted upon gene expression at the RNA level and which are imposed on the developmental system by RNA.

9.1. Maternal RNA

The average life-span of a molecule of bacterial messenger RNA is about three minutes, after this it becomes degraded by ribonucleases. In contrast, in the human cell line known as HeLa, about one third of the mRNA lasts for about seven hours, while the remaining two thirds may survive for up to 24 hours. Protection of RNA from attack by ribonucleases is one of the means by which differences in protein synthetic patterns arise. Some protection is achieved by masking the RNA with protein. As we have seen (Chapter 4), the cytoplasm of the egg also contains mRNA carrying ribosomes in 'blocked' inactive form known as an informosome. The maternal mRNA in the informosome is unmasked and

182

translated to provide the proteins required to carry the embryo through the early developmental stages, so that the embryo's own genes do not need to come into operation to a large extent until around gastrulation.

A theory put forward by Paul to explain the origins of differential gene expression in different parts of the embryo suggests that regional differences in the conditions to which cells are exposed cause selective unmasking of the informosomes. As we saw in Chapter 2, maternal RNA plays a crucial role in the early development of some species. According to this theory, one set of maternal messengers is translated in one part of the embryo, or in one tissue type, while another part sees the translation of a different set. Differential unmasking of informosomes could occur by partitioning of the other cytoplasmic constituents, for example under gravity, or by the establishment of different types of metabolism due, for example, to the availability of oxygen in superficial, compared with buried, tissues (Figure 9.1). In the ascidians, or sea squirts, maternal mRNA coding for alkaline phosphatase is translated in the embryonic endoderm, but nowhere else. If proteins coded by maternally derived messengers are capable of switching on the transcription of the embryo's own genes, then new, possibly self-perpetuating, regionally specific patterns of gene expression could become

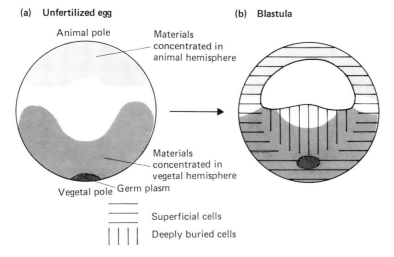

Figure 9.1. Theory of selective unmasking of maternal mRNA.
(a) Distribution of cytoplasmic materials in an amphibian egg. Some materials are concentrated near the animal or vegetal pole, masked maternal mRNA molecules (informosomes) are distributed throughout the cytoplasm. The germplasm is shown as a heavily shaded zone at the vegetal pole. (b) At the blastula stage the cells which have now formed have different cytoplasmic compositions. Different sets of informosomes may be unmasked and expressed at the poles, in the future ectoderm and endoderm respectively, due to the effects of unequal distribution of other materials. Another set may be expressed in superficial cells and yet another in deeply buried cells, due to the diverse effects of external and internal influences on cell metabolism. The future germ cells remain isolated from these events.

established. Such patterns would need to be initiated very early in development, however, as in some species maternal mRNA does not seem to survive much longer then a day or so once development has started.

In many species the mRNA accumulated in the oocyte is known to be synthesized on modified chromosomes known as lampbrush chromosomes, named after the brushes which were used for cleaning the glass chimneys of old-fashioned

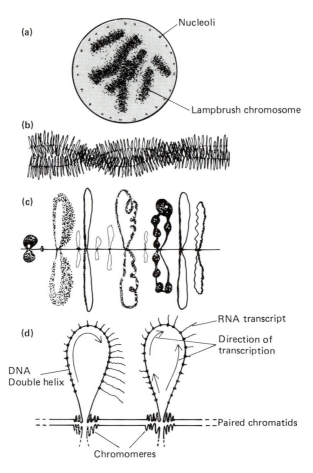

Figure 9.2. Lampbrush chromosomes.
(a) A mammalian secondary oocyte, there are about 1000 nucleoli at the periphery (after Grant, 1978). (b) Paired lampbrush chromosomes showing two chiasmata. (c) A region of a single lampbrush chromosome from a salamander oocyte, showing the variety of lampbrush loops which can be seen (after Waddington, 1966). (d) Details of lampbrush loop structure showing transcription taking place. One loop consists of a single transcription unit, the other includes three transcriptional units. For the sake of clarity, apart from RNA polymerase, other (chromosomal) proteins associated with the DNA and RNA have been omitted.

paraffin lamps. In lampbrush chromosomes portions of the DNA are extended into lateral loops away from the main axis of the chromosome, this conformation facilitating the transcription of RNA copies. Each loop consists of an extended part of one chromatid, the chromosomes being in diplotene (i.e., each consisting of two chromatids) at this stage. The RNA transcribed from these loops becomes packaged with proteins to produce structures of characteristic appearance (see Figure 9.2), which suggest some specificity in the association between the RNA and the protein packaging. Each loop can contain several transcriptional units which may be on either strand of the DNA double helix (Figure 9.2).

Ribosomal RNA also accumulates in enormous quantities within the oocytes. It is synthesized on the chromosomes (see Chapter 10) and also in bodies in the nucleus called nucleoli, which contain multiple copies of the gene sequences corresponding to the 18S, 5.8S and 28S rRNA subunits. The term 'gene amplification' is used to describe the multiplication of these ribosomal RNA genes in this extra-chromosomal DNA that occurs in some species (see Chapter 10). Normal vertebrate somatic cells have two nucleoli, each containing about 800 copies of each ribosomal RNA gene. Typically there are very many more nucleoli in the oocytes (see Figure 9.2), for example, in *Xenopus laevis*, the African clawed toad, there are up to 1200 per oocyte, so that each contains something like 1 000 000 copies of the rRNA genes.

9.2. Polytene chromosomes

Mass production of RNA is facilitated in certain tissues of the two-winged flies, the Diptera, by another modified form of chromosome known as a polytene chromosome. These are truly gigantic, measuring about 1 mm long (i.e., 100 times the normal length) by up to 50 μm in diameter and consisting of 256, 512 or 1024 chromatids (DNA double helices) packed side by side. Polytene chromosomes are normally linked together at the centromeres, producing a mass of heterochromatic material called the 'chromocentre' from which they radiate. Figure 9.3 shows the polytene chromosomes in *Drosophila melanogaster* salivary gland cells alongside a more conventional representation of *Drosophila* chromosomes; note that the size differential between polytene and normal chromosomes has been *reduced* by a factor of 10.

Each polytene chromosome contains equal multiples of both the homologous chromosome partners, and since these copies are lined up precisely beside one another features of the chromosome are presented side by side, giving the appearance of bands across the chromosome. Bands which stain darkly with chromosome stains represent the structural gene sequences, while the lighter-staining interband regions seem to represent non-coding regions of the DNA.

At intervals, some of the bands are swollen into what are known as 'chromosome puffs' where RNA synthesis occurs most actively. This can be demonstrated by incubating the tissues with the radioactive RNA precursor

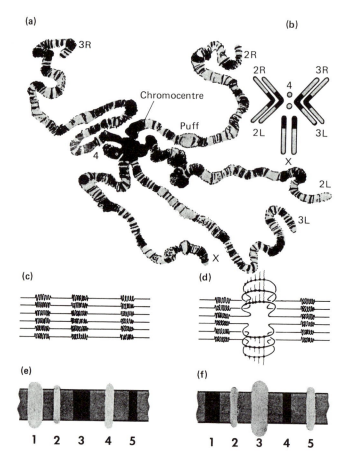

Figure 9.3. Polytene chromosomes.
(a) An impression of the structure of the chromosomes in the salivary glands of a female
Drosophila fly, stained with Feulgen stain. Homologous arms are fused and all the
chromosomes are joined at the chromocentre. Structural gene sequences stain as dark bands
when not puffed. Some puffs are shown. These chromosomes are represented as about 10
times the length of those in general somatic cells shown in (b), in reality they are 100 times
as long. (Drawn from a photograph by B. P. Kaufmann.) (b) The conventional representation of *Drosophila* chromosomes in general somatic cells of a female fly. Large areas of
heterochromatin and the adjacent centromeres are shown in black. (c) A representation of
the structure of the DNA in bands and interbands in a region without puffing. (d) A
chromosome puff showing the transcription of messenger RNA. For the sake of clarity,
apart from RNA polymerase, the protein normally associated with the DNA and RNA have
been omitted. (e), (f) Tissue differences in puffing activity. The same chromosomal region
is shown in salivary gland cells (e) and Malpighian tubules (f). Bands 1 and 4 are puffed
only in salivary glands, bands 3 and 5 only in Malpighian tubules, showing tissue-specific
differences in transcription. Band 2 is puffed in both tissues, showing transcription of a
common product.

[5-^{3}H] uridine. The regions which have incorporated radioactive label can then be revealed by autoradiography — by allowing the radioactive discharge to form an image on photographic film or emulsion. An impression of the differences in structure of DNA in the bands, interbands and puffs is represented in Figure 9.3, but it should be recognized that the DNA at this stage is also complexed with proteins (Chapter 10) which, apart from DNA polymerase, are not represented. By using antibodies directed against RNA polymerase, its presence can be revealed in both puffed and non-puffed bands, in accordance with the theory that transcription in differentiated cells occurs at a very low rate throughout the genome (see below). However, RNA synthesis and the concentration of RNA polymerase are enormously increased at the puffs.

Polytene chromosomes occur in several tissues of Dipterans, including the salivary glands, the cells of the mid gut and the Malpighian tubules. If patterns of puffing in different tissues are compared, they reveal that RNA synthesis varies between tissues (Figure 9.3(e),(f)). This illustrates the very important point that *some tissue-specific controls upon protein production operate at transcription of RNA* (see Chapter 12).

If the same tissue is examined at different developmental stages, patterns of puffing are found to follow a standard sequence, showing that the pattern of transcription changes in a standard fashion during development. Figure 9.4 shows the transient nature of some of this transcriptional activity. Some puffs, however, are neither stage- nor tissue-specific, indicating the synthesis of RNA for constitutive or 'housekeeping' proteins (see Chapter 8). Puffing is not an intrinsic property of the polytene chromosomes, but occurs in response to external stimuli. For example, when the moulting hormone ecdysone (see Chapter 4) is injected into

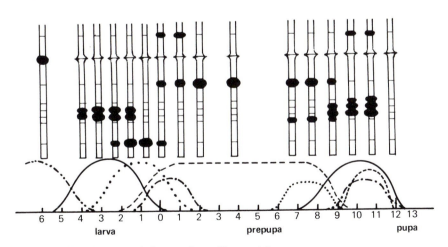

Figure 9.4. Developmental changes in puffing activity.
Puff changes in arm 3L of *Drosophila* salivary glands are shown in relation to the time in hours before and after puparium formation. Puffing is seen to be a transient phenomenon. (From Becker, 1979.)

a *Drosophila* larva, this induces the specific puffing pattern normally shown before moulting (Figure 9.7).

One species of the gnat *Chironomus*, known as *C. pallidivittatus*, has characteristic proteinaceous granules in what are known as 'special cells' in the salivary glands. Comparison of puffing patterns in the special cells with adjacent cells which do not produce granules reveals that the locus responsible for the granules is at the end of the fourth chromosome. A related species, *C. tentans*, produces no secretion in the equivalent cells. The two species are inter-fertile and

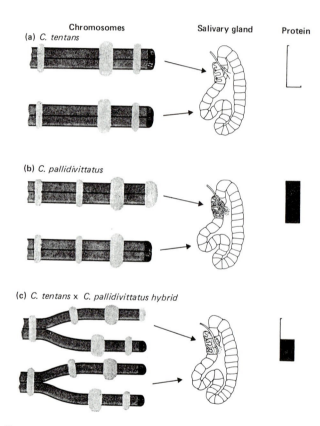

Figure 9.5. Puffing and gene activity in *Chironomus* salivary glands.
(a) The end of chromosome 4 in *C. tentans*. The terminal band is not puffed in any part of the salivary gland and no protein is secreted. (b) In *C. pallidivittatus* the terminal band is puffed in certain 'special' cells which visibly secrete a large amount of protein in the form of granules. Neighbouring cells which do not contribute to this secretion show no puffing at this region. (c) The terminal part of chromosome 4 does not synapse in the inter-specific hybrid, due to a chromosomal inversion. In the special cells, only the *pallidivittatus* chromosome puffs at the terminal band. In adjacent salivary gland cells neither homologue puffs at the terminal band, The amount of salivary protein produced is intermediate between that produced by the two parental species.

F1 hybrids secrete an intermediate amount of the protein (Figure 9.5). The F2 generation, produced by mating two of the hybrids, falls into three classes which resemble the *pallidivittatus* parent, the F1 hybrid and the *tentans* parent in the ratio 1 : 2 : 1, as would be expected if secretion of granules is due to a single gene. Detailed analysis of chromosome 4 in the salivary glands of the F2 generation revealed normal puffs in individuals secreting large amounts of granule protein and no puffs in those which secreted no protein. In the F2 individuals that produced an intermediate amount, the copy of chromosome 4 which came from the *pallidivittatus* parent displayed a puff, but the *tentans* chromosome did not. This is perhaps the first instance in which a gene identified by Mendelian criteria was actually observed in the act of producing a phenotypic character.

Control of puffing

In the *Chironomus* hybrids the *pallidivittatus* chromosomes puffed at this site while the *tentans* chromosomes did not. Normally homologous polytene chromosomes are in close contact, or in what is called synapsis, but in these hybrids they are not. This is because the species are distinguished by an inversion at the end of chromosome 4, which inhibits synapsis in the hybrid. Several mutant dipterans exist which, like *C. tentans*, are unable to form puffs at certain sites.

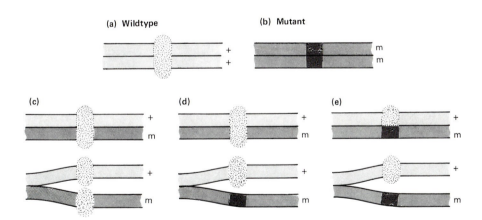

Figure 9.6. Puffing in relation to synapsis.
(a), (b) The same region of polytene chromosomes in a wild-type fly (a) and several puff-deficient mutants (b). (c) In this heterozygote both homologues puff irrespective of synapsis. This shows that the defect in the mutant is not closely linked to the locus represented by the band under observation. (d) In this heterozygote the chromosome from the puff-deficient parent puffs when in synapsis with the wild-type homologue, but not otherwise. In this case the mutant site is at or closely linked to the band observed. (e) In this heterozygote the chromosome from the puff-deficient parent does not puff in either situation. In this case also the mutant site is at or closely linked to the observed band.

With one type of mutation, puffs will occur on both mutant and normal homologues in flies heterozygous for the mutation, whether the chromosomes are in synapsis or not (Figure 9.6(c)). With another type, both homologues puff in the heterozygote only if they are in synaptic contact (Figure 9.6(d)); This is sometimes called 'infectious puffing'. With a third type (Figure 9.6(e)) the failure to puff is maintained autonomously by the affected homologue: it fails to puff in the heterozygote regardless of whether the chromosomes are synapsed. In the first example the mutation is probably not linked to the site which puffs, in the others it is either at the puffing site or is closely linked to it.

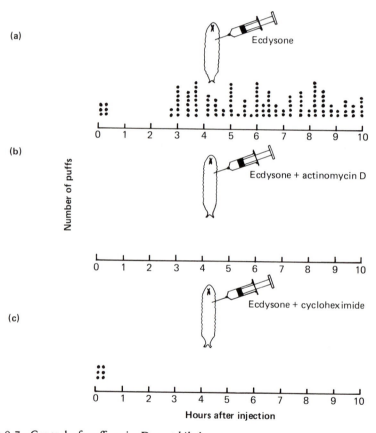

Figure 9.7. Control of puffing in *Drosophila* larvae.
(a) When a third-instar larva is injected with the moulting hormone ecdysone, six bands puff immediately in the salivary gland chromosomes, followed by a large number after 3–10 hours. (b) When actinomycin D is injected at high concentration with the ecdysone no puffs appear, as this inhibits RNA synthesis. (c) When the inhibitor of protein synthesis, cycloheximide, is injected with the ecdysone the first six puffs appear, but no more, showing that protein synthesis is required for stimulation of the second round of puffing.

It has been suggested that infectious puffing could involve contact stimulation of an open chromatin structure, rather like the formation of crystals stimulated by seeding a solution with crystal fragments. The repression of gene activity by heterochromatin could occur by the reverse of this effect (see Chapter 10).

If the larva of a Dipteran, such as *Drosophila*, is injected with the moulting hormone ecdysone, normally within about five minutes six bands will puff in the salivary gland cells, the puffs lasting for about four hours. After 3–10 hours, about 100 further puffs appear (see Figure 9.7(a)). If actinomycin-D is injected at high concentration with the ecdysone no puffs are produced, since actinomycin-D blocks RNA synthesis (see Chapter 12). If an inhibitor of *protein* synthesis, such as cycloheximide, is injected with the ecdysone, the first set of six puffs appear but there is no further puffing. This indicates that it is a *protein* product of the first puffing activity which initiates the later round of puffs.

This simple experiment confirms the generally accepted theory that *patterns of gene expression are directed by the protein products of previously expressed genes.* This idea is really of very far-reaching importance, since it means that cell types will differentiate more or less as a matter of course once initial patterns of protein synthesis have become established. A stimulus which affects cells in one region of the body and not another, and which initiates the synthesis of a gene-switching protein, can thus set off a cascade of metabolic events which, once started, may be virtually unstoppable without the death of the cell line. This is one way in which the canalization of development and cell determination (see Chapters 2 and 14) operate at a molecular level. It will be appreciated that such cascades of events could be initiated by the unmasking of maternal mRNA, as described above.

9.3. *The RNA populations of cells*

RNA–DNA hybridization

Tissue-specific puffing patterns give rise to different populations of mRNA in the different tissues. Vertebrates do not have polytene chromosomes, but tissue-specific differences in RNA populations can be demonstrated by a more refined technique, based upon the specific pairing relationships between the sequence of bases in the RNA and their complementary sequences in DNA. In this approach, double-stranded DNA is dissociated into small single-stranded fragments and incubated with a preparation of RNA. If the reaction is allowed to continue for long enough, unlikely as it may seem, the RNA molecules will bind to the very same DNA strands from which they were transcribed.

If the hybridization reaction occurs in suspension, each RNA sequence must compete with the complementary non-codogenic DNA strand for its specific coding sequence, and the reaction will not go to completion. This competition can be avoided, however, by binding the dissociated DNA on to nitrocellulose, so that the non-codogenic strand is rendered immobile along with its codogenic partner.

If the RNA preparation is made radioactive, the proportion which binds to the codogenic strand can be quantified by measuring the amount of radioactivity accumulated by the nitrocellulose.

This reaction, known as RNA–DNA hybridization, can be used for assessing similarities in the RNA populations of different tissues. In one experiment a

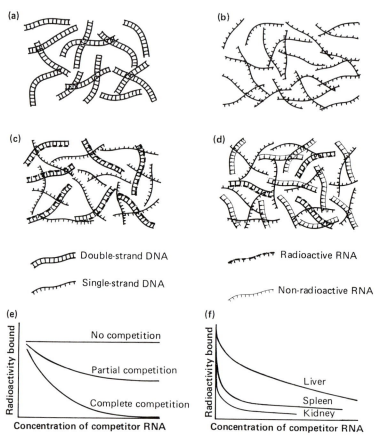

Figure 9.8. RNA–DNA hybridization competition experiment.
(a) Double strand DNA fragments. (b) DNA fragments dissociated into single strands and bound to a nitrocellulose filter. (c) A filter incubated with radioactive mRNA, which forms RNA–DNA hybrid molecules with the DNA codogenic strands. Unbound RNA has been removed by washing. (d) Non-radioactive mRNA is mixed with the radioactive RNA preparation and the mixture hybridized with bound DNA. The non-radioactive RNA competes for binding sites on the DNA codogenic strands. Unbound RNA has been removed by washing. (e) Theoretical plot to illustrate amount of radioactivity bound by filter under different degrees of competition. (f) Radioactive mouse kidney RNA was hybridized to mouse DNA in the presence of various amounts of non-radioactive RNA from liver, spleen and kidney. Competition occurs in the order: kidney, spleen, liver. ((e) And (f) after Markert and Ursprung, 1971.)

mouse was injected with radioactive uridine (which becomes incorporated predominantly into RNA) and the total RNA of the kidney was collected and purified. This radioactive kidney RNA was then mixed with different amounts of non-radioactive RNA extracted from the kidneys, livers and spleens of control mice. The RNA mixtures were incubated with nitrocellulose filters carrying dissociated mouse DNA and the amount of radioactive RNA which bound to the DNA on the filters was plotted against the amount of non-radioactive competitor RNA added to the hybridization mixtures (see Figure 9.8).

It will be appreciated that if there had been *no* competition between the added non-radioactive RNA and the radioactive preparation, accumulation of label by the filter should not have been inhibited. On the other hand, if the non-radioactive competitor RNA was identical in sequence to the radioactive preparation, accumulation of label should have decreased with increasing amounts of competitor (Figure 9.8(e)). Strong inhibition was in fact shown by the mixtures which included non-radioactive kidney RNA. RNA from the liver showed a relatively small degree of competition, while that from the spleen indicated values intermediate between kidney and liver. This experiment demonstrates differences between the RNA populations of the different tissues, and shows that some tissues are more similar in this respect than others.

Types and abundance of RNA molecules

Up to this point, the experiments described refer to the total RNA populations of cells and tissues. As we shall see below, the RNA of the nucleus is very different from that of the cytoplasm where messenger is translated into protein.

A typical mammalian cell contains in its cytoplasm around 350 000 mRNA molecules, and of these perhaps four species occupy the 'abundant class', each

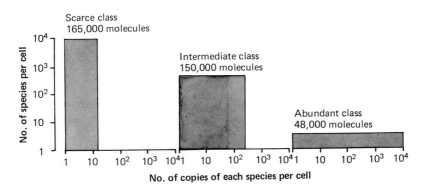

Figure 9.9. The cytoplasmic population of mRNA molecules in a typical mammalian cell. Very few mRNA species fall into the abundant class and very many species are represented by only a few molecules. Between these extremes is an intermediate class containing a large number of species with many representatives. (Data from Alberts *et al.*, 1983.)

being present in as many as 12 000 copies (Figure 9.9). These code for the 'luxury' proteins characteristic of that cell type and they differ between cell types (Chapter 8). There are also perhaps 500 other RNA species in an 'intermediate abundance class', with perhaps 300 copies of each. This class includes those RNA species responsible for the cell's 'household functions' (Chapter 8). There is also a 'scarce' class, composed of about 15 molecules each of 10 000 or so additional mRNA species. These probably include molecules that are really characteristic of other tissue types, since cytodifferentiation involves selective stimulation of production of functional mRNA transcripts of genes which in other cell types are expressed only at low rates. At this level, differential gene expression involves selective increases or decreases, rather like turning a dripping tap full on (or the reverse), instead of the popular image of a switch between 'all-or-none' expression.

However, the majority of cytoplasmic RNA is ribosomal, mRNA accounting for only 3–5% of the total. Calculation of the relative masses of the different RNAs shows that there are about 10 ribosomes for each messenger RNA molecule in the cytoplasm.

9.4. Split genes

One of the most elegant modern approaches towards studying many aspects of the genome involves cutting a gene out of the rest of the DNA and inserting it into a loop of bacterial DNA called a plasmid. The plasmid is then allowed to become incorporated into a bacterium which proliferates so that many copies of the modified plasmid are produced. The plasmid DNA is then isolated and the inserted gene cut out. This procedure for producing multiple copies of a single gene sequence is known as 'gene cloning'.

The rabbit β-globin gene was cloned in this way and, much to everyone's surprise, the DNA sequence corresponding to the rabbit β-globin gene was found to be considerably longer than the mRNA transcribed from it. This observation led to what is certainly the most significant recent advance in eukaryote genetics — the discovery that most eukaryote genes are 'split' and have non-coding intervening sequences, or introns, within the coding sequence, which are not represented in the messenger (see Figure 9.10). In order to distinguish the coding

---

Figure 9.10. Critical genetic features in the synthesis of mouse β-globin.
(a) A simplified diagram of the β-major globin DNA *non-codogenic* sequence is shown. (For further details, see Figure 12.3.) (b) The hnRNA primary transcript equipped with a methylguanosine cap at the 5′ end and a poly-A tail at the 3′ terminus. Note there is considerable base sequence homology at the exon/intron junctions. (c) The mRNA produced after excision, splicing and transfer to the cytoplasm. Note the non-translated leader sequence ends in AUG, (coding for the translation 'start' signal, formylmethionine) and the trailer sequence begins with the translation 'stop' codon, UAA. (d) The polypeptide produced after ribosomal translation. Note the orientation of N and C ends and that the AUG and UAA codons have no amino acid equivalents. (Drawn from data in Konkel, Maizel and Leder, 1979.)

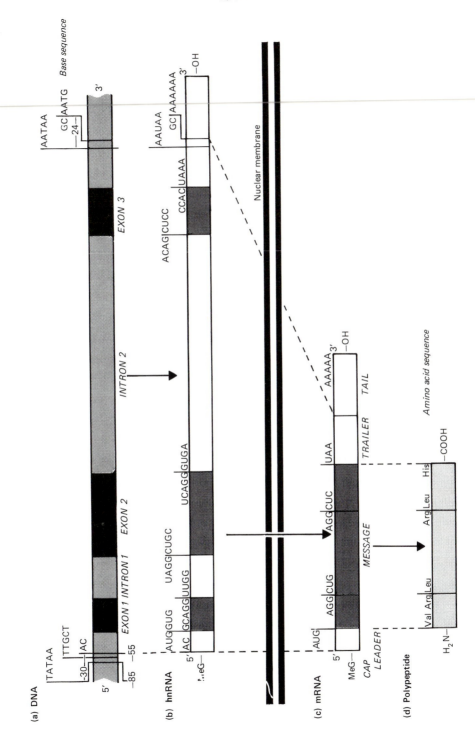

part of the gene from the introns, the former have been given the rather confusing name of exons or extrons. The β-globin genes of rabbits and mice consist of three exons with two introns.

Figure 9.10 illustrates the relationship between the mouse β-major globin gene and its RNA and protein products. This was the first eukaryote gene to be sequenced completely and illustrates many of the important features of eukaryote genes. We will return to it again in Chapter 12.

Introns have been found in most structural genes of higher organisms, apart from those coding for the histones. On average, the combined length of the introns is about double that of the exons, although in some genes they amount to 10 times the length. The average structural gene is therefore about three times as long as the messenger RNA molecule for which it codes.

It is currently accepted that the entire length of a transcriptional unit is transcribed as RNA, including a leader and a trailer sequence, to produce what is known as heterogeneous nuclear RNA (hnRNA). This hnRNA is then processed to produce the true messenger RNA, which is transported through the nuclear membrane into the cytoplasm. This contrasts markedly with the situation in prokaryotes, in which functional mRNA is produced directly by transcription. It is a matter of some discussion whether the prokaryote system is primitive or advanced in this respect (see Chapter 1). It could be that the elimination of intron DNA in prokaryotes and the development of RNA processing and a nuclear membrane in eukaryotes are alternative solutions to the problem of preventing the translation of the irrelevant 'information' in introns.

9.5. RNA processing

Processing of hnRNA involves first the 'capping' of the 5′ end with a methylated guanosine triphosphate (GTP) molecule in reverse polarity, at the beginning of transcription. Normally the first one or two adenines after the cap are also methylated.

Following the release of the RNA from the RNA polymerase, a 'tail' of adenosine molecules is built up through stepwise addition of adenylate residues to the 3′ terminus by the enzyme poly-A polymerase. Whether a suitable poly-A attachment site requires to be exposed by trimming back the transcript is a possibility, but this has not been established. The introns are then removed by a process of excision and splicing. The messenger RNA molecule, as it emerges from the nuclear membrane into the cytoplasm, thus has a guanosine molecule carrying a methyl group at its 5′ end, followed by a leader sequence, then the coding sequences (now adjacent to one another), followed by a trailer sequence, then a 'poly-A tail' of 150–250 adenosine molecules. So far as we know practically all eukaryote mRNA is produced in this way, but histone RNA is an exception to the general pattern, having no introns or poly-A tails. Bacterial mRNAs are also different, having no cap or poly-A tail.

In Chapter 8 we discussed the way in which the genetic information encoded in mRNA is translated into a sequence of amino acids, constituting a polypeptide, which then becomes modified to form a functional protein. Translation is carried out by transfer RNA within the ribosomes and occurs in accordance with the genetic code illustrated in Figure 8.3. The operation of the protein synthetic system also requires instructions about 'punctuation': where transcription should start and stop, which portions of the hnRNA initial transcript should be deleted and which part of the processed message should be translated. For the most part these instructions are also contained in the base sequences of the DNA and RNA. We will deal with the signals concerned with transcription and its control in Chapter 12.

Instructions for RNA processing

The 5' end of the hnRNA acts as an attachment point for the methylguanosine cap. The signal for polyadenylation is the sequence AAUAA, adenylate residues being attached one at a time usually to GC in the 3' position, 16–25 bases downstream from the AAUAA sequence. A clue to the specification of introns is

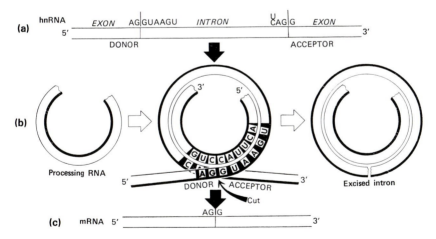

Figure 9.11. Removal of introns during RNA processing.
(a) Part of a typical hnRNA molecule containing one intron. Typical base sequences at the donor and acceptor junctions are shown. (b) The process of intron removal by excision and splicing. In this model the hnRNA is thrown into a loop, so that the sequences AGG at donor and acceptor junctions overlap. This conformation seems more realistic in terms of molecular torsion than the conventional side-loop model. The structure is maintained by the complementary sequence on 'processing RNA' in a specific small molecular ribonucleoprotein particle known as U1. The hnRNA is cut at the overlapping AGG sequence. (c) The resultant mRNA.

provided by homology in the sequences that flank them (Figure 9.10). This is not the whole story, however, as these regions are not completely homologous, while more distant regions are also conserved. Comparison of many flanking sequences led to the suggestion that all those in hnRNA may be derived from the ancestral sequences UCAGGU at the 'donor' 5′ end and CAGG at the 3′ 'acceptor' end. Transfer RNA also has introns but their flanking sequences are quite different.

The excision of introns and splicing together of the exons is thought to be mediated by small nuclear ribonucleoprotein (snRNP) particles, containing small nuclear RNA (snRNA) or processing RNA, with base sequences complementary to those at the intron ends. One such RNA species is illustrated in Figure 9.11. This occurs in an snRNP particle called U1, which is thought to mediate splicing of introns with AGGUAAGU at the donor and C (or U) AGG at the acceptor ends. Figure 9.11 illustrates how the sequence UCC in the processing RNA could bind to AGG at both the donor and acceptor termini. This overlap of complementarity is possibly necessary for the splicing operation.

Another condition for intron excision is its length. It has been found that deletions which reduce an intron to less than 15 bases prevent its removal. The excision of introns is linked to transport of RNA through the nuclear membrane in some way we do not yet understand. This is demonstrated by inserting DNA copies of mRNA (i.e., without introns) into the nuclear genomes of cultured cells. Such sequences are transcribed, but not exported, to the cytoplasm. Histone RNA is exceptional in that it is successfully transported through the nuclear membrane even though it contains no introns.

Instructions for translation

Translation starts at the first AUG codon following the 5′ cap (Figure 9.10). In prokaryotes there is no cap and AUG is translated as formylmethionine, but in eukaryotes it codes for methionine, which is usually later cleaved off the polypeptide. The translation 'stop' signals are the triplet codons UAA, UAG and UGA (see Chapter 8).

The functions of non-coding sequences in mRNA

The 5′ cap is essential for initiation of translation and probably also for protection from ribonuclease attack. The leader also contains bases that may be necessary for binding to ribosomal RNA. The sequence 3′–UAGGAAGGCGU–5′ is conserved at the 3′ end of the 18S ribosomal RNA subunit in several evolutionarily distant eukaryotes. Although the entire sequence complementary to this has not yet been found in mRNA leaders, certain short parts of it are found sufficiently frequently to support the theory that they are probably involved in ribosome binding. Some of these are shown in Table 9.1. The degree of complementarity of the leaders may be important in controlling translation rates, and so is their length, since messengers with short leaders are translated most rapidly.

Table 9.1. Homology between 18S rRNA sequences and leader sequences in eukaryote mRNA molecules.

18S rRNA conserved sequence in mouse, wheat and silkworm:	3′–U	A	G	G	A	A	G	G	C	G	U–5′
Theoretical ideal complementary sequence:	A	U	C	C	U	U	C	C	G	C	A
Complementary sequences in mRNA leaders:											
1. Rabbit α-globin				C	U	U	C				
2. Rabbit α-globin						U	C	C	G		
3. Rabbit β-globin				C	U	U	–	–	G		
4. Rabbit β-globin	A	U	C	C							
5. Human α-globin				C	U	U	C	–	G		
6. Human β-globin				C	U	U	C	–	G		
7. Sea urchin histone H4									C	G	C

The function of the poly-A tail has been the subject of much speculation. With time the poly-A tail shortens, and this prompted the suggestion that release of an adenosine molecule from the tail may be necessary for termination of translation and shedding of the ribosome from its 3′ end. If this is the case, the length of the tail could dictate the number of times the messenger would be translated. Gene expression may therefore be controlled at the level of translation by variable polyadenylation before export of the messenger from the nucleus.

There is some evidence that histone mRNA is polyadenylated in oocytes. In somatic cells it is not, yet it seems to be translated by the same ribosomes as other messengers. This suggests that the poly-A tail may have some function other than in translation. A clue to this is provided by a comparison between relative concentrations of 'abundant' and 'scarce' RNA sequences (Figure 9.9) in a line of cells known as 'Friend' cells. Among the RNA molecules engaged in translation in Friend cells, the ratio of abundant to scarce sequences was found to be 13 : 1. Among the polyadenylated hnRNA of the nucleus the ratio of the same sequences was 9 : 1, but among the non-polyadenylated transcripts within the nucleus it was only 3 : 1. This means that those RNA molecules which will contribute to the cytoplasmic abundant class are three times as likely to be polyadenylated as those that make up the rare class. Polyadenylation could therefore be a feature of a selective step at which gene transcripts are chosen for translation.

During the final stages of erythrocyte differentiation the majority of protein synthetic activity is devoted to the creation of globin chains. This apparently results from preferential conservation of globin mRNA and a few other mRNA species, while the majority are degraded. In breast tissue, stimulation of casein synthesis by prolactin involves a 20-fold increase in the half-life of mRNA, but only a two- to three-fold rise in transcription of casein mRNA. In these examples it seems therefore that mass production of a specialized protein product is a feature of the survival capacity of the mRNA which codes for it. The means by which certain messengers are preferentially conserved is not known, but one possible mechanism would involve differential polyadenylation, since messengers with the shortest tails are degraded most rapidly. In this connection there is an interesting, but unconfirmed suggestion by Brawerman (1974) that messengers may have their tails resynthesized within the cytoplasm. If this does occur it might throw some light on messenger longevity, although how certain messengers might be selected for preferential treatment remains a mystery.

9.6. *Cytodifferentiation at the RNA level*

As we saw above, the range of species of RNA molecules in differentiated mammalian cells shows some degree of overlap between cell types. This was revealed by RNA–DNA hybridization experiments carried out with tissue extracts and it accords with the tissue distribution of puffing patterns that we see in Dipteran polytene chromosomes. The discovery that primary RNA transcripts require processing before they can become functional as messenger RNA complicates this simple comparison and reinforces the possibility that cytodifferentiation may involve differential RNA processing in addition to the well-recognized differences in transcription. For example, it has been estimated that the distinction between 'scarce' and 'abundant' cytoplasmic messengers represented in Figure 9.9 is created in some cases by 50-fold differences in rates of transcription of different genes (see Chapter 12), by 10-fold differences in the relative rates of degradation of the RNA transcribed from them, as well as five-fold variations in the efficiency of post-transcriptional processing.

When the RNA compositions of eukaryote *nuclei* are compared, as expected a considerable degree of overlap is found, but what is particularly interesting is that the nuclear hnRNA populations fall into inclusive sets. In the rat, 30% of non-repetitive DNA sequences are transcribed in the brain, 20% in the liver and 10% in the kidney, but the brain set *includes* that in the other two tissues and the liver set *includes* that in kidney nuclei. This is a fairly recent discovery, the basis of which has not yet been established, but such an inclusive pattern of synthetic activity could result if transcription of the different sequences were triggered or limited by the concentration of a metabolite which varies quantitatively between different tissues, rather as postulated for segment determination in insects (see Chapter 5).

In this example, kidney-specific sequences are transcribed at all levels of RNA synthesis, liver-specific sequences at intermediate and high levels, and brain sequences only at the highest level. These tissues are respectively mesodermal, endodermal and ectodermal in embryonic origin (see Figure 2.11), and it is tempting to speculate that basic distinctions between derivatives of the embryonic germinal layers could derive from graded differences in metabolism, or the supply of some limiting factor which is distributed at different concentrations in the early embryo (see Chapter 4). However, as yet there is insufficient evidence to support a detailed theory of this sort.

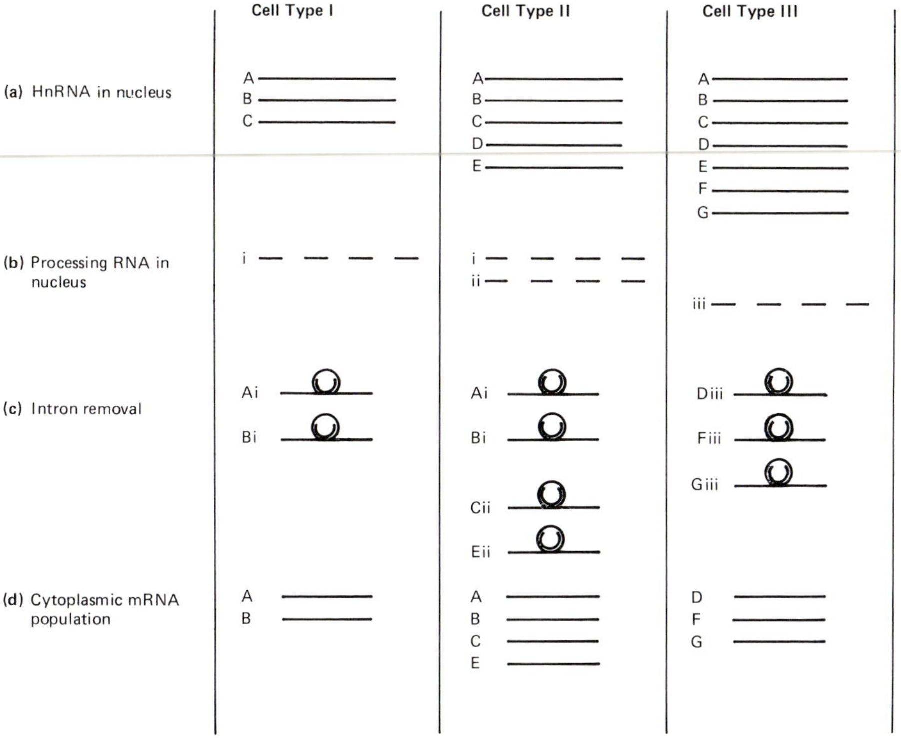

Cell Type I | Cell Type II | Cell Type III

(a) HnRNA in nucleus
A, B, C (Cell Type I)
A, B, C, D, E (Cell Type II)
A, B, C, D, E, F, G (Cell Type III)

(b) Processing RNA in nucleus
i (Cell Type I)
i, ii (Cell Type II)
iii (Cell Type III)

(c) Intron removal
Ai, Bi (Cell Type I)
Ai, Bi, Cii, Eii (Cell Type II)
Diii, Fiii, Giii (Cell Type III)

(d) Cytoplasmic mRNA population
A, B (Cell Type I)
A, B, C, E (Cell Type II)
D, F, G (Cell Type III)

Figure 9.12. A postulated scheme for the establishment of distinct populations of mRNA molecules in different cell types.
(a) Single-copy DNA sequences are transcribed as hnRNA, largely in an inclusive set. (b) Processing RNA is transcribed from repetitive DNA sequences. Each cell type has a distinctly different population of small nuclear RNA species, although there is some overlap between cell types. (c) hnRNA molecules become associated with the processing RNA by virtue of complementary sequences at the donor and acceptor sites. (d) Messenger RNA species in the cytoplasm. The different cell types contain different populations of mRNA, but there is overlap between the cell types.

The small nuclear RNA species that are thought to be involved in intron excision and exon splicing are derived by transcription of repetitive DNA sequences and there seem to be different kinds of snRNA in the different tissue types. Davidson and Britten (1979) have suggested that synthesis of different members of this class of RNA could be controlled by extra-cellular influences. If this is the case, by influencing the selective processing of hnRNA into translatable messengers, processing RNA could form a link between the extra-cellular milieu and the RNA population of the cell cytoplasm. The experimental establishment of such a relationship could pave the way to a new understanding of the molecular basis of embryonic induction (see Chapter 3).

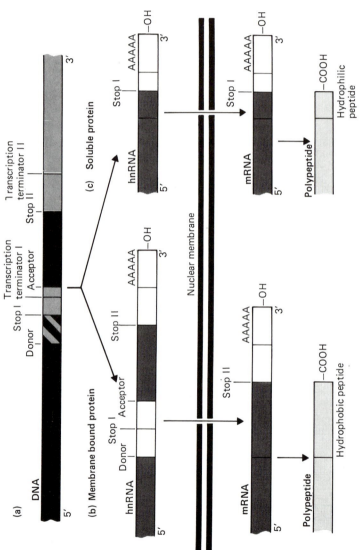

Figure 9.13. Synthesis of two closely related immunoglobulin proteins from a single DNA sequence. (a) The DNA sequence of the immunoglobulin μ heavy chain contains an intron in which are a translation 'stop' codon and a transcription terminator sequence. A second 'stop' codon is situated downstream, in the next exon and beyond it a second transcription terminator. (b) A membrane-bound protein is produced by transcription of a long hnRNA molecule, followed by excision of the intron coding for a hydrophilic series of amino acids, leaving a hydrophobic end to the protein. (c) A water-soluble molecule is coded when transcription terminates within the intron. Translation then produces a shorter protein with a concentration of hydrophilic amino acids near the end.

A scheme for the differentiation of cytoplasmic mRNA population, which incorporates this idea and the concept of inclusive patterns of hnRNA synthesis is shown in Figure 9.12.

Control of gene expression by differential RNA processing

Another way in which gene expression is controlled is by differential processing of hnRNA transcribed from a single gene sequence. This occurs in lymphocytes following stimulation with antigen, when they change from production of membrane-bound to secreted antibody.

The immunoglobulin heavy chain μ gene contains an intron, the nucleotides of which code for a series of hydrophilic amino acids, followed by a translation stop codon then a transcription terminator. The structure of this gene is shown in Figure 9.13. Membrane-bound immunoglobulin is produced in cells which transcribe *through* this terminator, to produce a long hnRNA molecule that includes the whole of the intron, plus the following exon coding for a series of hydrophobic amino acids, then another stop codon and a second transcription terminator. Subsequent splicing of the long transcript removes the intron and translation of the resultant messenger produces a protein which terminates in a string of hydrophobic amino acids. These allow the immunoglobulin μ chains to become embedded in the lipoprotein of the plasma membrane and so produce a display of antibody on the lymphocyte outer surface. When this cell is stimulated by antigen, transcription of the Hμ gene stops at the transcription terminator within the intron. As a result the intron is not processed out and translation produces a protein with a terminal peptide which is hydrophilic. This change causes the antibody to be water-soluble and instead of binding to the membrane it diffuses out into the bloodstream.

This is reminiscent of 'read-through translation' described in Chapter 8, but in the case of immunoglobulin synthesis the 'read through' occurs at the transcriptional level.

9.7. Summary and conclusions

The early embryo contains sufficient messenger and ribosomal RNA for synthesis of the proteins necessary to carry out its early functions. These are manufactured on lampbrush chromosomes and amplified ribosomal genes in the nucleoli of the oocyte. Selective unmasking and translation of masked maternal mRNA in different parts of the embryo may well play a part in the differentiation of the early tissue types.

Some of the somatic cells of Dipteran flies contain giant polytene chromosomes, transcription of which can be observed directly as chromosome puffs. Puffing patterns indicate similarities and differences in patterns of RNA synthesis in the different tissues. Puffing occurs in response to hormones and other external stimuli as well as to proteins produced in the same cell. An initial external stimulus can thus establish a cascade of transcriptional activity.

Differences between RNA populations can be demonstrated by RNA–DNA hybridization, which reveals varying degrees of overlap between the different tissues. There seem to be gross controls on transcription, which may be of a quantitative rather than a qualitative nature, as well as an elaborate system of processing of the primary (hnRNA) transcript formed within the nucleus.

RNA processing occurs in three stages:

1. *Capping*. The 5′ end of the hnRNA is capped with methylguanosine which at translation acts as an attachment site for the ribosomes.

2. *Polyadenylation*. A poly-A tail is attached to the 3′ end. This may protect the RNA from ribonuclease attack, act as a counter for episodes of translation, or be concerned with the recruitment of primary transcripts into the abundant class of mRNA molecules.

3. *Splicing*. This is essential for the removal of superfluous information contained in the introns, or for flexibility in the inclusion, or exclusion of key sectors of genetic information. In lymphocytes a modification of the splicing pattern allows synthesis of essentially the same immunoglobulin heavy chain, but with either hydrophobic or hydrophilic properties. Excision and splicing may be essential for transport of the RNA through the nuclear membrane.

Instructions for transcription, processing and translation of RNA are included in the nucleotide sequences of the DNA. Sequences flanking some introns are complementary to sequences in a ribonucleoprotein particle called U1, which could be instrumental in controlling intron excision and exon splicing.

Bibliography

Alberts, B., Bray, D., Lewis, J., Raff, M., Roberts, K. and Watson, J., *Molecular Biology of the Cell*. Garland, New York (1983).

Apirion, D., *Processing of RNA*. C.R.C. Press, Boca Raton, Florida (1983).

Balmain, A., Minty, A. J. and Birnie, G. D., *Nucleic Acids Res.*, **8**: 1643 (1980).

Becker, H. J., Die Puffs Der Speicheldrusen-chromosomen von *Drosophila melanogaster*. 1: Beobachtungen zum Verhalten des Puffmusters in Normalstamm und bei zwei Mutanten, giant und lethal-giant Larvae. *Chromosoma*, **10**: 656 (1979).

Browder, L. W., *Developmental Biology*. Saunders College/Holt, Rinehart & Winston, Philadelphia (1980).

Busch, H., Reddy, R., Rothblum, L. and Choi, Y. C., SnRNAs, SnRNPs and RNA processing. *Ann. Rev. Biochem.*, **51**: 617 (1982).

Chikaraishi, D. N., Deeb, S. S. and Sueoka, N., Sequence complexity of nuclear RNAs in adult rat tissues. *Cell*, **13**: 111 (1978).

Crick, F., Split genes and RNA splicing. *Science*, **204**: 266 (1979).

Davidson, E. H. and Britten, R. J., Regulation of gene expression: possible role of repetitive sequences. *Science*, **204**: 1052 (1979).

Doolittle, W. F., Revolutionary concepts in evolutionary cell biology. *Trends Biochem. Sci.*, **5**: 146 (1980).

Goldstein, L. and Prescott D. M., *Gene Expression: The Production of RNAs. Vol. 3, Cell Biology. A Comprehensive Treatise*. Academic Press, New York (1980).

Grant, P., *Biology of Developing Systems*. Holt, Rinehart and Winston, New York (1978).

Haggenbüchle, O., Sauter, M., Steitz, J. A. and Mans, R. J., Conservation of the primary structure at the 3′ end of the 18S rRNA from eucaryotic cells. *Cell*, **13**: 551 (1978).

Konkel, D. A., Maizel, J. V. Jr. and Leder, P., The evolution and sequence comparison of two recently diverged mouse chromosomal β-globin genes. *Cell*, **18**: 865 (1979).

Lerner, M. R., Boyle, J. A., Mount, S. M., Wolin, S. L. and Steitz, J. R., Are snRNPs involved in splicing? *Nature (Lond.)*, **283**: 221 (1980).

Lewin, B., *Gene Expression, Vol. 2. Eukaryotic Chromosomes*, 2nd edn. Wiley, Chichester (1980).

Markert, C. L. and Ursprung, H., *Developmental Genetics*. Prentice-Hall, Englewood Cliffs, New Jersey (1971).

Nevins, J. R., The pathway of enkaryotic mRNA formation. *Ann. Rev. Biochem.*, **52**: 441 (1983).

Nishioka, Y. and Leder, P., The complete sequence of a chromosomal mouse β-globin gene reveals elements conserved throughout evolution. *Cell*, **18**: 875 (1979).

Paul, R., *Transfer RNA: Structure, Properties and Recognition*. Cold Spring Harbour Laboratory, New York (1979).

Tobin, A. J., Evaluating the contribution of post-transcriptional processing to differential gene expression. *Dev. Biol.*, **68**: 47 (1979).

Waddington, C. H., *New Patterns in Genetics and Development*. Columbia University Press, New York (1962).

Whittaker, J. R., Cytoplasmic determinants of tissue differentiation in the ascidian egg. In *Determinants of Spatial Organisation*, edited by S. Subtelny. Academic Press, New York, pp. 29–57 (1979).

Chapter 10 Chromosomal proteins

10.1. The structure of chromatin

In prokaryotes the DNA is essentially naked, but the chromosomal DNA of eukaryotes is invariably bound up with a lot of protein and RNA to form the substance known to histologists as chromatin. Some chromatin is particularly dense and stains darkly with chromosome stains, and is known as heterochromatin to distinguish it from the weakly staining euchromatin. This difference in staining properties reflects a difference in the organization and protein content of the chromatin. The proteins are of two major types, basic histones and acidic non-histone chromosomal (NHC) proteins. Although a lot of RNA is also present, we do not yet know whether it is involved in transcriptional control. On the other hand, the chromosomal proteins have a very definite role in this respect.

The major species of histones are grouped into aggregates, or core particles, containing two molecules each of histones H2A, H2B, H3 and H4, arranged more or less as in Figure 10.1 and associated with 200 base pairs of DNA. The first turn of the DNA around the aggregate is held by a tetramer of histones H3 and H4, while a dimer of H2A and H2B lies on each side, fixing the ends of the second

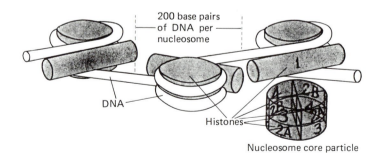

Figure 10.1 The arrangement of DNA and histone in extended chromatin.
The complex of 200 base pairs of DNA, two molecules each of histones H2A, H2B, H3 and H4, and one of H1 is called a nucleosome. H1 binds to the free ends of the DNA as it protrudes from the core particle. (After Kornberg and Klug, 1981.)

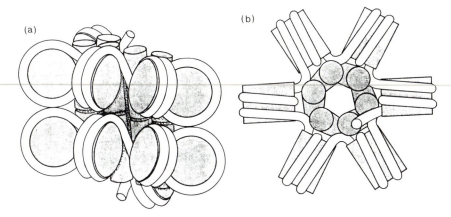

Figure 10.2. A possible arrangement of DNA and histone in chromatin in the solenoid form. H1 molecules link adjacent nucleosomes together and stabilize the solenoid structure.
(a) Two turns of the solenoid in side view. (b) The solenoid, end view.

turn. A single molecule of H1, or H5, seals off the package, binding to both free ends of DNA. This whole unit is known as a nucleosome. Under certain ionic conditions adjacent nucleosomes seem to be linked through the H1 molecules, with the nucleosome cores protruding on opposite sides of the main axis of the chromatin strand. The extended DNA–histone complex thus looks like a zigzag string of beads, but with the DNA on the outside (see Figure 10.1). If prokaryote DNA is allowed to interact with histones extracted from eukaryote chromosomes, it adopts a similar pattern to the beaded string, with 200 base pairs of DNA per nucleosome. So this structure seems to derive naturally from the properties of the constituents.

In transcriptionally inactive eukaryote chromatin the beaded string is wound into a coil or solenoid, which relaxes to allow transcription to take place. A diagrammatic impression of the possible relationships between the molecules in the solenoid is shown in Figure 10.2, which emphasizes the role of the H1 molecules in maintaining this structure.

Until recently, many biologists have considered transcriptional control to be the *only* control upon gene expression. This, of course, is not the case, but since no gene can act unless it is first transcribed into RNA, control of transcription must be one of the most basic aspects of gene expression (see Chapter 12).

10.2. Control of transcription by chromosomal proteins

The histones are basically very similar in all tissues and all species (see Chapter 1), although they are reversibly modified in what are probably tissue-specific patterns.

In contrast, the NHC proteins vary a great deal between tissue types and between species.

In one experiment to test the relative roles of the two, intact chromatin was used as a template for a transcription system *in vitro*. A quantity of RNA was produced, which we will give the value of unity, 1.0 (Figure 10.3). In the same system a quantity of RNA equivalent to 3.5 was transcribed from free DNA. When free DNA was mixed with histone prepared from the same chromatin, no RNA was transcribed, but if free DNA was mixed with the NHC protein from the same chromatin, transcription occurred to double the extent of that with intact chromatin. The degree of transcription was reduced to unity again when free DNA was allowed to associate with a mixture of histones and NHC proteins. This experiment indicates that histones inhibit transcription, while NHC proteins modify that inhibition.

The specificity of this control was assessed by hybridizing the RNA produced in such experiments with radioactive complementary DNA (cDNA) prepared by the action of the enzyme reverse transcriptase upon purified messenger RNA (see Chapter 11). In one such experiment, represented diagrammatically in Figure 10.4, chromatin was prepared from both mouse thymus and mouse bone marrow nuclei and fractionated into DNA, histone and NHC proteins. The chromatin was reconstituted by mixing these fractions in different combinations and was then transcribed into RNA in an *in vitro* system. When chromatin was reformed from mixed DNA, plus mixed histones, together with NHC protein

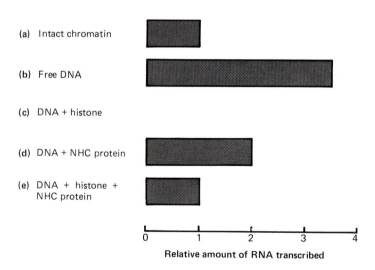

Figure 10.3. The amount of RNA transcribed from different fractions of chromatin. Histone (c) entirely inhibits transcription of DNA (b); NHC proteins inhibit transcription to a lesser extent (d); a mixture of NHC protein and histone (e) restores transcription to the level shown by intact chromatin (a).

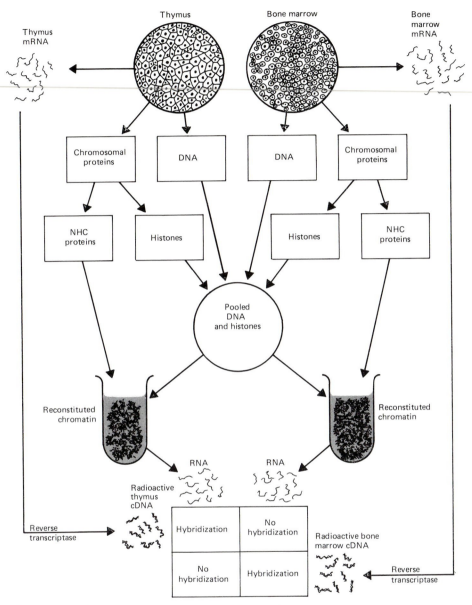

Figure 10.4. Control of transcription by NHC proteins.
Chromatin was extracted from mouse thymus and bone marrow nuclei and fractionated into DNA, histones and NHC proteins. The DNA and histones were then pooled and aliquots were mixed with NHC protein from thymus and bone marrow. The reconstituted samples of chromatin were then used as primers for transcription *in vitro*. The RNA produced was examined by hybridization with radioactive cDNA produced by the action of reverse transcriptase on the RNA products of the original tissues. The pattern of transcription of the reconstituted chromatin is defined by the NHC protein preparation.

from *either* mouse thymus *or* mouse bone marrow nuclei, the RNA transcribed was specific to *either* thymus *or* bone marrow, depending on the source of the NHC proteins. This specificity was demonstrated by the accumulation of radio-active label from one cDNA sample, but not the other, in the hybridization reaction between the cDNA preparations and the RNA synthesized from the reconstituted chromatin.

This is a very important finding, which has since been confirmed with chromatin prepared from other sources, although it should be pointed out that chromatin reconstruction experiments are regarded by some biochemists to be very suspect. If valid, the experiment demonstrates that tissue specificity in RNA synthesis can be controlled by the co-operative influence of basic histone, which acts in a non-specific fashion, and acidic nuclear protein, which confers the specificity. The reader will recall that a similar deduction was made about the control of synthesis of 'luxury' proteins in somatic cell hybrids (see Chapter 7) — that patterns of gene expression become established through the co-operative influences of 'expressive' and 'suppressive factors'. However, it should be stressed that the identity of the molecules involved in the cell hybrid systems has not been established, nor have the levels at which this control is imposed.

Analysis of the proteins in transcriptionally active chromatin reveals some very interesting features which provide some of the most important clues to the molecular bases of the determined and differentiated states of embryonic tissues. The histones in active chromatin are less highly charged, due to attachment of

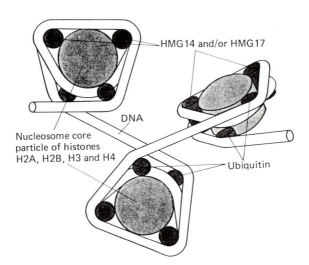

Figure 10.5. Unmasking of DNA by the NHC proteins ubiquitin, HMG14 and HMG17. This figure should be compared with Figure 10.1. Note the absence of histone H1. Ubiquitin becomes linked to H2A molecules. Two molecules of HMG14 or HMG17 are also frequently associated with the core particles in relaxed chromatin at transcriptionally active or potentially active sites.

acetyl groups to their lysine molecules and phosphate residues to their serines. This causes the nucleosome cores to be less tightly bound to one another, histone H1 being particularly loosely bound. The result of these modifications is that the chromatin solenoid (Figure 10.2) relaxes into the beaded string structure (Figure 10.1) allowing access of RNA polymerase and commencement of transcription.

Three specific NHC proteins are very frequently found in active chromatin. These are ubiquitin, which is covalently linked to H2A during interphase, and HMG14 and HMG17, which also bind to the nucleosome core proteins. The association of these three acidic proteins with the histones probably partially releases the DNA, allowing transcription to proceed. The concept of DNA 'unmasking' by NHC proteins is illustrated in Figure 10.5.

In polytene chromosomes the histone content of puffs is the same as that of unpuffed regions, but in contrast the puffs contain a higher concentration of NHC proteins compared with the non-puffed stretches (see Chapter 9). Observations on polytene chromosomes thus tend to confirm the general conclusions based on fractionated and reconstituted chromatin.

10.3. Developmental changes in histone synthesis

Although it is generally accepted that histones are basically insufficiently variable to exert a significant influence on tissue-specific patterns of transcription, there are some intriguing indications that the repertoire of histone species synthesized expands during development. These observations arose with recent improvements in the usual electrophoretic techniques used for examination of the histones. In sea urchins rather different forms of H1, H2A and H2B appear after the blastula stage, although those forms synthesized during cleavage continue to be produced throughout life. Little is known about the functions of the different variant forms, but they could obviously have an important role in the regulation of development changes in gene expression.

10.4. Experiments with mammalian deoxyribonucleases

The examination of patterns of gene expression in higher organisms is currently undergoing something of a revolution. This is due to a large extent to the discovery of enzymes with the capacity to cut DNA at, or close beside, regions which are being actively transcribed. Two such enzymes that are proving very useful in this respect are known as DNAase-1 and DNAase-2. DNAase-1 is prepared from mammalian pancreas and DNAase-2 from mammalian spleen.

As we have seen, much of the work on gene expression in higher organisms has been performed with red blood cells and their specific protein products, the globins. Another particularly useful cell type is the tubular gland cell which produces the characteristic egg protein, ovalbumin, in the chicken oviduct (see Chapter 4). Chromatin was extracted from chicken oviduct cells and exposed to DNAase-1 until 10% of the DNA was solubilized. The remaining 90% was then tested for the presence of ovalbumin gene sequences by hybridization with a radioactive cDNA preparation corresponding to the ovalbumin gene. This showed that two thirds of the ovalbumin genes had been destroyed by the DNAase-1 treatment (Figure 10.6).

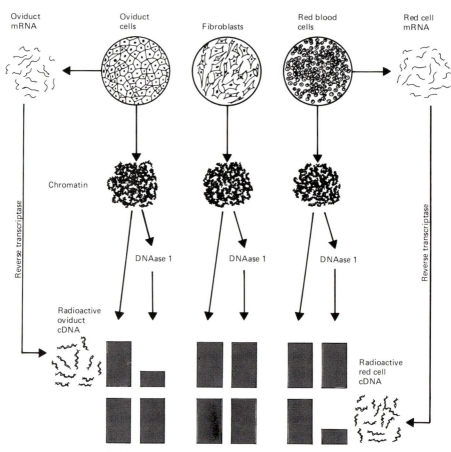

Figure 10.6. Effect of DNAase-1 digestion on patterns of transcription.
Chromatin was extracted from chicken oviduct cells, fibroblasts and red blood cells. A sample of each was then hybridized with radioactive cDNA, prepared by the action of reverse transcriptase on ovalbumin mRNA, isolated from oviduct cells and globin mRNA isolated from red blood cells. A second sample of each chromatin preparation was digested with DNAase-1 and the non-solubilized fraction was hybridized with the same preparation of cDNA. DNAase-1 digestion of fibroblast chromatin had no effect on either hybridization. This experiment shows that DNAase-1 preferentially destroys genes which are being actively transcribed.

The same preparation was also hybridized with cDNA corresponding to globin sequences, but, in contrast to the finding with the ovalbumin cDNA, DNAase-1 digestion had caused no reduction in globin genes. The reverse experiment was also performed, chromatin from chicken red cells being digested, and in this case it was the globin sequences that were found to have been preferentially destroyed. When the DNAase-1 digestion was repeated with chromatin extracted from chicken fibroblasts, neither globin nor ovalbumin sequences were damaged. It was concluded from such experiments that transcriptionally active chromatin is particularly accessible to enzyme attack and that DNAase-1 cuts chromatin at these sites and destroys them. Sites hypersensitive to DNAase-1 also extend 'upstream' of active genes. Chromatin that contains a high density of acetylated histone is cut 10 times as efficiently as that in regions which are not highly acetylated.

In contrast, DNAase-2 acts on genes which are under active transcription without destroying them. When the chromatin from chicken reticulocytes was treated with DNAase-2, the soluble fraction was found to be enriched with globin genes, whereas when chromatin from chicken liver was treated in the same way there was no globin-gene enrichment of the soluble fraction. From this and similar experiments, it is concluded that DNAase-2 recognizes actively transcribed DNA sequences and removes them by cutting the DNA on either side.

It is by means of this enzyme that preparations of transcriptionally active chromatin can be made for analysis of its protein content and comparison with that of inactive chromatin. Active genes cut out of chromatin by DNAase-2 contain DNA, histones and NHC proteins, whereas the insoluble fraction remaining after digestion contains DNA and histones, but has a much lower content of acidic protein.

An important feature of the action of these deoxyribonucleases is that they also act on chromatin which is *primed* for transcription, even though transcription is not currently taking place. This observation supports the theory that chromatin must attain a certain physical structure, or density, before transcription can be initiated.

10.5. Heterochromatin

There are two main types of heterochromatin, constitutive and facultative. Constitutive heterochromatin includes that at the centromeres and telomeres which hybridizes with satellite DNA (see Chapter 11), and its distribution is similar in all cells of the body. The term 'facultative heterochromatin' refers to chromatin which becomes heterochromatized in some selective fashion during development, the best-known example being the Barr body, derived by heterochromatization of one of the X chromosomes in female mammals.

In normal male mammals there is one X and one Y chromosome, whereas normal females have two X chromosomes, one of these being inactivated and condensed into a darkly staining or 'heterochromatic' body called sex chromatin, or the Barr body (see also Chapter 11). The sex chromatin can be revealed in female somatic cells by appropriate staining techniques, this being the basis of part of the sex test considered necessary by the Olympic Games authorities because of the occasional appearance of athletes whose bodies have an outwardly female appearance although genotypically they are male. Chromosome inactivation occurs at the blastula stage, when one X in each normal female mammalian somatic cell becomes condensed. The selection of which X is apparently random in most species, but in marsupials and in the extra-embryonic membranes of mice the one that came from the father is preferentially inactivated. In inter-specific female hybrids, such as mules, there is also often preferential X inactivation.

The inactivated chromosome replicates at mitosis, but it and its copy remain inactive in daughter cells, so that cell clones are set up expressing either the maternal or the paternal X. These clones are still detectable in mature adults and remain stable throughout life. If a female mammal is heterozygous for an X-linked gene, she thus becomes a mosaic for expression of the two alleles. A common example of this is the tortoiseshell cat, which is heterozygous for a colour gene on the X chromosome (Figure 10.7). One X-linked allele codes for black fur, the other for orange. Tortoiseshell cats are black–orange heterozygotes and since a normal male cannot be heterozygous for an X-linked gene, all tortoiseshell cats are female. Tortoiseshell cats with masculine characters should be derivable only from embryos carrying an X-autosome translocation, or where there is an additional X, as in Klinefelter's syndrome in humans.

Confirmation that it is the heterochromatic X which becomes inactive was provided by examination of mice heterozygous for Cattanach's translocation. In these mice, part of an autosome carrying dominant alleles affecting coat colour has

Figure 10.7. Tortoiseshell and white kittens showing different mosaic patterning of black and orange but standard patterning of white. The former is due to random inactivation of alternative X-borne alleles.

been translocated on to an X chromosome, the resultant chromosome being recognizable by its large size. In female mice heterozygous for the translocation, the autosomal coat colour alleles are expressed only in those parts of the skin where the large X remains uncondensed.

Heterochromatin seems to be a state alternative to euchromatin, and it is possible that small-scale facultative heterochromatization may be a means by which differential expression of genes is controlled. Active genes rarely map to heterochromatic regions and active genes which become translocated to positions adjacent to heterochromatin tend to become inactive, although their degree of inactivation can vary from cell to cell.

An illustration of the control of gene expression by heterochromatin is provided by *Drosophila* eyes, as shown in Figure 10.8. The eyes of wild *Drosophila* are dark red, but the recessive *white* allele (*w*) causes them to be white, due to a defect in the binding of pigment. Flies which are heterozygous for *w* and the wild-type allele w^+ have red eyes, since *w* is recessive. Translocation heterozygotes can be bred in which the wild-type allele has been translocated from euchromatin on chromosome I, close to heterochromatin on chromosome IV. These flies have eyes that are mottled white and red. This is because the wild-type allele is inactivated in some cell clones, allowing the *white* gene to be manifest, whereas in others the wild-type allele remains fully expressed.

Another recessive which affects the eyes, known as *split facets* (*spl*), is closely linked to *w*. When the two *wild-type* alleles of *w* and *spl* are translocated together (in coupling) close to heterochromatin, it is found that the allele closest to the heterochromatin is switched off more frequently than the other. The allele distal to the heterochromatin is repressed only when the proximal one is also, so that the eye contains patches which are normal in both respects, or red with split facets, or white with split facets, but never white with normal facets. This phenomenon is one of several examples that show how the repressive influences of heterochromatin spreads away along the chromosome. It is convenient to visualize this effect as a local tightening or coiling of the chromatin solenoid, but the way it works has not yet been elucidated.

Once a piece of chromatin becomes heterochromatized, the same region is usually heterochromatic in all progeny cells. This means that the structure and/or chemical composition of the modified chromatin is reproduced in newly synthesized chromosomes. In vertebrate tissue cultures the pattern can, however, be disrupted by agents which interfere with methylation of cytosine in the DNA (see Chapter 11).

We do not know whether stable patterns of gene expression are normally established by processes akin to heterochromatization, but cytosine methylation is important in the control of gene expression, so it is possible that tightening and relaxation of chromatin gross structure, on a restricted scale, may close down or open up short stretches of the chromosome and allow transcription to take place. When larger stretches are closed down in a similar fashion this may produce the conformation we recognize microscopically as heterochromatin.

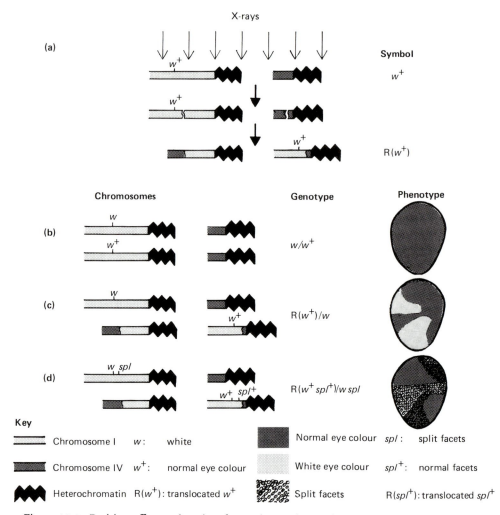

Key

Chromosome I	w :	white	
Chromosome IV	w^+ :	normal eye colour	
Heterochromatin	$R(w^+)$:	translocated w^+	

Normal eye colour spl : split facets
White eye colour spl^+ : normal facets
Split facets $R(spl^+)$: translocated spl^+

Figure 10.8. Position effect variegation due to heterochromatin
(a) X-irradiation can induce chromosome breaks, promoting translocations and formation of chromosomes of novel structure. The illustration shows a reciprocal (balanced) translocation in which the *normal* allele for *white* eyes in *Drosophila melanogaster* is translocated from chromosome I to a situation on chromosome IV close to heterochromatin. (b) Chromosomal constitution, genotype and appearance of the red eye of a normal heterozygote. (c) A translocation heterozygote carrying the *white* allele on chromosome I, plus the wild-type allele in the translocated situation ($R(w^+)$). This produces a mottled red and white eye, due to non-expression of w^+ in some cells but not in others. (d) A translocation heterozygote carrying the closely linked alleles w and spl on chromosome I, plus the normal alleles of both genes translocated close to heterochromatin on chromosome IV. This produces a mottled eye, with a variegated pattern of split facets. The spl^+ allele is represented independently of w^+ in some cells, but w^+ is never repressed without concomitant repression of spl^+ because spl^+ is closest to the heterochromatin. (From Markert and Ursprung, 1971).

10.6. Acquisition and retention of DNA programmes in relation to mitosis

The chromosomal DNA sheds its protein during mitosis and reacquires it after the S phase, when new DNA is synthesized. It is probably during late S and G2 phases that DNA becomes 'reprogrammed' by attachment of protein, so establishing new patterns of gene expression. In some systems, inductive stimuli seem to operate only if the cells undergo one or two (or more) cycles of division during or after exposure to the stimulus. An example is the progressive induction of proliferating mesenchyme cells by the ectoderm of the apical ectodermal ridge (AER) during the development of vertebrate limbs (see Chapter 6). The inference is that some inductive stimuli may provide conditions which favour specific patterns of histone modification or association of NHC proteins with the newly synthesized DNA. These could then cause new combinations of genes to be expressed and the establishment of new cell phenotypes. However, mitosis is not an invariant requirement. Hydra will regenerate a new hypostome and tentacles without cell division and tadpoles can undergo extensive morphogenesis and tissue development even when cell division is blocked by drugs.

Heterochromatin is replicated at the end of S phase in the division cycle, and it is interesting to note that DNA synthesis during early and mid-S phase is by no means synchronous in euchromatin. It is possible, therefore, that the precise stage when a particular piece of DNA becomes replicated and then reprogrammed is a function of its initial state of supercoiling. If the availability of NHC proteins, or the conditions appropriate to acetylation or phosphorylation of histones vary during S phase, this would provide a means by which patterns of DNA programming could be maintained through mitosis. This attractive possibility does not seem to have been investigated.

10.7. Summary and conclusions

Unlike prokaryote DNA, that of eukaryotes is complexed with proteins which play an important role in the control of transcription. Histones occur in aggregates containing two molecules each of H2A, H2B, H3 and H4, and either H1 or H5, associated with 200 base pairs of DNA, to form a 'nucleosome' of characteristic structure. Histones exert a non-specific repressive effect, while the non-histone chromosomal proteins modify this repression in a tissue-specific fashion. In transcriptionally active areas the histones are acetylated and phosphorylated and there are developmental changes in the repertoire of histone synthesis. The acidic proteins ubiquitin, HMG14 and HMG17 are frequently associated with active genes, in addition to other acidic proteins concerned with specificity in response. Parts of the genome can be 'opened up' or 'primed' for transcription, by changes in the association of the DNA with these proteins. The sequence of changes in chromatin structure and composition leading up to transcription are summarized

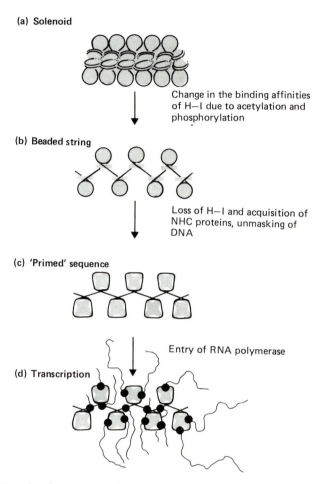

(a) Solenoid

Change in the binding affinities
of H–I due to acetylation and
phosphorylation

(b) Beaded string

Loss of H–I and acquisition of
NHC proteins, unmasking of
DNA

(c) 'Primed' sequence

Entry of RNA polymerase

(d) Transcription

Figure 10.9. Postulated sequence of major changes in chromatin organization during expression of a gene.
The tightly coiled chromatin solenoid (a) relaxes into a beaded string stucture (b). This is associated with modifications of the biochemical properties of the histones, particularly H1. As the structure becomes loosened, H1 is lost and acidic proteins enter the chromatin (c), facilitating access by RNA polymerase and allowing transcription to take place (d).

in Figure 10.9. It is suggested that some inductive stimuli may operate by influencing the patterns of histone modification and association of NHC proteins with DNA, following its replication during mitosis.

Gene expression can also be repressed more or less irreversibly by heterochromatization. This probably involves a change in chromatin supercoiling and seems to be related to methylation of cytosine residues in the DNA. Evidence for the relationship between gene expression and cytosine methylation is provided in Chapter 11.

Bibliography

Alberts, B., Bray, D., Lewis, J., Raff, M., Roberts, K. and Watson, J., *Molecular Biology of the Cell*. Garland, New York (1983).

Browder, L. W., *Developmental Biology*. Saunders College/Holt, Rinehart and Winston, Philadelphia (1980).

Cooke, J., Morphogenesis and regulation in spite of continued mitotic inhibition in *Xenopus* embryos. *Nature (Lond.)*, 242: 5392 (1973).

Gurdon, J. B., *The Control of Gene Expression in Animal Development*. Harvard University Press, Cambridge/Oxford University Press, London (1974).

Gurdon, J. B., *Gene Expression during Cell Differentiation*, 2nd edn, Carolina Biology Readers, edited by J. J. Head. Oxford University Press/Carolina Biology Supply Company, Burlington, North Carolina (1978).

Hnilica, L. S., *Chromosomal Nonhistone Proteins*, C.R.C. Press, Boca Raton, Florida (1983).

Holtzer, H., Mitosis and transformation. In *General Physiology of Cell Specialization*, edited by D. Mazia and A. Tyler. McGraw-Hill, New York, p. 80 (1963).

Igo-Kemenes, T., Hörz, W. and Zachau, N. G., Chromatin. *Ann. Rev. Biochem.*, 51: 89 (1982).

Johns, E. W., *The HMG Chromosomal Proteins*. Academic Press, New York (1982).

Kornberg, R. D. and Klug, A., The nucleosome. *Sci. Am.*, 244: (2): 48 (1981).

Lewin, B., *Gene Expression, Vol. 2: Eucaryote Chromosomes*, 2nd edn. Wiley, Chichester (1980).

Lyon, M. F., Mechanisms and evolutionary origins of variable X-chromosome activity in mammals. *Proc. R. Soc. Lond. [Biol.]*, 187: 243 (1974).

Markert, C. L. and Ursprung, H., *Developmental Genetics*. Prentice-Hall, New Jersey (1971).

Strickberger, M. W., *Genetics*, 2nd edn. Macmillan, New York/Collier-Macmillan, London (1976).

Suzuki, D. T., Griffiths, A. J. F. and Lewontin, R. C., *An Introduction to Genetic Analysis*, 2nd edn. Freeman, San Francisco (1981).

Wolpert, L., Hicklin, J. and Hornbruch, A., Positional information and pattern regulation in regeneration of *Hydra*. *Symp. Soc. Exp. Biol.*, 25: 391 (1971).

Chapter 11 DNA

The haploid DNA content of a diploid organism is known as its C value, and since it is DNA which carries the major burden of information required for the structure and functioning of an organism, it would be expected that C values should increase in proportion to organismic complexity. In fact, there is no simple relationship between the two. For example, among the Amphibia we find a 100-fold variation in C values, without any notable difference in complexity. This is known as the 'C-value paradox'. The paradox was resolved by the finding that a lot of the DNA in higher organisms has no coding function and is either included in introns and spacer regions, or has a role in the structure of chromosomes, or may even be quite redundant.

The study of eukaryotic DNA emphasizes the enormous gulf that has formed between eukaryote and prokaryote systems and, as new features emerge, we begin to see the elements of gene patterning, analogous in some ways to the patterns of phenotypic features that have arisen through the agency of the same evolutionary forces (see Chapter 5). What is particularly interesting is that, as suggested in the context of the control of body segmentation in insects, the patterns of the genes seem to constitute an important feature in the integration of their expression. Before we examine these relationships we need to look at some of the earlier more general findings on eukaryote DNA.

11.1. DNA complexity and renaturation

In Chapter 9 we considered RNA–DNA hybridization studies, the starting-point of which involved denaturation of DNA into small single-strand fragments. This can be achieved by passing a DNA preparation through a small orifice or by shearing in a high-speed blender, before heating briefly at about 100°C. If this denatured DNA is allowed to cool, it begins to 'renature' or re-anneal — the single-strand structures begin to pair up with their partners to form double helices again.

If a sample of native or denatured DNA is examined in a spectrophotometer, it is found to absorb light maximally in the region of 260 nm in the u.v. range, single-stranded DNA absorbing 40% more than double-stranded DNA. This property can be used to estimate the rate of re-annealment of denatured DNA. Alternatively, the rate of re-annealment can be assessed by a technique similar to that described in relation to RNA–DNA hybridization. The denatured DNA is bound to a nitrocellulose filter which is then coated with albumin. This prevents further binding to the nitrocellulose, but not to the exposed surfaces of the DNA. A sample of radioactive DNA is then prepared from a similar source, denatured and incubated with the prepared filter. The radioactive DNA cannot bind directly to the filter, due to the presence of the albumin, but it can and does bind to complementary DNA sequences, which themselves are bound to the filter. The accumulation of radioactive label by the filter is then assessed at intervals and the percentage of the total radioactive bound is plotted against time. In practice, the product of the initial concentration of single-strand DNA and time in seconds, or C_0t, is calculated and plotted on a logarithmic scale.

Figure 11.1(a) shows the type of plot produced with DNA from the bacterium *Escherichia coli* and two bacteriophages. In these organisms essentially all genes are present as single copies and simple patterns of renaturation are produced. The value of log C_0t corresponding to 50% renaturation (log $C_0t_{\frac{1}{2}}$) depends on the size of the genome examined, renaturation occurring relatively rapidly with a small genome but much more slowly with a large one. This is because the probability of chance encounter of complementary sequences is inversely related to the number of different sequences present.

When the same experiment is done with DNA from a eukaryote, the pattern produced is very much more complicated. Figure 11.1(b) shows the type of pattern produced when denatured mammalian DNA is allowed to renature and is assessed in this fashion. The curve has a raised base-line and three shoulders, corresponding approximately to 10%, 30% and 100% renaturation. This indicates four fractions, representing perhaps 1 or 2% and about 10%, 20% and 70% of the DNA. The raised base-line is caused by almost instantaneous formation of double strands from pieces of DNA which contain inverted repeats — sequences that are the same when one strand is read left to right ($5'-3'$) or the other is read right to left (also $5'-3'$). These single-strand pieces are self-complementary and can fold back on themselves to form hairpin loops as shown in Figure 11.2. This is called foldback DNA.

The next 10% to renature does so in the standard fashion, by formation of double-strand duplexes. This also occurs relatively rapidly because these sequences are present in very many copies. This first shoulder therefore represents highly repetitive sequences, each sequence being represented about 1 000 000 times.

The second shoulder, caused by renaturation of the next 20%, represents DNA sequences which are also repeated many times, but less frequently than those in the first fraction. This intermediate-repetitive group consists of sequences present as 1 000–100 000 copies.

The last shoulder represents the renaturation of single-copy or unique-sequence DNA — gene or non-coding sequences which are present only once or a very small number of times in the haploid genome. The number of different single-copy

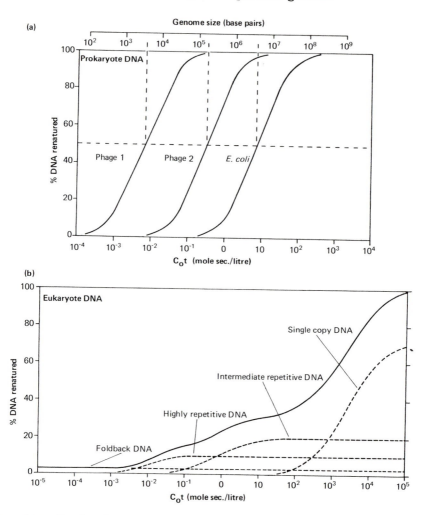

Figure 11.1. Graphs of renaturation of denatured DNA.
(a) Renaturation of DNA from three non-eukaryote species. In separate experiments DNA from the bacterium *Escherichia coli* and two phages were denatured into single strands, which were then allowed to reassociate into double helices. The percentage of DNA renatured is plotted against the product of the initial concentration (C_0) of single strand DNA and the time (t) since the beginning of the reaction. Speed of renaturation decreases with genome size and the patterns of renaturation follow simple S-shaped curves, as expected when there are equal numbers of each sequence present. Genome size can be estimated from the $C_0t_{\frac{1}{2}}$ value when the DNA is 50% renatured. (b) Idealized renaturation curve for eukaryote DNA. The experiment is conducted as with prokaryote DNA. One fraction (foldback DNA) renatures instantaneously, followed some time later by the highly repetitive fraction. This is followed by renaturation of the intermediate repetitive fraction and lastly by single-copy DNA. The combined effect is to produce an irregular curve with two shoulders. (Redrawn from Ford, 1976.)

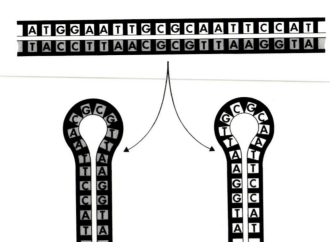

Figure 11.2. Renaturation of foldback DNA.
Spontaneous renaturation of denatured DNA is thought to occur by self-complementation, due to the formation of hairpin loops by both strands of DNA within inverted repeat sequences.

sequences can be estimated as around 10% by reference to the scale of genome size in Figure 11.1(a).

It is uncertain how much of the DNA in each fraction is functionless, but the single-copy fraction includes most of the structural genes, the intermediate-repetitive fraction contains the genes for tRNA, rRNA and the histones, while the highly repetitive sequences include the DNA of the constitutive heterochromatin near the centromeres. Foldback DNA is present in the trailer sequences of some genes (see below and Chapter 1), and in the genes for transfer RNA. Small inverted repeats are sites that can be recognized by bacterial restriction endo-nucleases (see below). It is quite possible that eukaryotic proteins may also utilize the special properties of inverted repeats by binding to the DNA at these sites during gene-switching operations.

Repetitive DNA is not found in bacteria. It increases in quantity approximately in proportion to evolutionary complexity and most of it does not code for protein. RNA is transcribed from some of it although, apart from tRNA and rRNA, most of this remains within the nucleus. An important point is that there are claimed to be tissue-specific differences in the pattern of transcription of the repetitive sequences. As we saw in Chapter 9, RNA in this class may be concerned with the processing of hnRNA transcripts of single-copy, structural genes into functional mRNA.

11.2. RNA–DNA *hybridization* in situ

The highly repetitive fraction of eukaryote DNA can be isolated by collecting the DNA which renatures after the spontaneous formation of foldback hairpins. Alternatively, since it has an atypical ratio of guanine and cytosine to adenine and thymine (GC : AT), it has a different buoyant density compared with typical DNA and can be isolated by centrifugation in a gradient of a dense solution such as caesium chloride. When separated in the latter fashion it is known as satellite DNA.

Guinea-pig satellite DNA consists of the sequence TTAGGG repeated very many times. Hermit crab satellite DNA has the sequence TAGC, and that from *Drosophila* ATAAACT. In humans there are three satellite sequences, the ratio of which varies from one chromosome to another. Mouse satellite DNA consists of a more complex mixture.

The location of satellite DNA on the chromosomes has been deduced by what is known as *in situ* RNA–DNA hybridization (Figure 11.3). This involves producing a preparation of chromosomes on a microscope slide and treating them in such a way that the double helices split apart, but remain in position on the slide. If the preparation is then incubated with a radioactive RNA copy of satellite DNA, the latter will bind to those sites on the chromosomes which contain matching complementary sequences. The unbound radioactive RNA is then washed off and the slide is coated with photographic emulsion. This operation is performed in the dark and the coated slide is then maintained in darkness to allow the radioactive atoms to cause crystallization of silver in the emulsion. The emulsion is then developed as if it were a photographic negative and the location of the radioactive atoms becomes revealed as a photographic image. This technique is known as autoradiography or radioautography, and it reveals that satellite DNA occurs mainly in the heterochromatin on either side of the centromeres of the chromosomes and also to some extent near the distal ends, at the telomeres (Figure 11.3).

It has been suggested that satellite DNA could be concerned with gross functions of the chromosomes, such as homologous pairing at meiosis, but its true function is still unknown.

11.3. *Gene amplification*

Cells which are undergoing intensive protein synthesis require large numbers of ribosomes. For this reason there are multiple copies of the 18S, 5.8S and 28S ribosomal RNA genes, which occur in tandem repeats at five sites in the human genome. Bulk synthesis of ribosomal components is further facilitated by amplification of these genes into extra copies, which are shed from the chromosomes and stored in the nucleoli (see Figure 9.2). Amplification of the

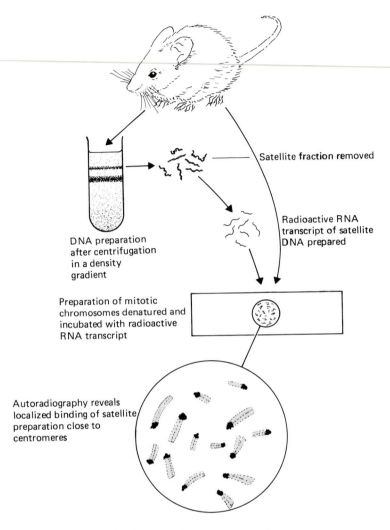

Figure 11.3. Localization of satellite DNA sequences on chromosomes.
A preparation of mouse DNA is centrifuged in a density gradient of caesium chloride solution and the less dense satellite band removed. A radioactive RNA copy is prepared by transcription *in vitro*. Actively dividing cells are arrested at mitosis, fixed and their DNA denatured to expose the base sequence. This preparation is then incubated with the radioactive RNA preparation. The unbound RNA is washed off and the slide coated with photographic emulsion. On development several days later silver grains produced by the action of the radiation on the emulsion are found to be concentrated over the heterochromatin beside the centromeres.

ribosomal genes is particularly important in the oocytes, which typically contain about 1000 extra nucleoli (see Chapter 9).

Following the discovery of ribosomal gene amplification, the Specific Gene Amplification Theory was put forward as a suggested explanation for the capacity of some cell types to produce large quantities of tissue-specific proteins. The theory was that cells which specialize in the production of massive quantities of one or two proteins may do so by amplifying just the genes concerned with production and functioning of those proteins. The theory was put to the test by the following type of experiment, illustrated in Figure 11.4, using avian red blood cells which, unlike those of mammals, retain their nuclei.

DNA was extracted from the erythrocytes, denatured and bound to a nitrocellulose filter. Another filter was coated with denatured DNA taken from the rest of the bird's body. Both filters were then incubated with a radioactive preparation of avian reticulocyte mRNA, which should bind specifically to globin gene sequences in the DNA (reticulocytes being the precursors of erythrocytes). Unbound RNA was washed off the filters and the amount of radioactivity remaining was assessed in a scintillation counter. If there are more globin sequences in the DNA of avian red cells than in the other tissues, the filter coated with red-cell DNA should have accumulated more radioactive label than the control filter, carrying DNA from the rest of the body. On the other hand, if the globin genes are not amplified in red cells there should be no difference between the two. The results of this and similar experiments seemed to disprove the concept of specific gene amplification as an aspect of cytodifferentiation (Figure 11.4). Furthermore, this conclusion fell into line with the experiments on nuclear transplantation (Chapter 7) which established the now well-accepted theory that differentiated somatic cells generally contain a complete and essentially unmodified genome.

The theory of specific gene amplification was therefore rejected, but in recent years the idea has again come to the fore, with a very exciting report of transient amplification of actin genes during muscle differentiation and indications of a modest degree of permanent amplification in other rather special situations. Of the latter, one of the best known is a 30-fold amplification of the egg chorion protein genes in *Drosophila* egg follicle cells. This seems to involve a kind of localized polytenization, producing multiple copies 'in parallel', as distinct from the reiterated ribosomal sequences situated 'in series' at the nucleolar-organizer sites (see Chapters 9 and 12)

The actin story is at the time of writing very new and unique, although it is sure to spark off a lot of experimentation on similar lines. During muscle development in chickens the early muscle cells, or myoblasts, proliferate then fuse at 17–19 days to form multinucleated myotubes. At this time the concentration of α-actin mRNA molecules increases some 270-fold, following an 8- to 10-fold increase in detectable actin-coding sequences present as DNA (Figure 11.5). A day or so later, a second α-actin DNA species appears and increases until 85 times the normal number of actin genes and some 20 000 actin mRNA molecules are present per nucleus. The DNA copies then disappear, although the RNA remains

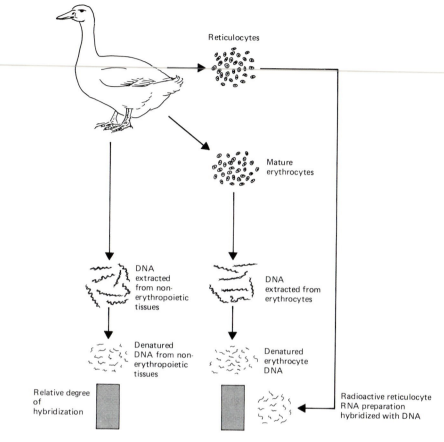

Figure 11.4. A test of the specific gene amplification theory.
DNA was extracted from avian erythrocytes and also from the rest of the bird's body. This DNA was denatured and hybridized with a radioactive preparation of RNA from reticulocytes, i.e., erythrocyte precursors containing mainly globin mRNA. There was no difference in the proportion of reticulocyte–specific RNA bound by the two DNA samples, indicating that the genes for red cell function are not amplified in the DNA of red blood cells.

at high titre for a prolonged period, dropping eventually to 20% of its former peak value.

What is particularly interesting about this observation is that the gene coding for a tissue-specific product is amplified only during the short period when that tissue differentiates. Any later investigation into the number of gene copies present in the mature tissue, such as that described above, could well draw a blank.

The means by which these extra genes are produced is also of great interest. Could they be formed by reverse transcription of messenger RNA? As yet we do not

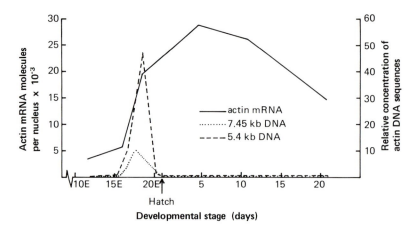

Figure 11.5. Temporary amplification of actin genes during muscle differentiation. The concentrations of two fractions of chicken embryo DNA, produced by digestion with restriction endonuclease and containing α-actin gene material, are plotted against time since the commencement of development. They are described as the 7.45 and 5.4 kilobase fractions. These fractions are amplified, reaching peak values at 17–19 days. Actin mRNA reaches a peak around five days after hatching. (After Zimmer and Schwartz, 1982.)

know, but this observation could well revolutionize our understanding of the way in which cytodifferentiation occurs.

Vertebrate cells cultured in the presence of methotrexate have also been shown to produce resistant strains containing multiple copies of the gene coding for dihydrofolate reductase, which is inhibited by methotrexate. However, there is no evidence that forced gene amplification of this type has any role in normal development.

11.4. Gene deletion

The converse of gene amplification is gene deletion. As mentioned in Chapter 7, there is no good evidence that deletion of DNA plays a major part in cytodifferentiation or any other aspect of development in most organisms, although there are a few unusual instances, like the maturation of lens fibre cells in the eye, where it is important. Several invertebrates, such as the Copepods, are unusual in that about half the DNA is lost from all their somatic cells, the full complement being retained only in germ cells. In one species, *Cyclops strenuous*, the heterochromatic DNA which is lost is interspersed among euchromatin all along the chromosomes, so that its removal requires many episodes of cutting and rejoining.

An even more extreme situation exists in the protozoan *Oxytricha*. In this species, as in many protozoans, genetic information is transmitted between the generations in a compacted form within a micronucleus. The DNA which is required to be expressed during the lifetime of the individual amounts to only 3% of that in the micronucleus and this is cut out and spliced together in a less compact form to produce a larger macronucleus. The specificity of cutting and splicing is apparently guided by a set of repeat sequences situated at each end of the excised pieces. The repeats are all the same and they amount to an average of about 1000 copies for *each* of the 17 000 different structural gene sequences. Repetitious DNA can therefore apparently be the result of gene amplification or the instrument of gene deletion. As we shall see below, repeat sequences also constitute 'switch sites' for gene rearrangements.

11.5. Multi-gene families

In Chapter 5 we considered the control of body segmentation in *Drosophila* by the genes of the bithorax complex. This set of genes probably arose by tandem duplication of a single gene, followed by modification of the individual replicates. This seems to be a common means by which the genotypes of organisms become extended and diversified to produce what are known as multi-gene families. Well-known examples are the three immunoglobulin families and the 18S–5.8S–28S ribosomal RNA family (see Chapter 12). Some of the tRNA genes also constitute mixed multi-gene families, as do the genes for the histones. In sea urchins one copy of each of the H2A, H2B, H3, H4 and H1 genes are linked together in units separated by spacer regions (see Chapter 12) and there are 400–1000 copies of this family per haploid genome.

Developmental systems have capitalized on the close proximity of such genes of related function, by co-ordinating the control of their expression. As we saw in Chapter 1, the prokaryote approach is to transcribe linked sequences of related function as one polycistronic messenger RNA molecule, but this does not seem to occur in eukaryotes. The bithorax complex (Chapter 5) is an example of a eukaryotic application of the multi-gene family in the establishment of body pattern. Below we deal with another application, in the rearrangements which occur in the immune system, but another intriguing instance occurs among those molecules beloved by teachers of genetics, the haemoglobins.

Five different β-type globin genes have been described in humans, which are closely linked and on the basis of their amino acid sequences are considered to have evolved progressively from an ancestor shared with α-globin and myoglobin. Their evolution diverged from that of α-globin about 400 million years ago, and the β-globin cluster is not linked to the α-globin genes (Figures 11.6(a), (b), 15.1). Synthesis of α-globin chains begins very early in development and continues throughout life. On the other hand, synthesis of the members of the β group is regulated in relation to developmental stage: first epsilon (ε), then the two gammas (γ), then beta (β) and lastly delta (δ). The different forms are synthesized in the different tissues according to the time-schedule shown in Figure 11.6(c), (d).

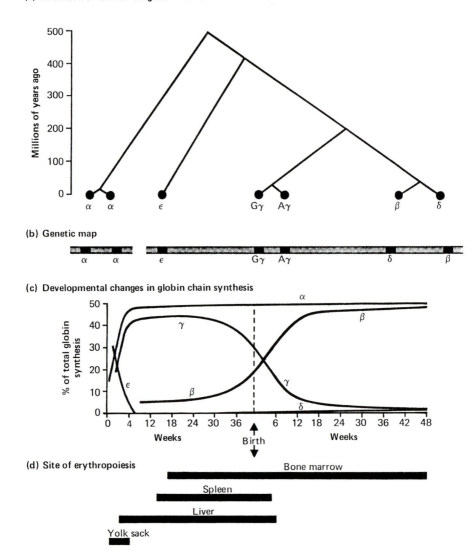

(a) Evolution of the human globin chains

(b) Genetic map

(c) Developmental changes in globin chain synthesis

(d) Site of erythropoiesis

Figure 11.6. The relationship between the evolution, genetic map, time of expression during development and site of expression, of the members of the α- and β-globin families in man.

(a) The postulated sequence of times of divergence of the human globin genes, as deduced by their amino acid sequences. (b) The genetic map of the human globin sequences. The β-globin multi-gene family includes the ε, Gγ, Aγ, δ and β structural sequences in a linked group. Apart from the β and δ, the order of these sequences along the chromosomes is the same as the order in which they arose during evolution. The α-globin sequences are also linked together, but situated elsewhere in the genome. (There are indications that the human genome contains two further α-globin sequences, but their location is not known.)

What is particularly interesting about this system is that the order in which the different globin genes are expressed during development more or less parallels both the order in which they are thought to have evolved and their order along the chromosome. Their relative positions on the chromosomes are generally considered to be the natural result of gene duplication, due to illegitimate chromosome pairing and erroneous crossing over at meiosis, but the relationship with time of expression is another matter. This could be merely coincidental, but it suggests a physiological version of the recapitulation of morphological stages that occurs to some extent during development and which has been summarized by the expression 'ontogeny repeats phylogeny' (see Chapter 15). We will see below that regulation of expression of these genes occurs partly by demethylation and methylation of cytosine residues in the DNA (see also Chapter 8).

11.6. Gene rearrangement in the immune system

The vertebrate immune system contains three multi-gene families and represents a special case of cytodifferentiation since it includes a large collection of cell clones derived from B lymphocytes, each secreting a different species of antibody protein produced (in quantity) in response to exposure to different foreign antigens. In Chapter 9 we saw how differential processing of the hnRNA coding for immunoglobulin heavy chains leads to production of first membrane-bound, then water-soluble antibody. This is one of several extraordinary events that occur during the generation of antibody diversity.

A healthy mammal or bird can construct an antibody of unique structure to match the shape and charge distribution of practically every large molecule in existence. This facility is necessary to counter the invasion of the body by diverse parasites and disease organisms which would otherwise rapidly kill the host. In addition, there are classes of antibodies sharing the same specificity, but with variation in their other properties, such as capacity to cross the placenta or to bind to macrophages (Table 11.1). This enormous diversity in protein species synthesized is achieved by rearrangement of a relatively small number of segments of genetic information, carried out to some extent by the cutting and splicing of the genomic DNA. This gene rearrangement occurs in the chromosomes of individual B-cell precursors, which mature when exposed to the appropriate antigen and proliferate to produce the new gene product in large amounts.

(c) Developmental changes in globin production. Synthesis of α-globin commences early in development and continues throughout life. Production of the other globins occurs approximately in the order of their structural genes on the chromosome and the order in which they evolved. (d) The site of synthesis of haemoglobin changes during development. Before birth the yolk egg sac, spleen and particularly the liver are the major sites of erythropoiesis. After birth this function is taken over by the bone marrow. The temporal change in globin synthesis is thus based in a change in the cell types that are expressing the different globin genes. (Based on Wood, 1976 and Zuckerkandl, 1965.)

Table 11.1. The classes, composition and special properties of the immunoglobulins in man.

Class	Percentage of total Ig in blood	Heavy chain	Light chain	Special properties and functions
IgG	80	γ	$\varkappa$ or λ	Predominant serum antibody Activities complement Crosses placenta Binds to surface of phagocytes and assists their function
IgM	5	μ	$\varkappa$ or λ	Cell-surface receptor Serum antibody early in immune response Activates complement Exists as a pentameric molecule
IgA	15	α	$\varkappa$ or λ	Predominant antibody in saliva, colostrum and intestinal fluids Exists as a monomer or dimer
IgD	< 1	δ	$\varkappa$ or λ	Cell-surface receptor on immature B lymphocytes
IgE	< 1	ε	$\varkappa$ or λ	Releases histamine from mast cells Involved in anti-parasite immune response

From Alberts *et al.* (1983) and Gough (1981).

The structure of antibodies

Antibody, or immunoglobulin, molecules are each composed of two identical light polypeptide chains and two identical heavy chains linked by disulphide bonds (Figure 11.7). In the mammals there are five major classes of immunoglobulin (IgG, IgM, IgA, IgD and IgE) which carry out rather different functions. Each class and subclass is distinguished by the nature of its heavy chain, gamma (γ), mu (μ), alpha (α), delta (δ) or epsilon (ε) respectively (see Table 11.1), but the light chains, which are of two types, kappa ($\varkappa$) and lambda (λ), occur in all classes. All types of antibodies can be produced in membrane-bound or water-soluble forms by differential RNA processing, as described for the μ chain in Chapter 9.

The structure of each chain is maintained by intra-chain disulphide bridges, which cause consecutive portions to form into discrete globular sections or 'domains' connected by short segments of more extended polypeptide. There are close similarities in the amino acid sequences of these domains, suggesting that they probably arose in evolution by a process of gene duplication. The immunoglobulins have carbohydrate groups attached to their heavy chains which vary between the five major classes.

Each heavy and each light chain has a variable (V) region which shows great diversity, and a constant (C) region. Antigen-binding specificity is determined by

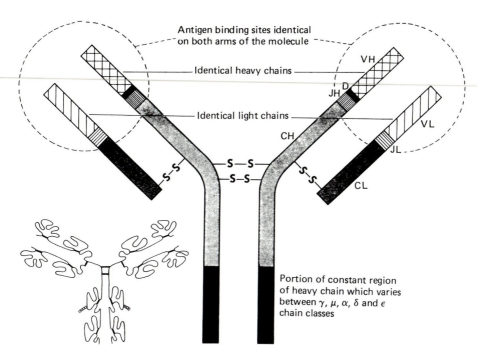

Figure 11.7. An immunoglobulin molecule.
The molecule is symmetrical and consists of an identical pair of heavy chains and an identical pair of light chains, linked though cystine residues by disulphide bonds. The light
chains are of either λ or ϰ type and consist of a constant (CL), a joining (JL) and a variable
(VL) section. Each heavy chain consists of a long constant (CH) region, short joining (JH),
and diversity (D) sections and a variable (VH) region. The terminal portion of the CH chain
differs in the different classes of immunoglobulin. Inset: Each heavy chain is built up of
four 'domains', and each light chain of two 'domains' of broadly similar structure stabilized by intra-chain disulphide bonds. The heavy chains each carry a carbohydrate moiety
which varies in structure between the immunoglobulin classes.

the amino acid sequences of the variable regions, and membrane binding or
other properties by the terminal parts of the C regions of the heavy chains. The
V and C parts of the protein are linked by a variable joining (J) region of 13 amino
acids in the light chains, and by a J plus a variable D (diversity) sequence in the
heavy chains (Figure 11.7).

The genetic basis of antibody diversity

The lambda, V, J and C, the kappa, V, J and C, and the heavy chain, V, D, J
and C regions are each coded by non-adjacent gene sequences grouped in separate
kappa, lambda and H multi-gene families located on different chromosomes
(Figure 11.8). In the mouse each C region of the five classes of heavy chain and

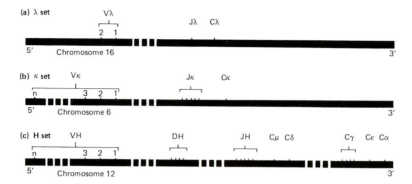

Figure 11.8. Genetic map of the immune system in the mouse.
The genes for the different immunoglobulin chains fall into three multi-gene families situated on different chromosomes. The λ chain set consists of two V genes, one J and one C on chromosome 16. (b) The $\varkappa$ chain set is made up of a large number of V genes, five J genes and one C gene situated on chromosome 6. (c) The heavy chain set includes many V genes, four D and five J genes. There are several CH, one each for μ, δ, ε, and α and four of Cγ, all on chromosome 12.

the kappa and lambda light chains is encoded by a unique C gene, except for the gamma heavy chain for which there are four alternative sequences. There are several different J genes in the kappa and H sets, and also several different D genes in the H set. In each diploid set of mouse chromosomes there are 100–300 V kappa and VH gene sets, but probably only two V lambda. Variability in antibody properties is brought about by ringing the changes on different combinations of these gene sequences, by rearrangement of the DNA within individual cells.

During differentiation of an antibody-producing B lymphocyte two different types of gene rearrangement take place. The first occurs before exposure to antigen and involves association of V, J and C genes in either the kappa or lambda set, and V, J, D and C μ genes in the heavy-chain set, to produce a functional immunoglobulin with a defined antigen specificity. There is unlikely to be any 'attempt' by the genome to undergo rearrangements in order to match a particular antigen. The different specificities are almost certainly produced by random events and selected by their capacity to bind with antigen.

The second rearrangement occurs after antigen exposure. In this case the heavy-chain C μ gene is exchanged for another class of heavy-chain C gene, with a consequent change in the biological activity of the product, but without loss of antigen specificity.

During the first rearrangement, particular V and J light-chain genes become linked and the intervening stretch of DNA is deleted (Figure 11.9(c)). This occurs by some mechanism that is not yet understood, but it depends on the presence of repetitive sequences at the switch sites where the DNA is spliced. The V–J–C sequence is then transcribed as hnRNA, commencing at the beginning of the V

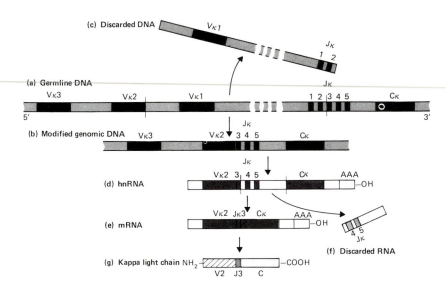

Figure 11.9. Generation of diversity by rearrangement in the *x*-chain gene.
(a) The arrangement of the *x*-chain genes in mouse germ-line DNA is shown (see also Figure 11.7). A section of DNA between the V and J regions is excised and discarded (c), bringing one of the V genes (V*x*2 in this instance) adjacent to one of the J genes (J*x*3) (b). (d) Transcription commences at the start of the V*x*2 gene and continues through all the J genes and the C, plus a trailer region. Processing of the hnRNA transcript involves discarding the transcripts of all the J genes except J*x*3, to produce an mRNA molecule in which V*x*2, J*x*3 and C*x* are adjacent (e). Translation of this molecule produces a *x* light chain which is just one of many which could be coded by the *x* family. (Redrawn from Alberts *et al.*, 1983.)

gene adjacent to the J section, and terminating after the C *x* gene. The spacer section between J and C is then removed during processing into mRNA.

In the heavy-chain system a similar sequence of events take place, but with one extra step. A V gene joins to a D, then the V–D sequence joins to the J–C *μ* section. Intervening pieces of DNA are deleted and the remaining non-coding spacers are processed out of the RNA transcript.

Heavy-chain class switching

The first type of heavy chain produced is the *membrane-bound* form of the *μ* chain, as IgM, since the C *μ* gene is closest to the JH section. Cells undergoing this type of synthesis can switch to *secretion* of water-soluble IgM by a change in RNA processing, as described in Chapter 9. Alternatively they can switch to simultaneous production of the membrane-bound form of any of the other classes of immunoglobulin distinguished by its heavy-chain C region. This the cell does by transcribing a long hnRNA molecule that includes a transcript of several, or

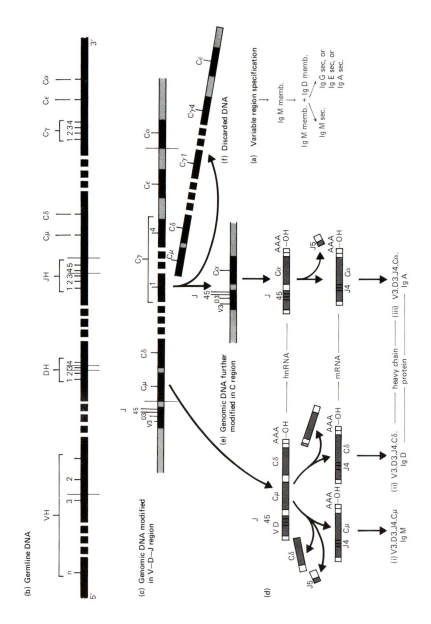

all of, the C regions. This long molecule is then processed differentially to produce *two* membrane-bound Ig types simultaneously, for example IgM from the C μ region and IgA using the C α region. If such a cell is stimulated with its appropriate antigen it proliferates to establish a clone which *secretes* one Ig species only, for example, IgA, and this it does by deleting from the DNA all the intervening CH genes between C α and the JH regions (see Figure 11.10).

The ability of an assembled heavy-chain variable region to associate with any of the different classes of constant chains has important functional implications, since this allows a particular antigen binding site to become distributed through all the different classes of immunoglobulins, with all their disparate properties.

Allelic exclusion

Loss of genomic DNA from the antibody-producing cells represents a determinative step that cannot be reversed. It might be expected that each cell should still have an alternative option open to it however, since each lymphocyte would normally contain two chromosome sets. This is prevented by a phenomenon known as allelic exclusion, which ensures that each clone of cells produces only one species of immunoglobulin. The mechanism of allelic exclusion is unknown, but it ensures that each antibody molecule synthesized is mono-specific and can participate in the formation of large lattices of cross-linked antigens which are probably essential for antibody function.

Figure 11.10. Heavy-chain class switching. This diagram illustrates how the same antibody specificity can become associated with different classes of heavy chain in the mouse.
(a) The sequence of changes which can take place in a B lymphocyte lineage. The variable region is first specified by gene rearrangement as in Figure 11.9. This produces a membrane-bound IgM molecule containing a pair of heavy chains with the terminal portions coded by the C μ gene. This cell can switch to secretion of IgM by differential RNA processing (Figure 9.9), or to simultaneous production of the membrane-bound form of IgM, plus that of any of the other Ig classes, while retaining the same antibody specificity. On exposure to antigen which matches the antibody site on the immunoglobulin protein, the cell proliferates to produce a clone with the same antibody specificity, which secretes a class of Ig molecule originally bound to the surface membrane of its parent cell. (b) The heavy-chain germ-line gene map in the mouse (cf. Figure 11.8). (c) The genomic DNA in B lymphocytes after gene rearrangement in the V–D–J region (cf. Figure 11.9). (d) Both IgM (i) and IgD (ii) can be produced by transcription of the long hnRNA molecule which is differentially processed to form either a C μ or a C δ type of mRNA. In both cases J5 is eliminated during RNA processing. (e) Formation of the other classes of heavy chain is accomplished by excision of genomic DNA 3' to the commencement of the C μ gene. In the example shown, the region including C μ, C δ, C γ 1–4 and C ε is discarded leaving C α adjacent to J5. The J5 section is removed during RNA processing to form an IgA molecule of the same antibody specificity as before. (After Gough, 1981.)

11.7. *Gene rearrangement in cytodifferentiation*

The existence of a mechanism for gene rearrangement suggests a model for control of the differentiation of other body cells based on reassortment of a few gene sequences which might then code, for example, for a range of non-histone chromosomal (NHC) proteins (see Chapter 9). A range of mutually exclusive gene-switching NHC proteins could then be produced in different cell types. Alternatively, some of the gene products could become bound to the plasma membrane and embryonic inducer molecules could act like antigens, causing selective proliferation of cells with particular gene rearrangements. Whether or not such models have any basis in reality is as yet unknown.

Another model involves rearrangement of structural genes with respect to controlling sequences. This is known to occur in a rather random fashion in maize, *Drosophila* and several other species. In man it is an important feature in the development of some cancers, such as chronic myelogenous leukaemia, following translocation between chromosomes 22 and 9 to produce the 'Philadelphia chromosome', a slightly smaller than usual copy of 22. The fluctuating display of surface antigens in trypanosome parasites and the mating types of yeasts are also controlled by rearrangement of structural genes with respect to sites that control transcription.

One cytodifferentiative system of higher organisms has been described which seems to include a rearrangement of this sort. This involves the α-amylase gene *Amy-1*[a] in mice. The enzyme is present in the liver, and also the salivary gland at 100 times the activity in the liver. The mRNA concentrations show a similar discrepancy, suggesting control at transcription or RNA processing. The leader sequences of the α-amylase mRNA found in the two tissues are different, although the remainder of the transcripts are identical. This observation inspired the suggestion that tissue-specific control could be exerted at transcription through rearrangement of transcription rate modulators (see Chapter 12) with respect to the structural gene. A powerful set of upstream controlling sequences seems to operate in the salivary glands exclusively, while a weak set apparently operates in both tissues. Exactly how this is achieved has not yet been elucidated, but if these features are found in other systems they would obviously explain much of the mystery surrounding tissue-specific differences in gene expression.

11.8. *Chemical modification of DNA*

A group of enzymes which is proving to be of immense importance in current investigations into DNA structure is the bacterial restriction endonucleases. These enzymes have the capacity to recognize and cut certain base sequences with great specificity, each endonuclease recognizing a specific inverted repeat. For example, the enzyme Hpa II cuts the base sequence 5'-CCGG-3', which has the identical and complementary sequence on the other strand, when read in the same direction relative to the molecular bonding between the nucleotides.

Another restriction endonuclease, Msp1, cuts the CCGG sequence and will do so even if the second cytosine carries an extra methyl (-CH₃) group, whereas Hpa II will not cut the sequence if the second C is methylated. When these two enzymes are allowed to digest samples of the same DNA preparation they reveal characteristic tissue-specific patterns of cytosine methylation at CCGG sequences. This elegant technique has opened up what is currently one of the most exciting areas of investigation into the control of gene expression. In combination with the technique of gene cloning it promises to reveal some very valuable information about the earlier controls upon tissue-specific patterns of gene expression.

Cytosine methylation in the β-globin cluster

In one experiment to test for methylation of the CCGG sequence in the region of the β-globin gene, chromatin was isolated from several different adult rabbit tissues, cut into small pieces using another restriction endonuclease (EcoRI) and then divided into two fractions, one being exposed to MSP1 and the other to Hpa II. The DNA fragments were then separated according to size by electrophoresis in a gel, and the fragments containing the β-globin sequence were identified by autoradiography, following hybridization with a radioactive molecular probe corresponding to the β-globin sequence.

As expected, preparations which had been treated with Msp1 produced β-globin gene fragments of the same size from all tissues, since Msp1 does not discriminate between different degrees of methylation (Figure 11.11). DNA from a line of cultured rabbit cells produced identical patterns when treated with the two enzymes, indicating no methylation of the CCGG site in the region of the β-globin gene. In contrast, however, some of the DNA prepared from tissues and treated with HpaII contained large β-globin gene fragments, while other samples contained small fragments, because this site is methylated in some tissues but not in others. The experiment indicates that in DNA from rabbit sperm the β-globin site is invariably methylated, in adult liver DNA about 50% of the cells have methylated sequences in this region, while adult rabbit brain DNA is 80% methylated at the β-globin sequence (see Figure 11.11).

A similar approach was applied to human material and extended to cover a wider range of tissues and a broader section of the β-globin gene cluster. Figure 11.12 shows the positions of the sites considered relative to the globin genes and the degree of under-methylation of those sites. In sperm and foetal brain DNA all the sites are completely methylated, but in placenta all are highly demethylated. The liver is one of the principle foetal organs of erythropoiesis, but after birth this function is largely taken over by the bone marrow (see Figure 11.6). This change is reflected by a general increase in methylation levels in liver during development.

Interestingly, despite the fact that foetal liver does not synthesize β-globin, the two sites in the β-globin region were as demethylated in foetal liver as they were in adult bone marrow. This shows that additional controls must regulate production of β-globin (see Chapters 8 and 12).

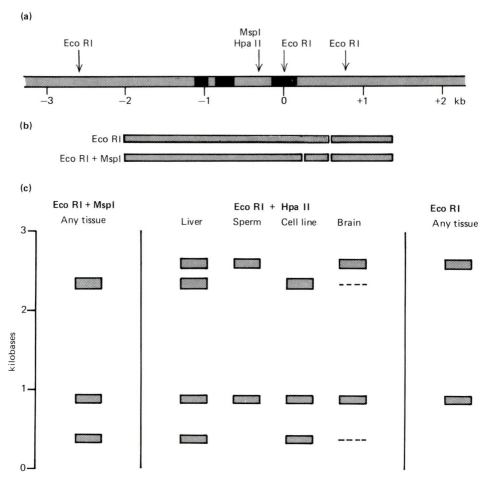

Figure 11.11. Methylation of a CG sequence within the rabbit β-globin second intron in relation to tissue type.

(a) Map of the rabbit β-globin gene region. The coding sequences, or exons, are shown in black (cf. Figures 9.13, 12.3) and restriction enzyme cutting sites are indicated. (b) The fragments produced by digestion of the DNA with Eco R1 and a mixture of Eco R1 and Msp1. The latter enzyme cuts a CG site within the second β-globin intron whether or not that site is methylated. (c) The pattern of DNA fragments containing the β-globin gene after electrophoresis in agarose gel. In this system mobility of the fragments is inversely proportional to molecular weight. The positions of fragments are located by transfer to a sheet of nitrocellulose paper (Southern blotting) and hybridization with a radioactive copy of the β-globin gene, followed by autoradiography. Since the enzyme Hpa II cuts the CG sequence only if it is unmethylated, comparison of the density of labelling in the different bands reveals the proportion of cells in the tissue in which the sequence is methylated. The liver sample was found to be 50% methylated, sperm 100%, a cell line 0% and the brain sample 80% methylated at this site. In the brain sample the fragments represented by the dotted line were not actually detected, but are indicated here for logical clarity. (Redrawn from Razin and Riggs, 1980; after Waalwijk and Flavell, 1978.)

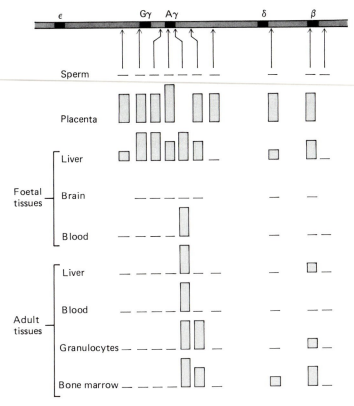

Figure 11.12. Demethylation of cytosine residues in the human β-globin cluster in relation to tissue type and developmental stage.
The diagram relates the digestibility of the DNA at the indicated sites by enzymes which cut only methylated sequences. These sites are largely unmethylated in foetal liver and placenta and completely methylated in the sperm and brain DNA. There is a dramatic change in the pattern shown by the liver during development as it ceases to synthesize haemoglobin (cf. Figure 11.6). (Drawn from data in Van der Ploeg and Flavell, 1980.)

Cytosine methylation at other sites

Similar results have been obtained with the ovalbumin gene, which is expressed in secretory cells in the oviducts of laying hens (see below). In this case, sequences in the region of the ovalbumin gene were found to be much less frequently methylated in oviduct tissue than in liver cells and erythrocytes and, like the β-globin sequence, are apparently always methylated in sperm.

A most interesting feature of this work is the observation that the variably methylated sites are not necessarily only in the coding sequences, but also for some distance on either side of them. The variably methylated site in the vicinity of the rabbit β-globin gene is actually within the first intron, in other cases these sites are in flanking regions.

In the chicken there are two α-globin genes, some 4 kilobases apart, together with a related sequence called the U gene, which probably codes for an embryonic type of α-globin. As the erythroid cells mature, expression of the U gene is switched off, while that of the α-genes is switched on and this correlates with methylation of the U region and demethylation of the α region (see Figure 11.13).

In mature erythrocytes the demethylated region occupies a much broader stretch of the chromosome, about 6.5 kilobases long, spanning the two α-globin genes. This region selectively binds the NHC proteins, HMG14 and HMG17, is digestible by DNAase-1 and contains RNA polymerase II, all features characteristic of transcriptionally active chromatin (see Chapter 10). For 10–20 kilobases on either side of this unmethylated zone is a domain of intermediate sensitivity to

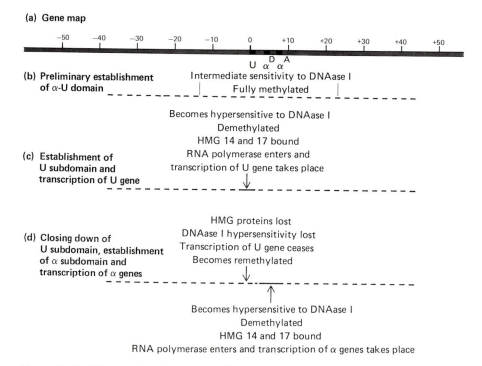

Figure 11.13. The postulated sequence of changes that take place in the chromatin around the α-globin cluster in the chicken in relation to expression of the α-globin genes.
(a) The map of the α-globin cluster. There are two α-globin genes and a third gene, named U, which probably codes for an embryonic type of α-globin. The chromatin is initially highly methylated in all tissues and insensitive to DNAase-1. (b) A domain of intermediate sensitivity to DNAase-1, extending to 20–30 kilobases on either side of the α-globin cluster, becomes established in erythropoietic tissues. (c) A subdomain around the U gene becomes hypersensitive to DNAase-1 and demethylated. It binds HMG14 and HMG17. RNA polymerase II enters the U-gene region and the U gene is transcribed. (d) The U-gene subdomain reverts to its former state, as indicated in (b). A new subdomain becomes established around the α-globin genes as in (c).

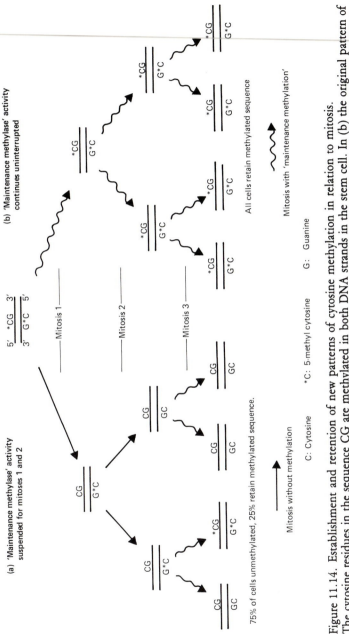

Figure 11.14. Establishment and retention of new patterns of cytosine methylation in relation to mitosis. The cytosine residues in the sequence CG are methylated in both DNA strands in the stem cell. In (b) the original pattern of methylation is retained through three cell divisions, due to methylation of the sequence CG base-paired to 5-methyl CG. In (a) the activity of maintenance methylase is inhibited at this site for two cell divisions. This allows transmission of the methylated sequence to daughter cells, but the partner strands remain unmethylated. When maintenance methylase recommences its action, one in four cells acquire the same methylation pattern as the stem cells, but the other three remain unmethylated. Scheme (a) shows how a mixed population of cells could arise under the influence of an embryonic inducer which temporarily blocks the activity of maintenance methylase in a tissue of proliferating cells.

DNAase-1, but which has none of the other characteristics of transcriptionally active chromatin. The actively transcribed region therefore occupies a sub-domain within the domain defined (so far) only by its intermediate degree of sensitivity to DNAase-1.

Considering all the systems examined, several important points emerge:

1. The DNA of most organisms contains modified bases, but 5-methylcytosine is the only one found in vertebrates, and occurs mainly in the sequenced CC and CG. According to some authors, 50–70% of all CG sequences are methylated in avian and mammalian DNA, although most authorities quote lower figures.

2. Repetitive sequences, including satellite DNA, are generally highly methylated.

3. Some CG sequences remain methylated and some remain unmethylated in all tissues, while others are variably methylated in relation to tissue type or developmental stage. In general, gene activity is associated with under-methylation.

4. The variably methylated sites extend across a fairly extensive length of chromosome which encompasses the gene or gene complex of particular importance.

5. Patterns of methylation tend to be maintained in daughter cells. It has been suggested that this could be due to the activity of a 'maintenance methylase' enzyme which acts only on CG sequences base paired with methylated partners (see Figure 11.14).

6. Whereas methylation occurs actively, unmethylated DNA results from suspension of methylation of newly synthesized DNA, not by removal of methyl groups from methylated sequences. Establishment of 'homozygous', new patterns of non-methylation therefore requires two rounds of cell division (Figure 11.14).

7. Apart from situations of 100% methylation as in sperm, it is usually only a proportion of the cells in a tissue that displays a defined pattern of DNA methylation, the observed pattern in a tissue being the consensus of a range of individual patterns (see Figure 11.14). The unmodified cells could act as a reserve, utilized for example in regeneration (see Chapters 6 and 7).

Cytosine demethylation in relation to development

A theory which is gradually gaining acceptance is that the DNA throughout the body becomes highly methylated at CG or CC sites during very early development and that selective gene expression in the different tissue types is accompanied by selective demethylation of specific regions of the chromosomes. Whether demethylation is the result or the cause of gene activity has not yet been properly established. One approach to this problem is to incubate cells with an analogue of cytidine, 5-azacitidine (5-aza C), which becomes incorporated into DNA, but cannot be methylated, as the carbon atom which would carry the methyl group is replaced by nitrogen. When cells of the embryonic mouse mesoderm line $10T\frac{1}{2}$

are allowed to incorporate 5-aza C for several mitotic cycles, they differentiate into muscle, chondrocytes and adipocytes, suggesting that tissue-specific patterns of gene expression can become established following demethylation and hence that demethylation is a causative factor in gene expression. Analysis of the proteins in these cultures suggests that 5-aza C causes a stable change which is maintained indefinitely despite proliferation, but expression of characteristic differentiative proteins becomes activated only when culture conditions become appropriate. Methylation and demethylation therefore seem to control cytodifferentiation at the determinative level.

Heterochromatization is a gross means by which gene expression can be controlled. For example, during early mammalian development one X chromosome in female cells is inactivated in this way, forming the Barr body. Active loci translocated close to heterochromatin can become inactivated in a rather imprecise manner, by an influence that seems to spread from heterochromatin (see Chapter 10). It is interesting to note that heterochromatin contains a high proportion of methylcytosine. This is certainly the case with satellite DNA heterochromatin situated beside the centromeres (see above) and seems to apply also to the Barr body, since incubation of female mammalian cells with 5-aza C can cause reactivation of genes in the heterochromatic X chromosome.

A theory that links these observations is that methylcytosine may catalyse conversion of DNA from the familiar right-handed double helix, or B form, to a left-handed zig-zag helix or Z form (Figure 12.6). This is the case in synthetic oligonucleotides, although it awaits demonstration in natural DNA. The total net charge and the charge distribution on the phosphodiester backbone are quite different in the two conformations and they would be expected to have quite distinct binding properties with respect to protein. It is possible therefore that interconversion between B and Z forms, in association with demethylation and methylation of cytosines, could control patterns of gene expression through differential protein binding.

Antibodies have been prepared against Z-form DNA and these bind specifically to the inter-band regions of *Drosophila* chromosomes, that is between the structural genes (see Chapter 9).

11.9. A general model for control of gene expression in vertebrates

In Chapter 10 we considered whether those inductive influences which operate during cell proliferation (such as the action of the AER on limb bud mesenchyme, described in Chapter 6) could do so by modifying the association of NHC proteins with DNA. In that chapter it was suggested that this might be due to biochemical changes in the proteins, but an alternative would be that such inductions might initially cause modified patterns of DNA demethylation. This could cause localized changes in DNA conformation and modified patterns of binding of chromosomal proteins such as histone H1. This in turn could cause loosening of the chromatin solenoid, allowing entry of the HMG proteins and RNA polymerase, and enabling transcription of previously unexpressed genes to take place.

The big unknown in this argument relates to the establishment of new patterns of methylation due to selective masking of regions of the genome from the activities of 'maintenance methylase'. One explanation would be that chromosome-binding proteins which are available, or in an appropriate state only under conditions of embryonic induction, may bond to specific nucleotide sequences, so blocking access of the methylating enzyme.

It would be attractive to accept this speculative model as a possible central feature of eukaryote developmental genetics. Unfortunately things are not that simple, since the DNA of some species, such as *Drosophila*, does not contain 5-methylcytosine although some of it is in the Z conformation. In these species the function performed in vertebrates by this modified base is presumably carried out by some other means.

A particularly interesting point to emerge from this study is that some controls upon gene expression are apparently exerted over extensive tracts of chromosome on either side of the gene. It is possible to envisage the whole genome being divided into many domains, some of which initially become rendered accessible to proteins of the size of DNAase-1 before selective demethylation takes place. Transcriptional control could therefore be seen as a multi-step process involving 'homing in' on the gene in question.

We would expect the locations at which these events take place to be written in the DNA, with key controlling sequences within or beside structural genes and also some distance away from them. As we shall see in the next chapter, controlling sequences within and close beside the genes are now being recognized, but we know next to nothing about the more distant signals. Perhaps the widely scattered highly repetitive sequences so characteristic of eukaryote DNA perform a role in this respect. Support for this view comes from the ovalbumin system. The ovalbumin gene, together with two related genes, X and Y, are located within a 100 kilobase domain, which shows increased susceptibility to DNAase-1 in oviduct tissue. This domain contains several repetitive sequences, especially at the 'upstream' end, including two of a type called CRI near the upstream boundary. There is also a CRI sequence in reverse orientation near the downstream limit, beside a high density of CCGG sites. This suggests that the ovalbumin domain may be *defined* by the CRI sequences and its operation *controlled* by variable methylation at the 'downstream' end.

11.10. Summary and conclusions

DNA renaturation studies indicate that much of the genome of eukaryotes consists of sequences which are repeated very many times. This includes the histone and tRNA genes and satellite DNA, the latter located mainly in the heterochromatin near the centromeres. Some repetitive sequences have a role in co-ordination of related events that occur in widely separated parts of the genome and that involve neighbouring structural genes. The genes which code for transfer and ribosomal RNA are also repeated many times in the chromosomes and further amplified as extra-chromosomal copies, and stored in the nucleoli. Specific gene amplification is not a widespread permanent feature of eukaryote genomes, but transient gene amplification seems to occur during muscle differentiation and may well take place in other systems also.

The immune system of the vertebrates reveals a remarkable capacity for gene rearrangement, which increases the coding capacity of the genome many fold. There is evidence from other eukaryotes that gene rearrangement is probably a widespread phenomenon and there are indications that rearrangement of structural genes with respect to controlling sequences may take place during normal cytodifferentiation in vertebrates.

During the early development of vertebrates many of the cytosines in the sequence CG, as well as in other sequences, carry methyl groups. There is some evidence that the development of diverse patterns of gene expression in the different tissues involves selective demethylation of these sites. This may change the conformation and/or the degree of supercoiling of the chromatin and probably constitutes an important primary feature in the control of gene expression in vertebrates at the level of determination.

Gene activity is associated with undermethylation, sensitivity to DNAase-1 and binding of the protein HMG14 and HMG17 throughout a subdomain stretching for thousands of bases on either side of transcriptionally active regions. These are within much larger domains characterized by intermediate sensitivity to DNAase-1. Demethylation is thought to occur by inhibition of an enzyme responsible for maintenance of methylation patterns, which comes into operation following synthesis of new DNA. This could be an important aspect of embryonic induction and might operate through physical blocking of specific sites on the DNA by other agents which become active under the conditions of induction. The pattern of events that takes place is likely to be guided by specific sequences in the DNA, which may be within or quite distant from the gene expressed. In the next chapter we will explore the signals carried by the DNA which control the transcription of specific gene sequences.

Bibliography

Alberts, B., Bray, D., Lewis, J., Raff, M., Roberts, K. and Watson, J., *Molecular Biology of the Cell*. Garland, New York (1983).

Behe, M. and Felsenfeld, G., Effects of methylation on a synthetic polynucleotide: the B–Z transition in poly(dG-m^{5}dC).poly(dG-m^{5}dC). *Proc. Natl. Acad. Sci. USA*, **78**(3): 1619 (1981).

Bird, A. P. and Southern, E. M., Use of restriction enzymes to study eukaryotic DNA methylation: I. The methylation pattern in ribosomal DNA from *Xenopus laevis*. *J. Molec. Biol.*, **118**: 27 (1978).

Brown, D. D., Gene expression in eukaryotes. *Science*, **211**: 667 (1981).

Burdon, R. H. and Adams, R. L. P., Eukaryotic DNA methylation. *Trends Biochem. Sci.*, **5**: 294 (1980).

Chater, K., Cullis, C., Hopwood, D., Johnston, A. and Woolhouse, H., *Genetic Rearrangement. The Fifth John Innes Symposium*. Croom Helm, Beckenham (1983).

Cold Spring Harbour Symposia on Quantitative Biology, *Structures of DNA*. Cold Spring Harbour Laboratory, New York (1983).

Doerfler, W., DNA methylation — a regulatory signal in eukaryotic gene expression. *J. Gen. Virol.*, **57**: 1 (1981).

Ford, P. J., Control of gene expression during differentiation and development. In *The Developmental Biology of Plants and Animals*, edited by C. F. Graham and P. F. Waring. Blackwell, Oxford (1976).

Gally, J. A. and Edelman, G. M., The genetic control of immunoglobulin synthesis, *Ann. Rev. Genet.*, **6**: 1 (1972).

Gough, N., The rearrangements of immunoglobulin genes. *Trends Biochem. Sci.*, **6**(8): 203 (1981).

Herrick, G. and Wesley, R. D., Isolation and characterization of a highly repetitious inverted terminal repeat sequence from *Oxytricha* macronuclear DNA. *Proc. Natl. Acad. Sci. USA*, **75**: 2626 (1978).

Holliday, R. and Pugh, J. E., DNA modification mechanisms and gene activity during development. *Science*, **187**: 226 (1975).

Jelinek, W. R. and Schmid, C. W., Repetitive sequences in eukaryotic DNA and their expression. *Ann. Rev. Biochem.*, **51**: 813 (1982).

Konieczny, S. F. and Emerson, C. P, 5-Azacytidine induction of stable mesodermal stem cell lineages from $10T\frac{1}{2}$ cells: evidence for regulatory genes controlling determination. *Cell*, **38** (3): 791 (1984).

Lewin, B., *Gene Expression, Vol. 2, Eucaryotic Chromosomes*, 2nd edn. Wiley, Chichester (1980).

Mandel, J. L. and Chambon, P., 1979. DNA methylation: organ-specific variations in the methylation pattern within and around ovalbumin and other chicken genes. *Nucleic Acids Res.*, **7**, 2081 (19   ).

Markert, C. L. and Ursprung, H., *Developmental Genetics*. Prentice-Hall, New Jersey (1971).

Mohandas, T., Sparkes, R. S. and Shapiro, L. J., Reactivation of an inactive human X chromosome: evidence for X inactivation by DNA methylation. *Science*, **211**: 393 (1981).

Pardue, M. L. and Gall, J. G., Chromosome structure studied by nucleic acid hybridisation in cytological preparations. *Chromosomes Today*, **3**: 47 (1972).

Peterson, P. A., The position hypothesis for controlling elements in maize. In *DNA: Insertions, Elements, Plasmids and Episomes*, edited by A. I. Bukhari, J. A. Shapiro and S. L. Adhya. Cold Spring Harbour Laboratory, New York, pp. 419–460 (1977).

Razin, A. and Riggs, A. D., DNA methylation and gene function. *Science*, **210**: 604 (1980).

Rich, A., Nordheim, A. and Wang, H.-J., The chemistry and biology of Z-DNA. *Ann. Rev. Biochem.*, **53**: 791 (1984).

Schibler, U., Hagenbüchle, O., Young, R. A., Tosi, M. and Wellauer, P. K., Tissue-specific expression in mouse α-amylase genes. *Adv. Exp. Med.*, **158**: 381 (1982).

Shapiro, J. A., *Mobile Genetic Elements*. Academic Press, London (1983).

Spradling, A. C. and Rubin, G. M., Drosophila genome organization: conserved and dynamic aspects. *Ann. Rev. Genet.*, **15**: 219 (1981).

Stumph, W. E., Baez, M., Lawson, G. M., Tsai, M.-J. and O'Malley, B. W., Chromatin structure of the ovalbumin gene domain. In *Gene Regulation* (UCLA Symposium on

Molecular and Cell Biology, Vol. xxvi), edited by B. W. O'Malley, C. F. Fox and S. Malone. Academic Press, London (1982).

Taylor, J. H., DNA methylation and cellular differentiation. *Cell Biol. Monographs*, 11: 1 (1984).

Taylor, S. M. and Jones, P. A., Multiple new phenotypes induced in 10T$\frac{1}{2}$ and 3T3 cells treated with 5-azacytidine. *Cell*, 17: 771 (1979).

Van der Ploeg, L. H. T. and Flavell, R. A., DNA methylation in the human $\gamma\delta\beta$-globin locus in erythroid and non-erythroid tissues. *Cell*, 19: 947 (1980).

Waalwijk, C. and Flavell, R. A., DNA methylation at a CCGG sequence in the large intron of the rabbit β-globin gene: tissue-specific variations. *Nucleic Acids Res.*, 5: 4631 (1978).

Weintraub, H., Larsen, A. and Groudine, M., α-Globin-gene switching during the development of chicken embryos: expression and chromosome structure. *Cell*, 24: 333 (1981).

Wood, W. G., Haemoglobin synthesis during human fetal development. *Br. Med. Bull.*, 32: 282 (1976).

Zuckerland, E., The evolution of haemoglobin. *Sci. Am.*, 212 (5): 110 (1965).

Zimmer, W. E. and Schwartz, R. J., Amplification of chicken actin genes during myogenesis. In *Gene Amplification*, edited by R. T. Schimke. Cold Spring Harbour Laboratory, New York (1982).

Chapter 12 Transcription and its control

We have seen in earlier chapters that before a gene can be transcribed, the chromatin in which it is contained must acquire a suitable state. This probably involves changes in the coiling and supercoiling of the DNA double helix, modifications in patterns of methylation of its contained cytosine residues, changes in the biochemistry of its associated histones and the acquisition of non-histone chromosomal proteins. These tissue-specific variations on the central theme carried in the germ-line genome are orchestrated by the DNA itself. On either side of the genes, within their own coding sequences and in the introns included within them, is the hidden 'score', which until very recently was read and interpreted only by molecules. Molecular biologists are now crossing the threshold of understanding into this new world of molecular dialogue, as they begin to decipher the intricate cues upon which the transcription of the genes depends. As always in such enquiries they are searching for unifying theories, but as usually happens these are hard to find. Even at the level of transcriptional control there seem to be several different kinds of instructions operating simultaneously.

12.1. RNA polymerase

Transcription of the DNA gene sequence into an RNA copy is carried out by enzymes called DNA-dependent RNA polymerases. These operate along stretches of DNA called transcriptional units, each of which probably corresponds in most cases to one structural gene plus introns, leader and trailer sequences (see Chapter 9). There are three known major species of RNA polymerase in eukaryotes. Polymerase I or A (Pol I) operates in the nucleolus and is concerned with synthesis of 18S, 5.8S and 28S ribosomal RNA (rRNA). Its activity requires magnesium (Mg^{2+}) ions and is particularly sensitive to the drug actinomycin D, but not to another transcription inhibitor called α-amanitin. RNA polymerase II or B (Pol II) operates in the nucleoplasm and it is this enzyme which is responsible for the synthesis of messenger RNA. This polymerase is activated by ammonium sulphate and manganese (Mn^{++}) ions, it is relatively resistant to actinomycin D, but particularly sensitive to α-amanitin. RNA polymerase III, or C (Pol III) is fairly resistant to both drugs, although high concentrations of actinomycin D will inhibit all three polymerases. Polymerase III is also present in the nucleoplasm and is concerned with the synthesis of small RNA species such as tRNA and 5S rRNA. A fourth RNA polymerase is responsible for transcribing the genes on the mitochondrial DNA. This is coded by a nuclear gene, it requires Mg^{2+} ions for its activity and is inhibited by manganese, but not by α-amanitin. The literature contains few reports on this enzyme.

Since the properties of the four polymerases are so clearly distinct, we can assume that they operate independently. The three nuclear RNA polymerases can be likened to three tired journalists representing newspapers with diverse interests, all listening to the same boring speech. The reporter from an economists' paper pricks up his ears at key phrases like 'inflation' and 'balance of payments', while the other two doze on. The one representing a paper dealing with defence comes to life in response to words like 'deterrent' and 'armed forces', while a third, who is interested in ecological issues, wakes up to words like 'environment' and 'pollution'. Each has heard the same speech, but has responded only to those remarks preceded by a key phrase that attracted his attention. Similarly, the three nuclear RNA polymerases can potentially transcribe any part of the genome, but actually come into operation only at regions that are signposted by specific key sequences with particular significance for each polymerase.

At the beginning of transcription, an RNA polymerase molecule attaches to specific base sequences (see below) at the 'upstream' end of a transcriptional unit, causing the double helix to unwind further and split into single strands (Figure 12.1). The enzyme then moves along the DNA, causing local unwinding and splitting, followed by reformation of the double helix as it proceeds. Starting at a defined point called the transcription initiator and using the DNA strand orientated in the 3'−5' direction, the codogenic strand, as a template, the polymerase molecule links ribonucleotides one by one to produce a complementary RNA sequence orientated with a reverse (5'−3') polarity. In other words it produces a copy of the 5'−3' DNA strand, using ribonucleic instead of deoxyribonucleic acid and substituting uridylic for thymidylic acid. An impression of the way in which the polymerase may split the DNA to expose the base sequences and assist the assembly of ribonucleotides into RNA is shown in Figure 12.1. When the enzyme reaches the transcription terminator at the 'downstream' end of the transcriptional unit, both it and the attached RNA are released.

Synthesis of RNA proceeds very rapidly, at 37°C about 30 ribonucleotides being added to each RNA chain per second. By the time the RNA strand is released, several other polymerase molecules may already be transcribing the same sequence, rather like a factory production line. In bacteria, translation of the RNA into polypeptides can begin even before the RNA is completely synthesized, but in eukaryotes there is an interval between to two events during which the RNA is processed and exported through the nuclear membrane into the cytoplasm (see Chapters 1 and 9).

It is perhaps worth restating at this point that RNA is translated in the 5'−3'

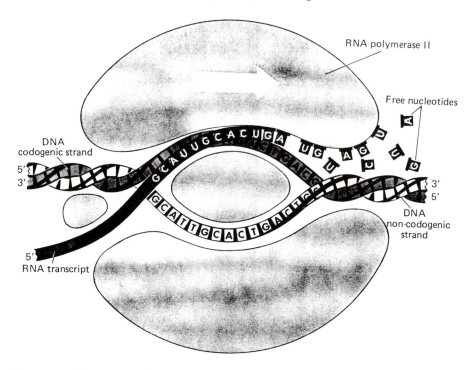

Figure 12.1. The process of transcription.
This diagram visualizes the molecular events which take place within a molecule of RNA polymerase. The DNA double helix is represented as being split open to expose the base sequence. An RNA copy is then produced by matching unbound ribonucleotides with exposed bases on one strand (the codogenic strand) and linking the free ribonucleotides together. These processes are assisted by a suitable distribution of charged groups in the polymerase. The polymerase molecule is indicated by pale shading and the direction of its movement by the white arrow.

direction, that is with the opposite molecular polarity compared to transcription but in the same sense with respect to the genetic information (see Chapter 8).

12.2. *The transcription punctuation code*

In Chapter 9 we introduced the concept of the other genetic code, which, instead of defining amino acids (Chapter 8), introduces punctuation marks, and stop and start signals into the relatively straightforward triplet-coded information contained in the base sequence of messenger RNA (Chapter 8). Perhaps the most important and intriguing aspects of this punctuation code are the signals contained in the DNA that control its transcription. As suggested in Chapter 11, we would expect to find such signals within or beside the structural genes and also some distance away from them. We might expect signals common to those sequences transcribed by each species of RNA polymerase and signals unique to each gene family, or each set of genes normally expressed in co-ordination in any cell type. Other signals would be expected to be unique to individual genes and we might expect variants of all of these due to the evolutionary divergence of species.

One approach to deciphering the punctuation code is to look for homologies in the sequences of nucleotides around the structural genes. Another approach is to prepare mutant versions of selected sections of DNA and to inject these into the nuclei of *Xenopus* oocytes where they may be transcribed. Examination of the RNA transcripts produced can then throw light on the importance of the particular section which was mutated or deleted.

A starting-point for these investigations was the knowledge that the bacterial promoter (see Chapter 1) to which the bacterial RNA polymerase becomes attached and which is indispensable for initiation of transcription *in bacteria*, contains the sequence 5′–TATAATG–3′. This is known as the Pribnow box and is situated about 10 bases 'upstream' from the RNA transcription initiation site. A second sequence, the recognition site, has also been noted in some prokaryote promoters, in a region centred about 35 bases upstream from the initiation site. This has the sequence 5′–TTGACA–3′ and is believed to act as an initial binding site for the polymerase.

Variants of these sequences are found upstream of most prokaryote structural genes. It should be noted that the sequences described are situated *in this form* in the *non*-codogenic strand of the DNA, that is the strand which is copied, not the one which acts as template.

12.3. *Transcription of ribosomal RNA by Pol I*

The ribosomes are the bodies in which the messenger RNA is translated into polypeptide (see Chapter 8). They are composed of proteins plus four RNA species known as 5S, 5.8S, 18S and 28S rRNA from their coefficients of sedimentation when centrifuged in dense media. The 18S, 5.8S and 28S structural genes are situated adjacent to one another, constituting a multi-gene family which is repeated very many times and in some species distributed about the genome. In man there are five groups of these sets, on chromosomes 13, 14, 15, 21 and 22, making a total of 200 copies per haploid genome. These are in what are known as the nucleolar organizer regions. In *Xenopus* there are 450 copies per haploid genome, but these are all located at one site. The 5S sequences are also highly repetitive, but are situated elsewhere in the genome (in man on chromosome 1) and are transcribed by a different enzyme, RNA polymerase III.

Each rRNA gene set constitutes a transcriptional unit and consists of a short spacer near the 3′ end which is transcribed, then the 18S sequence, another spacer, the 5.8S sequence, a third transcribed spacer, then the 28S sequence. This unit is repeated many times and the repeats are separated by spacers which are not transcribed. The primary RNA transcript illustrated in Figure 12.2 has the structure 18S–5.8S–28S. This long molecule is cut and trimmed to produce the separate 18S, 5.8S and 28S molecules which then become associated with the 5S transcript and a host of proteins to form the ribosome.

Little is known about the nucleotide sequences that specify the points on the DNA at which transcription by Pol I should start and stop, but there are indications that the sequence defined by nucleotides − 12 to + 16, with respect to the transcription initiation site, are particularly important in defining its position. It also seems to be a general rule for all the polymerases that transcription commences at a purine, that is an A or a G residue.

In bacteria termination of transcription occurs when the RNA polymerase traverses one of various inverted repeat sequences and then a run of T residues. The newly formed RNA transcript folds back on itself as a hairpin loop, due to self-complementarity (see 'foldback DNA', Chapter 11), and terminates in the poly-U sequence coded by the T residues. Termination of transcription by Pol I also seems to require the sequence of T residues, represented by U in the RNA.

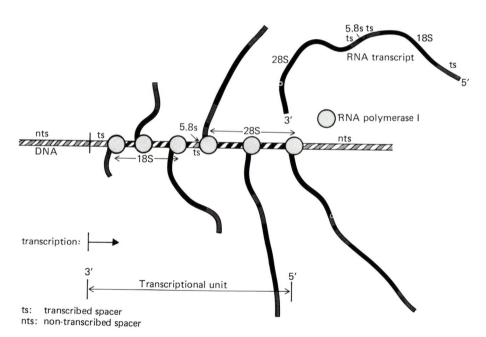

transcription: ⊢——▶

ts: transcribed spacer
nts: non-transcribed spacer

Figure 12.2. Transcription of ribosomal RNA by Pol I.
A molecule of RNA polymerase I attaches to the 3′ end of a transcriptional unit and transcribes a short spacer, then the 18S sequence, another spacer, the 5.8S sequence, another spacer, then the 28S sequence. Transcription terminates at a run of T residues. The RNA and polymerase are then released from the DNA. Several strands of RNA are being synthesized simultaneously. ts.

In fact the requirement for four or more adjacent T molecules in the transcription terminator seems to be a general one for all the RNA polymerases, although alone it is not a sufficient signal and there are other terminators (see below).

12.4. *Transcription of unique sequences by Pol II*

There is some evidence that *most* of the coding part of the genome is transcribed constitutively *at a very low rate* in most cell types, at least in the embryo (see Chapter 9). However, transcription of specific genes can proceed effectively only in relaxed chromatin, following changes in the proteins associated with the DNA in that region.

As they are synthesized, the RNA transcripts become complexed with protein and each unique sequence is given a 'cap' and a 'tail' as described in Chapter 9. These modifications are characteristics of this class of RNA, they are necessary for translation, they protect the mRNA from ribonuclease digestion and they may also be related to the functioning of RNA polymerase II as distinct from that of the other RNA polymerases.

Sequence homologies upstream of structural genes

In eukaryotes an analogous sequence to the prokaryote Pribnow box is present around -30 relative to the transcription initiation site. This is referred to as the TATA box or Hogness box and usually has the sequence $5'-TATAA_T^AA_T^A-3'$. This signal is present before every known mRNA coding gene transcribed by Pol II, but its precise position varies, in most cases the location of the first T being at -31 (± 2) relative to the mRNA transcription initiation site. The significance of this interval is that it corresponds to just three turns of the double helix. The Pribnow box in prokaryotes is only one turn before the initiation site.

The transcription initiation site itself is defined by an A residue surrounded by pyrimidines (Py, i.e., either C or T) in the general sequence Py---Py A Py Py Py Py Py (see Figure 12.3). This is sometimes known as the cAp site, since it represents the site in the RNA transcript which receives the methyl guanosine cap required later for ribosome attachment at translation. The cAp site is preceded by the pentanucleotide, TTGCT, in a variable position.

Within the region corresponding to the mRNA leader, at variable positions are four or more bases believed to define a sequence involved later in binding to the ribosome. The most common of these sequences is $5'-CTTC-G-3'$ (see Table 9.1).

Another sequence of homology in several genes, $5'-GC_T^CCAATCT-3'$, has been noticed at -70 to -80, relative to the cAp site. This is known as the CAAT box (see Figure 12.3).

(a) Mouse β-globin major controlling sequences

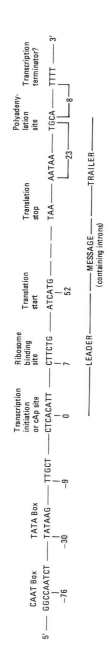

(b) *Psammechinus* **histone H2A controlling sequences**

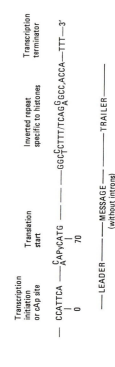

Control of transcription initiation by upstream sequences

One of the best-studied eukaryote genes is that coding for histone H2A in the sea urchin *Psammechinus miliaris*. The histones are unusual proteins in several respects: (a) they are very ancient in origin and very strongly conserved in structure (Chapter 1); (b) their structural genes are in repetitive multi-gene families (Chapter 11); (c) their RNA transcripts contain no introns and do not become polyadenylated (Chapter 9); (d) they are very intimately concerned with the functioning of eukaryote DNA (Chapter 10); (e) they are transcribed unusually early in development and (f) they are produced at stages of the cell cycle when most protein synthesis has stopped (see Chapter 8). For these reasons it is uncertain to what extent the molecular genetics of the histones can be considered to be typical of eukaryote proteins. Nevertheless, they have features in common with less unusual proteins and their genes are transcribed by Pol II. The primary transcript of the *Psammechinus* histone H2A gene carries a 70-base 5′ leader and a 52-base trailer. The cAp site and immediately adjacent sequences fall into the general patterns described above.

Since all eukaryotes have histones, each species has several classes of histone and the genes coding for the histones are repeated many times in each genome, it is possible to look for nucleotide sequence homologies between animal species, between the genes for the different types of histone in the same species and between the multiple gene copies coding for a single type of histone in a single species. In *Psammechinus* one of the most impressive homologies occurs in the latter category and constitutes a gene-specific upstream sequence present before all H2A genes and before no others. This is found between positions − 144 and − 173. It is rich in A and T, and contains the pentanucleotide ACAAT repeated many times, plus an inverted repeat with 10 bases in each half (see Figure 12.3(b)). By analogy with other inverted repeats, such as those that act as substrates for restriction endonucleases, it has been suggested that this site may be an attachment site for a regulatory protein.

Birnstiel and co-workers have elucidated the functions of this and other

---

Figure 12.3. Nucleotide sequences that control transcription by Pol II.
(a) The mouse β-globin major gene. The positions of significant bases are indicated relative to the transcription initiation (cAp) site. The positions of the CAAT and TATA boxes are probably critical. The point at which transcription terminates is not known for certain, but attention is drawn to the presence of four adjacent T residues downstream of the polyadenylation site. (Data from Konkel, Maizel and Leder, 1979.) (b) The *Psammechinus miliris* histone H2A gene. The CAAT and TATA boxes are at similar positions to those in β-globin, but there are additional features conserved in the multiple repeats of H2A (the upstream inverted repeat) and in histone genes in general (the inverted repeat in the trailer of the RNA transcript). Since histone mRNA is not polyadenylated, there is no polyadenylation site. In accordance with convention, the DNA strand is represented in which the nucleotide sequence resembles that in the RNA transcript (the non-codogenic strand). (Data from papers by Birnstiel *et al.*)

sequences in the region of the H2A gene by the *Xenopus* oocyte injection technique mentioned above. Deletion of the AT-rich sequence containing the ACAAT repeat enhanced the rate at which H2A messenger was produced. In other words, the presence of this ACAAT upstream sequence in this orientation *slows down* the transcription of H2A. Deletion of the larger spacer segment lying between positions − 110 and − 416 reduced transcription to nearly one twentieth of the original value, whereas inversion of the same segment increased transcription three- to four-fold. This region was given the name of modulator or enhancer.

In the previous section mention was made of a sequence of homology (the CAAT box) in several genes at positions − 70 to − 80. Deletion of this region had no appreciable effect on transcription of the histone H2A gene, but the same treatment given to the rabbit β-globin gene reduced its rate of transcription in cultured cells.

In bacterial systems deletion of the Pribnow box abolishes transcription. Surprisingly, deletion of the eukaryote homologue, the TATA box, from the *Psammechinus* H2A gene did not prevent transcription, but instead caused it to commence at the wrong place. Deletion of the equivalent sequence from the rabbit β-globin gene had a similar effect, while deletion of the DNA between this site and the cAp site caused transcription initiation to be displaced downstream. In eukaryotes the TATA motif therefore seems to assist in defining the location of the cAp site, or transcription initiator, and has consequently been named the selector.

Deletion of the cAp site itself also failed to abolish transcription, but instead caused it to begin at the wrong place. The definition of the correct site for commencement of transcription therefore seems to depend on both the TATA motif at − 30 and the initiator sequence itself. Deletion of the TATA box together with the CAAT box completely eliminated transcription of the rabbit β-globin gene.

Downstream sequences

As we saw above, transcription termination in bacteria occurs when a run of T residues follows an inverted repeat. In eukaryotes several signals may precede the T residues. In all histone genes examined so far, there is a 23-base conserved sequence at the tail end, which contains a 16-base hyphenated inverted repeat. This has the structure shown in Figure 12.3(b) and is followed by ACCA and then, in some histones, a run of T residues. This structure is therefore directly analogous to the bacterial terminator.

In non-histone genes, termination of transcription is signalled by AATAA, commencing 18–25 bases before GC, the latter corresponding to the polyadenylation site (Figure 12.3(a)). In non-histone genes there seems to be no requirement for an inverted repeat, but it is difficult to be sure of the exact termination point, since the primary transcripts are processed very rapidly. It is probably significant that runs of T residues are frequently reported just downstream of non-histone gene polyadenylation sites, although their actual implication in termination has not yet been demonstrated.

Histone mRNA is successfully transported through the nuclear membrane despite having no introns, although intron removal seems to be an essential feature of the transport of non-histone messengers (see Chapter 9). It is possible therefore that some other feature of the histone messengers could be involved in transport. Possibly the hairpin loop tail has this function instead of being involved in transcription. As we saw above, the TATA sequence has different meanings in prokaryotes and eukaryotes.

In addition to transcribing all the messenger RNA, Pol II also produces some of the small nuclear RNA (snRNA) molecules which are thought to be involved in the processing of mRNA (see Chapter 9).

12.5. *Transcription of repetitive sequences by Pol III*

The sequences transcribed by Pol III include those coding for transfer RNA, 5S ribosomal RNA and the remainder of the snRNA molecules thought to be involved in the processing of mRNA. Pol III is unusual in that initiation of its action is controlled by sequences within the gene itself. In the case of the 5S rRNA genes, this sequence binds regulatory proteins. This is reminiscent of the 'positive control' systems of bacteria mentioned in Chapter 1.

It has been suggested that the DNA coding for tRNA may adopt a 'side-stem-and-loop' conformation partially resembling its clover-leaf-shaped transcript (see Figure 12.4). This capacity depends on the existence of self-complementary sequences in these genes, which give the tRNA molecule its characteristic structure. Two of the sequences involved in establishing this structure are essential for transcription initiation of tRNA.

There are two types of 5S rRNA, one synthesized only in the oocyte, represented 20 000 times or more per haploid genome, and a somatic form coded by some 400 genes per haploid genome and expressed in both oocytes and somatic cells. Both sets of genes are regulated by a protein initiation factor called TF111A and possibly by an additional factor that discriminates between them. It is these factors which bind to the centre of the gene, 55–83 bases beyond the initiator site, to enable transcription to take place (Figure 12.4).

The 5S rRNA system provides one of the best understood examples of transcriptional control, and it may have a much broader relevance to other aspects of the eukaryote genome. This system operates through feedback inhibition, based on competition for TF111A, between the DNA non-codogenic sequence and the 5S RNA transcript itself. Production of 5S RNA inhibits further transcription because the transcript itself binds all surplus initiation factor so that transcription ceases. When the 5S RNA gets used up in ribosome formation, the initiation factor binds instead to the same sequences in the 5S gene so causing transcription to recommence. In addition to this control, the precise site of initiation of transcription

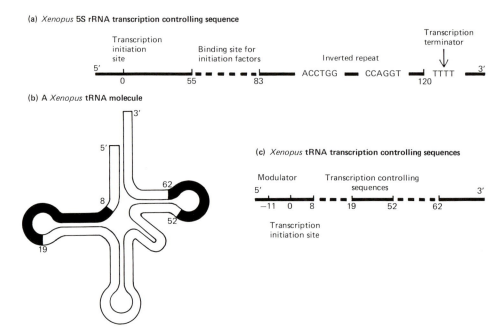

(a) *Xenopus* 5S rRNA transcription controlling sequence

(b) A *Xenopus* tRNA molecule

(c) *Xenopus* tRNA transcription controlling sequences

Figure 12.4. Nucleotide sequences that control transcription by Pol III.
(a) The *Xenopus* 5S rRNA gene sequences. Transcription is regulated by binding of an initiation factor to the region between bases 55 and 83. The function of the inverted repeat is not known. (b) A tRNA molecule from *Xenopus* showing the regions coded by transcription controlling sequences (in black). (c) The gene sequence corresponding to the tRNA molecule shown in (b). Transcription is dependent on the sequences in the regions 8–19 and 52–62.

of the 5S gene and the rate of synthesis of 5S RNA are further influenced by sequences at, or surrounding the transcription initiation site.

Transfer RNA is transcribed by the same class of RNA polymerase and a variety of different upstream sequences have been proposed as controlling the site of tRNA transcription initiation. One found in *Xenopus* is of particular interest as it consists of a 9-base alternating series of purines and pyrimidines with the potential for forming a left-handed helix. As pointed out in Chapter 11, in the context of cytosine methylation, transitions between left- and right-handed DNA helices (Figure 12.6) could constitute an important switch mechanism for transcription, as this change would radically modify the protein-binding properties of the DNA.

The termination signal for 5S rRNA synthesis in *Xenopus* also shows resemblance to the prokaryote signal in that it consists of a hyphenated inverted repeat, followed by a cluster of T residues in a region rich in G and C (Figure 12.4(a)). As in the other systems where termination is known to be regulated in this way, if the T cluster is reduced to below four T molecules in a row, RNA synthesis continues past this site.

12.6. *Transcription of the mitochondrial genome*

In Chapter 1 attention was drawn to the fact that the mitochondria show many features which suggest their origin in symbiotic aerobic bacteria that became incorporated into the cytoplasm of an anaerobic organism during the very earliest steps of eukaryote evolution. In keeping with this idea, the DNA of the mitochondrion exists as a closed loop without attached protein. In higher eukaryotes this loop has one transcription initiator site per strand of the double helix, to which mitochondrial RNA polymerase can attach. (In yeast mitochondria there are many initiator sites per strand.) Transcription therefore occurs along both strands, so that two giant RNA molecules can be produced, each containing a full-length copy of one strand of the DNA. These raw transcripts are, however, normally processed into separate mRNA, tRNA and rRNA species before completion of transcription. The mitochondrial transcripts are the only well-characterized truly polycistronic messenger RNA molecules in vertebrates (see Chapter 1).

12.7. *Co-ordination of gene expression*

To take up the analogy used in the introduction to this chapter, music makes sense only when notes follow or accompany one another in accordance with definable rules. Similarly, cell physiology is 'meaningful' only when certain combinations of enzymes are present in the cytoplasm at the same time, or succeed one another in a particular sequence. For physiological harmony, therefore, genes must be expressed with some degree of co-ordination. If several non-adjacent genes are flanked by similar regulatory sequences, they could be expressed co-ordinately in response to a single regulatory signal. For example, if the genes coding for several enzymes in a biochemical pathway were all controlled by a specific upstream sequence common to all the genes, they could all be transcribed in unison in response to the synthesis of several molecules of a single regulatory protein. There is evidence that such a system may exist with respect to the proteins produced by *Drosophila* in response to heat shock. When *Drosophila* larvae, or their excised salivary glands, are exposed to the unusually high environmental temperature of $37\,^\circ$C for half an hour or so, a characteristic pattern of puffs occurs on their polytene chromosomes as a recognized set of heat-shock protein genes is transcribed (see Chapters 4 and 9). The means by which this response is generated, through release of a subunit of a mitochondrial enzyme, is described in Chapter 4. All of these heat-shock protein genes are preceded by the inverted repeat $5'-CT-GAA--TTC-AG-3'$, which it is thought has a role in this co-ordination.

Co-ordinated systems of this sort are probably of great significance in higher organisms, and in 1969 Britten and Davidson put forward a model for the co-ordination of transcription based on this principle. This model was at one time very fashionable, but has since fallen somewhat into disfavour as it postulates transcriptional control exerted by RNA. The model, illustrated in Figure 12.5,

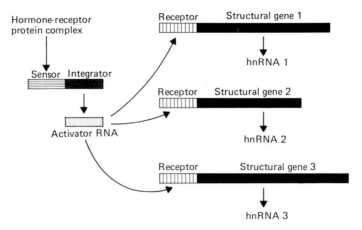

Figure 12.5. The Britten and Davidson (1969) model for co-ordinated control of transcription.

A hormone–receptor protein complex (or embryonic inducer molecule) interacts with a 'sensor' sequence linked to an 'integrator' gene. The latter codes for an 'activator' RNA molecule that interacts with receptor sequences linked to one or more structural genes. Binding of activator RNA to these common receptor sequences is considered to trigger transcription of the structural genes. Despite the general appeal of the model, there is little evidence that transcription can be controlled by RNA as the original concept required.

proposes the existence of sensor sites which are responsive to inductive or hormonal stimuli and linked to integrator sites. The latter it is suggested could code for activator RNA molecules produced as a result of the induction. The activator RNA is envisaged as interacting with a distant receptor site linked to the structural gene, which is transcribed as messenger RNA in response. Integration of several structural genes is achieved by the location of similar receptor sequences beside them.

There is as yet no evidence that transcription can be controlled directly by RNA, which would perhaps require the DNA to be split before specific DNA sequences could be recognized. The indications are that transcription is controlled by proteins, but the expression of different structural genes may well be co-ordinated by common adjacent sequences as Britten and Davidson suggested. (The same authors have since designed other models concerning control at RNA processing (see Chapter 9).)

Some function of the histone genes or their transcripts is presumably co-ordinated by the conserved sequence found in the trailers of their messengers, although whether this is concerned with transcription or transport is uncertain (see above). The Lewis model for the control of expression of the bithorax complex in insects is an adaptation of the same general principle of co-ordination by proximal sequences (see Chapter 5).

12.8. Torsion of the DNA double helix

In bacteria, an enzyme called DNA gyrase causes the DNA loop to become super-coiled — to resemble an elastic band twisted about itself like a string of figures of eight. This occurs in a negative sense relative to the twist in the DNA helix and the torsion produced causes the latter to unwind slightly, sometimes allowing entry of RNA polymerase. Another enzyme, topoisomerase I, has the opposite effect, relaxing the supercoil and tightening the helix. These two enzymes can therefore control transcription of some regions where RNA polymerase has difficulty in 'entering' the DNA. These enzymes are not known to occur in eukaryotes, but it is possible that eukaryote genes may also be opened up to RNA polymerase by slight untwisting of the DNA helix, through interaction of gene-switching proteins with upstream modulator sequences or by the formation of side loop structures as shown in Figure 12.6.

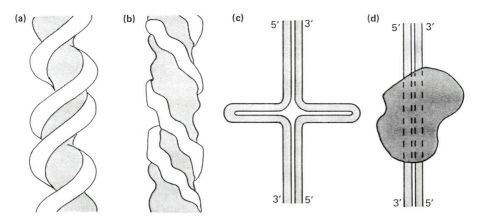

Figure 12.6. Structural modifications of DNA that could be related to gene expression. (a) Normal right-handed helical B-form DNA. (b) Left-handed helical Z-form DNA. This form acquires its name from the zig-zag form of the phosphodiester backbone. This form is promoted by methylated cytosine residues and has a different binding affinity for protein compared with the B-form. (c) Postulated side-loop structures which could form due to self-complementarity in each DNA strand at regions containing inverted repeat sequences. The two DNA strands are represented in an untwisted form for the sake of clarity. It is suggested that such structures, which occur within or beside structural genes, could act as transcription controlling sequences. More complicated structures could be formed in the case of the genes for tRNA. (d) This represents a transcription control signal conferred by binding of a protein to an inverted repeat or other unusual nucleotide sequence.

12.9. Summary and conclusions

A great deal still awaits discovery concerning the 'punctuation code' which directs the RNA polymerase molecules in their selective transcription of genetic information. Nevertheless, much has already been elucidated, and in a few years' time we will probably know enough of other species and other genes to see these findings in perspective. Something that emerges from this study is that each function, be it initiation, termination, or regulation of transcription, is generally controlled by more than one factor, or in more than one way at the level of each gene. The 'belt and braces' policy seems to be a general feature of development, another good example being lens induction in amphibia, in which several different stimuli contribute to the same effect (see Chapter 3). During evolution the various alternatives have become elaborated diversely by different species, so that generalization is a perpetual problem for the investigator and dramatic or well-described examples, which may be atypical, are the most frequently quoted. However, a rather unexpected feature which is now beginning to emerge is the possibility that the chromosomal DNA, instead of remaining as a static right-handed coil, may well on occasion unwind locally and even reform in a left-handed spiral. In places it may form stem-and-loop structures, sticking out like side coils in a partially untwisted rope. This would be most likely to occur in regions containing inverted repeats. Such regions seem to be recognizable by proteins and perhaps that is why. It is possible that contortions of the DNA could be facilitated by binding of protein or ribonucleoprotein (RNP) particles to specific sites on the DNA. Alternatively, the bound protein could itself produce a three-dimensional structure that makes the chromosomes recognizable to RNA polymerases or other proteins relevant to gene expression (Figure 12.6).

Among the sequences that control transcription, some apparently define which type of RNA polymerase will operate in that region. Pol I copes with most of the ribosomal RNA and Pol III with the remainder, plus transfer RNA and some of the small nuclear RNA species. The rest of the small nuclear species and messenger RNA are transcribed by Pol II. Other gene-flanking sequences may define a group of genes within the polymerase class, like the characteristic trailer sequence of histone genes. There are also gene-specific sequences, such as those upstream of each copy of the histone H2A gene. Other sequences define the points at which transcription should start and stop and the rate at which it should proceed, while others probably have no function at this stage, but come into operation during processing of the RNA transcript (Chapter 9) or translation (Chapter 8).

DNA sequencing alone tells us little about the way transcription is actually controlled. The operon concept (see Chapter 1) provides a useful start for model building, but although operons have similarities with the eukaryote systems there are also major differences. In bacteria, the binding of a regulatory protein to a specific DNA sequence is controlled by its allosteric response to an inducer molecule, co-repressor or other ligand which may come from inside or from outside the organism. An analogous event may occur in the activation of genes by steroid hormones (Chapter 4), following modification of proteins in the responses to hydrophilic hormones, or in the heat-shock response (see above and Chapter 4). In all these systems we still lack molecular details. The 5S rRNA system is perhaps the best described at a molecular level, and in this the RNA product competes directly with its DNA non-codogenic strand for the transcription initiation factor. This is quite a different kind of control compared with that normally seen in bacteria.

The nucleotide sequences that control transcription also show common features in prokaryotes and eukaryotes, but they are not necessarily utilized in the same way. The best example is TATA, which in prokaryotes is concerned with ensuring that transcription is initiated at all, whereas in eukaryotes it defines the site at which transcription commences.

With the 5S rRNA and tRNA systems, deletion of parts of the structural gene abolishes transcription of the remainder. This may turn out to be characteristic of genes transcribed by Pol III, but it is not shown by the more interesting class of mRNA transcribed by Pol II. We therefore still await a good illustration of the regulation of transcription of messenger RNA in higher eukaryotes.

Bibliography

Alberts, B., Bray, D., Lewis, J., Raff, M., Roberts, K. and Watson, J., *Molecular Biology of the Cell*. Garland, New York (1983).

Birchmeier, C., Grosschedl, R. and Birnstiel, M. L., Generation of authentic 3′ termini of an H2A mRNA *in vivo* is dependent on a short inverted DNA repeat and on spacer sequences. *Cell*, **28**: 739 (1982).

Birnstiel, M. L., Developmental control of histone gene expression. In *Progress in Clinical and Biological Research*, Vol. 85A. *Embryonic Development, Part A: Genetic Aspects*. Alan R. Liss, New York, pp. 1–12 (1982).

Breathnach, R. and Chambon, P., Organization and expression of eucaryotic split genes coding for proteins. *Ann. Rev. Biochem.*, **50**: 349 (1981).

Britten, R. J. and Davidson, E. H., Gene regulation for higher cells: a theory. *Science*, **165**: 349 (1969).

Brown, D. D., Gene expression in eukaryotes. *Science*, **211**: 667 (1981).

Busslinger, M., Portmann, R., Irminger, J. C. and Birnstiel, M. L., Ubiquitous and gene-specific regulatory 5′ sequences in a sea urchin histone DNA clone coding for histone protein variants. *Nucleic Acids Res.*, **8**: 957 (1980).

Cold Spring Harbour Symposia on Quantitative Biology, *Structures of DNA*. Cold Spring Harbour Laboratory, New York (1983).

Gluzman, Y., *Enhancers and Eukaryotic Gene Expression*. Cold Spring Harbour Laboratory, New York (1983).

Grosschedl, R. and Birnstiel, M. L., Identification of regulatory sequences in the prelude sequences of an H2A histone gene by the study of specific deletion mutants *in vivo*. *Proc. Natl. Acad. Sci. USA*, **77**: 1432 (1980).

Grosveld, F., Busslinger, M., Grosveld, G., Groffen, J., De Klein, A. and Flavell, R. A., The structure and expression of the haemoglobin genes. In *Stability and Switching in Cellular Differentiation*, edited by R. M. Clayton and D. E. S. Truman. Plenum, New York (1982).

Hagenbüchle, O., Sauter, M., Steitz, J. A. and Mans, R. J., Conservation of the primary structure at the 3′ end of 18s rRNA from eucaryotic cells. *Cell*, **13**: 551 (1978).

Hall, B. D., Clarkson, S. G. and Tocchini-Valentini, G., Transcription initiation of eucaryotic transfer RNA genes. *Cell*, **29**: 3 (1982).

Konkel, D. A., Maizel, J. V. Jr and Leder, P., The evolution and sequence comparison of two recently diverged mouse chromosomal β-globin genes. *Cell*, **18**: 865 (1979).

Korn, L. J., Transcription of *Xenopus* 5S ribosomal RNA genes. *Nature*, **295**: 101 (1982).

Levens, D., Luetig, A. and Robinowitz, M., Purification of mitochondrial RNA polymerase from *Saccharomyces cerevisiae. J. Biol. Chem.*, **256**: 1474 (1981).

Lewin, B., *Gene Expression Vol. 2, Eucaryotic Chromosomes*, 2nd edn. John Wiley, Chichester (1980).

Lewin, B., *Genes.* John Wiley, Chichester (1983).

Losick, R. and Chamberlain, M., *RNA Polymerase.* Cold Spring Harbour Press, Cold Spring Harbour, New York (1976).

Mainwaring, W. I. P., Parish, J. H., Pickering, J. D. and Mann, N. H., *Nucleic Acid Biochemistry and Molecular Biology.* Blackwell, Oxford (1982).

Pelham, H. R. B., A regulatory upstream promoter element in the *Drosophila* Hsp70 heat-shock gene. *Cell*, **30** (2): 517 (1982).

Schlesinger, M., Ashburner, M. and Tissières, A., *Heat Shock from Bacteria to Man.* Cold Spring Harbour Laboratories, New York (1982).

Smith, G. R., DNA supercoiling: another level for regulating gene expression. *Cell*, **24**: 599 (1981).

Chapter 13 Growth and morphogenesis

Animals increase in size by several means, but growth is largely ascribable to increase in cell numbers brought about by mitotic proliferation. Most growth regulation therefore depends on the control of mitosis by tissue interactions or on the effects of hormones secreted by distant endocrine glands. In addition, nerves penetrate most tissues and combine growth regulation with nervous integration (see Chapter 4), while neurosecretory cells act as communication links between the neural and endocrine systems. Although not entirely true, it is generally assumed that cells withdraw from the mitotic cycle when they differentiate, while some inductive influences operate only on proliferating cell populations. Thus there is an interrelation between growth and cytodifferentiation, such that control of one affects the other. In this chapter, however, we are not concerned with mitosis and its control, or its relationship to differentiation of cell types, but with the relevance of growth to morphogenesis.

Morphogenesis refers to the development of the shape of an organism or that of its parts, and it involves four main processes: directed cell movement, differential growth, localized cell death and the generation and application of mechanical forces. Many features of morphogenesis originate at the cellular level.

13.1. Cell shape and movement

The shape of individual cells depends very largely on their environment and their state of mobility. A cell's environment has three aspects, a substratum beneath it, a liquid medium around it and neighbouring cells beside it. The same cell may adopt a variety of shapes and have different capacities for movement depending on its environment, but cells of the different tissues tend to have characteristic forms. Cells packed tightly together may appear polygonal, while sedentary cells in loose connective tissue are frequently star-shaped with many processes. In contrast, mobile fibroblasts are elongated, their shape alternating between bipolar and pear-like as they move along. Changes in cell shape occur by streaming of cytoplasm and the operation of microtubules and microfilaments, which push and pull the cell from within. In an epithelium in contact with basement membrane this can result in the outward or inward curvature of the whole sheet in one or

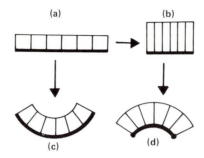

Figure 13.1. Effects of cell adhesion on epithelial morphology.
If the adhesions between cuboidal epithelial cells (a) increase in strength they can either become columnar if their basement membrane can fold or shrink (b) or bend convexly (c) if the membrane remains of fixed length. If the adhesions between cells of columnar epithelium (b) weaken while the basement membrane remains fixed at its ends then the epithelium becomes concave. (From Gustafson and Wolpert, 1967.)

two dimensions (see Figure 13.1). The neural tube, optic cup and lens cup are probably created by this means (see Chapter 3).

Independent cell movement requires first a loosening of inter-cellular attachment or breakdown of other physical barriers. Channels are required for cellular migration and cells also need a capacity to recognize their destination and to respond appropriately on reaching it. Other stimuli to movement may include changes in the concentrations of diffusible molecules or provision of new substrata. In some cases swelling of extra-cellular material creates passages for invasion of migratory cells, like those from the neural crest.

Cells will align beside or migrate along physical features like fibrils of collagen. This is called contact guidance and probably explains colonization by nerve cells along routes taken previously by blood vessels. Cells will also move from regions of low adhesivity to those of high and by chemotaxis up or down the concentration gradient of a chemical. Pigment cells in the skin modulate their tendencies to repel or attract one another, producing spotted and striped patterns. A study of cell behaviour during the formation of the extravagant colour patterns of tropical fishes would probably reveal a wealth of picturesque information on cell dynamics.

The classic and most dramatic example of positive chemotaxis occurs during aggregation of slime mould amoebae. This occurs at times of food shortage, presaging sexual reproduction, and involves the conversion of a community of independent unicellular animals into a multi-cellular colony, which behaves with the coherent attributes of a single creature. The stimulus to aggregation is an attractant called acrasin, which in *Dictyostelium discoideum* has been identified as cyclic AMP (see Chapter 4), although other species use different, so far unidentified, biochemicals. The formation of aggregates is initiated by cells that spontaneously emit cyclic AMP and this draws surrounding cells towards them. Responding cells also secrete cyclic AMP, which amplifies the signal and attracts still more cells. This physiological mimicry is reminiscent of the spread of pigmentation from leader cells in transdifferentiating neural retina cultures (see Chapter 7), but in the case of the slime moulds it is combined also with directed cell movement.

An isolated fibroblast moves by active ruffling of the anterior end as it seeks out regions of attachment. It then hauls its rear part forward and the process is repeated. If it contacts another cell the anterior ends of both immediately cease ruffling and this is known as contact inhibition of movement. After a while, another part of each cell becomes active, this becomes the new fore-end and the two cells move apart. This behaviour accounts for the tendency of migratory cells to occupy every space available to them and is responsible, for example, for the formation of dermatoglyphic patterns on the palms of the hands and soles of the feet of primates (see Chapter 14). There may be some inter-cellular specificity in contact inhibition since some migratory cells can apparently migrate over or among other cell types. The cranium of vertebrates forms from neural crest cells that invade and disperse between brain and epidermis before consolidating as bone.

Cells multiply as they migrate, but cease to divide when densely packed. This is called density dependent inhibition or contact inhibition of growth. In some regions, such as at the sites where bones form from mesenchyme in the limb, initially uniformly distributed cells become densely aggregated as part of the differentiation process (see Chapter 6).

13.2. *Cell-surface effects*

Cell reaggregation

An important factor in morphogenesis is the mutual but selective adhesivity of cell surfaces. If we take a tissue from an animal, disaggregate the cells, then cause them to reaggregate by swirling in a gyratory shaker, the cells will often rearrange themselves so as to reconstitute the structure of the original tissue. Cells isolated from two different tissues and aggregated together will sort themselves out according to the type of tissue from which they came, and if mixtures of amphibian ectoderm, mesoderm and endoderm are so treated the cells not only separate into types, but also tend to readopt the relative positions they originally held in the embryo (Figure 13.2). One interpretation of this behaviour is that each type of cell has a characteristic degree of cohesiveness, and the cells of the mixed population rearrange themselves to minimize the total cohesive free energy. They therefore adopt the most stable configurations possible, with the less cohesive enveloping the more cohesive.

It is difficult to find instances in which this sorting capacity is exploited as a major feature of development, although it must surely be relevant to the establishment of discrete boundaries between contiguous tissue masses. One of the

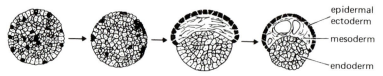

Figure 13.2. Reaggregation pattern of embryonic cells.
Epidermal ectoderm, mesoderm and endoderm tissues were taken from an amphibian neurula, disaggregated by trypsinization and reaggregated by swirling in a gyratory shaker. Following reaggregation the cell types move relative to one another so as to reacquire their original relative positions (see Figure 2.8). (After Waddington 1966.)

most interesting applications is shown by the so-called 'annual fishes' such as *Australofundulus myersi*, which live in water-holes and swamps that dry up seasonally. If conditions become adverse following cleavage, the cells separate within the egg case and this avoids the drastic disruptive effects of dehydration on gastrulation and neurulation. Later, when conditions improve, they reaggregate and normal embryonic processes resume.

The *T* locus

In the mouse, a series of mutations at the *T* locus affects cell-surface determinants of crucial import during early embryogenesis. The *T* allele is semidominant and lethal in homozygotes (*T/T*), the posterior half of the embryo failing to develop, so that the animal dies and is aborted at 11 days' gestation. In normal (+ / +) embryos the tail undergoes a growth spurt at 11 days. In *T*/ + heterozygotes a constriction appears half-way along the tail at this stage and distal tissues atrophy, producing a short-tailed, but otherwise normal mouse. In heterozygotes carrying the *T* allele plus another recessive at the same locus (*T/t*), the constriction develops at the tail base and the whole tail is lost.

At least six different recessive mutants are known and in the homozygous state most have lethal effects, although in heterozygous combinations complementation occurs and viable offspring are produced. The *T* allele acts relatively late in ontogeny, but the others come into operation at various earlier stages and embryos are aborted over the range 6–10 days. All the alleles affect axial organization, differentiation of ectoderm and inductive relations between ectoderm and notochord.

The *T* series is also expressed on the surface of sperm causing a most curious breach of Mendel's law. All alleles are transmitted in equal frequency by female heterozygotes, but sperm carrying dominant *T* alleles are much more successful at fertilization. Thus in some crosses 99% of the offspring of male *T*/ + heterozygotes are of mutant type. This effect is known as meiotic drive and is quite exceptional.

Specificity of neural connections

The most highly specific of all cell-surface interactions are those involved in the interlinking of peripheral sense organs with the central nervous system. For example, each of the optic tecta of the brain contains an array of cells that in their distribution precisely match the sensory cells (rods and cones) of the retina. Communication is made between the two organs by growth of neurones back from the retina and into the exact complementary area of the optic tectum on the opposite side. Contacts are made with such precision that each tiny group of light-sensitive retinal cells is linked specifically to its exact counterpart in the brain, enabling the visual field to be interpreted as a coherent whole, rather than like the scrambled pieces of a jigsaw puzzle.

Contact relations between nerve cells are believed to be governed by matching of similar or complementary molecules on their surfaces, and a great deal of ingenious experimentation has been directed at finding out how this is achieved. In some parts of the nervous system more neurones are initially generated than are found in the mature animal. Selective destruction of neurones could well be one means by which the specificity of cell matching is tidied up, only appropriate synaptic contacts being allowed to survive. In the retina large-scale cell death follows differentiation of the rods and cones.

It has been suggested that the basis of specificity of neuronal connections could depend on matching sets of gradients of cellular activity or biochemicals within optic tectum and retina, but their initial connection seems a somewhat haphazard affair. A lot of shuffling about of nerve endings occurs in the early stages, suggesting that cells are not uniquely labelled, but able to assess their relative positions and to break and re-make connections. In most organisms the fine tuning of visual acuity probably requires actual use of the eyes, although the domestic chicken, for example, can apparently see very well immediately on hatching.

13.3. The generation of mechanical forces

Apart from cleavage, the first major morphogenetic event in sexually reproducing eukaryotes is gastrulation. This involves conversion of a single-walled sphere, the blastula, into a double-walled sack, the gastrula (see Chapter 2). The details of gastrulation in insects, sea urchins, frogs and birds are well described and are all quite different, although the end result is essentially similar. In sea urchins the cells of the vegetal pole begin to pulsate, without detaching from one another, to produce a hemispherical inpushing. They then throw out long extensions, filopodia, across the central cavity. These have sticky ends, like miniature grappling hooks, that make contact with the junctions between ectodermal cells at the animal pole. The extensions then contract, pulling the base of the embryo upwards. At the point of contact between the walls a hole appears, penetrating both walls, and this later becomes the mouth, while the central tube so formed (the archenteron) becomes the future gut (see Figure 13.3). In this case then, the mechanical activities of a few individual cells cause a gross change in shape of the whole structure.

Glycosaminoglycans in the extra-cellular matrices around cells can absorb water and swell to occupy space 10 000 times that of the molecules themselves. Such swelling exerts considerable pressure on surrounding tissues and can cause general expansion of organs, or changes in their shape. This is the mechanism which causes the primitive brain and eyeballs to swell.

Fluid pressure can be generated in other ways, for example by the differential contraction of fluid-filled tubes. Such means seem to be employed in the distension of insect wings, for example. Evidence derived from surgically restructured blood systems suggests that the blood flow itself modifies the shapes of the vessels that create and contain it (see Chapter 15).

The development of insect wings illustrates the role of genes in the establishment and exploitation of hydrostatic forces, but in order to understand how they operate we need first to consider normal wing development.

Drosophila wings develop from small hollow bags called imaginal discs. As we saw in Chapters 5 and 7, stimulation of their development and differentiation depends on an increased ratio of ecdysone to juvenile hormone, which precipitates pupation. One side of the wing disc is thicker than the other and during pupation

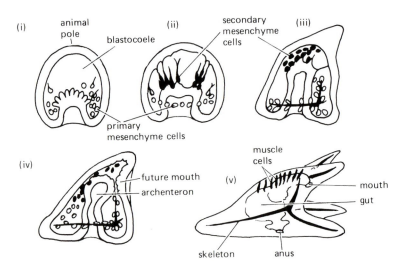

Figure 13.3. Gastrulation in the sea urchin.
Secondary mesenchyme cells (coloured black) throw out filopodia that adhere to the region of the animal pole and then contract, pulling the vegetal pole cells towards the future mouth region and creating the future gut. (From Gustafson and Wolpert, 1967.)

this thicker side folds in towards the centre of the bag. The thick wall grows rapidly and bursts through the thinner side, which then degenerates. The folded area of the thick wall then flattens into a small double-walled blade, which grows at different rates along its length and across its width, while its base becomes attached to the outer surface of the body. The wing blade is then inflated by body fluid, then the sides collapse together and the wing contracts as the fluid drains away. The two surfaces make close contact except for a few channels which form the 'veins' of the adult wing. Each cell then increases in area and the wing expands, but in doing so within the confined space of the puparium becomes crumpled into folds. When the fly emerges body fluid is again pumped into the veins which expands the wing to its full size, rather like an inflated-rib poleless tent. The wing finally dries out, its cells die, and it attains its final functional form.

Patterns of growth have been discerned by establishment of genetically marked cell clones produced by somatic chromosome cross-over following X-irradiation (see Figure 5.6). These suggest that proliferation is directed mainly along the axis of the wing, possibly by preferred orientation of cleavage spindles, as occurs in the early development of spirally cleaving embryos (see Chapter 5).

The mutant genes known to affect wing development act in a variety of ways. Some control growth in length, others control its width, some control inflation,

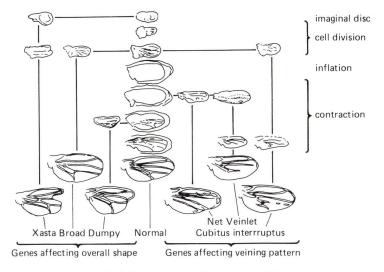

Figure 13.4. Genetic control of development of the *Drosophila* wing.
Stages in normal development are shown in the centre column. On the left are shown the effects of three mutant alleles that affect the general shape of the wing. *Xasta* affects the very earliest stages in the imaginal disc. *Broad* affects the first expansion by cell division. *Dumpy* causes increased contraction. On the right are three mutant alleles affecting the veins. In *net*, contraction is reduced and extra veins appear. In *veinlet* the tips of the veins are obliterated. In *Cubitus interruptus* there is a fault in the first appearance of veins during the growth phase. (Redrawn from Waddington, 1966.)

others contraction. The final product is an organ of defined shape, size and vein pattern. Each of the developmental stages, up to that of the complete wing, can be upset by the actions of mutant alleles, and the stages at which these come into play can be discovered by following the course of wing development in mutant stocks. The times of action of some are shown in Figure 13.4, together with the phenotypes they produce. A particularly important observation is that the (wild type) controlling genes act antagonistically to one another, so that the form adopted at any stage represents a balance between opposing influences. Thus if you understand the events that led to its formation, a mature wing can be seen as a record of the shifting states of balance between the different opposing forces that contributed to its formation.

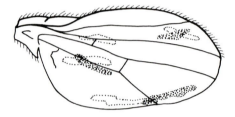

Figure 13.5. Cell lineages in *Drosophila* wing development.
Irradiation of embryos or larvae heterozygous for a recessive visible mutation, such as yellow body colour, can cause a cell to undergo chromosome rearrangement, so that one of its daughter cells becomes homozygous for the recessive allele (see Figure 5.6). After growth and metamorphosis the clone of homozygous cells produces a visible patch of distinctive phenotype that indicates the direction in which growth has occurred by cell proliferation. The diagram combines observations made on several individuals. (Based on Garcia-Bellido, Lawrence and Morata, 1979.)

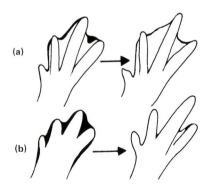

Figure 13.6. Programmed cell death in the separation of the digits.
Areas of high incidence of cell death in the embryo and the later foot forms in a duck (a) and a fowl (b). (From Newth, 1970, after J. W. Saunders.)

13.4. Programmed cell death

Detailed moulding of the body frequently results from localized patterns of cell destruction, arising from what is known as programmed cell death, the cells dying as if that is their differentiative fate. This occurs for example during the reduction of the tails of frog tadpoles in response to thyroxine. In the inter-digital regions of the hands and feet the mesenchymal cells also become committed to die, so separating the digits (Figure 13.6). In evolutionary terms the latter is a relatively new invention, since amphibians achieve the same result by elongation of the digits themselves. The split between radius and ulna also occurs by localized cell death.

13.5. Differential growth and timing

At the level of gross phenotype, the two most important processes underlying evolutionary change in body form are heterochrony and allometric growth. Heterochrony refers to changes in the timing of developmental events and allometry deals with the relative growth of parts, but the two concepts are interrelated.

Heterochronic change

Heterochronic change can involve displacement of onset signals for growth, displacement of offset signals or changes in growth rates. These can produce minor or major changes, but the most radical of all involves a shift in the relative rates of somatic and gonadal maturation. The terms used in this subject area have been applied in a somewhat idiosyncratic fashion by different authors, but one useful terminology is that adopted by Gould, who uses the term paedomorphosis to describe the retention of ancestral juvenile features at later developmental stages in descendants. Neoteny is the term for paedomorphosis due to retardation of somatic development, whereas paedomorphosis due to precocious sexual maturation is called progenesis.

The classic examples of paedomorphic species are members of the genus *Ambystoma*, or *Amblystoma*, the axolotls. These were unknown to European naturalists until the invasion of Mexico by Cortes and the Spanish conquistadores, but have been popular objects of study ever since. The axolotls are urodele amphibians related to newts, some of which reach sexual maturity without changing out of the larval form. The sexually mature larvae actually resemble somewhat streamlined frog tadpoles (see Figure 13.7). The scientific world was astonished to find that some species of axolotl will however transform into terrestrial adults if treated with the hormone thyroxine, increase of which relative to prolactin is normally required for amphibian metamorphosis (see Chapter 4). When this was done an archaic life-form, the adult *Ambystoma*, was recreated to be seen for the

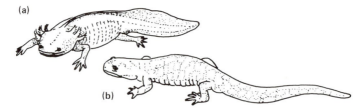

Figure 13.7. The salamander *Ambystoma*.
Many members of this genus attain sexual maturity while still in the larval form (a). On exposure to thyroxine or iodine some of these larval species metamorphose into a terrestrial adult form (b). (Redrawn from Young, 1958.)

first time by man. Apparently paedomorphosis occurred as a measure that counteracts either the failure of synthesis of thyroxine, or the failure of response to it, or the over production of prolactin, which may occur at the lower temperatures encountered at high altitude. This therefore seems to be an example of progenetic adaptation to enforced retardation of overall development. In some species of *Ambystoma* the paedomorphic phenotype has become completely assimilated by the genome, as these animals fail to metamorphose even when iodine or thyroxine are supplied. This form has probably lost the thyroxine receptors from its cell membranes (see Chapter 4).

Evolutionary implications of paedomorphosis

Paedomorphosis is one of the most important means by which new species are created, since one of the major obstacles to evolutionary progress is specialization. Species which have followed a particular way of life for a very long time become so well adapted phenotypically and genotypically that it becomes almost impossible to change should that ecological niche become closed to them. Paedomorphosis provides a way around this impasse as it allows sexual reproduction to occur among less-specialized juvenile forms. Some of the giant steps in evolution are thought to have resulted from this type of situation. For example, the insects with nearly a million species are the most successful invertebrates, and probably arose from a paedomorphic larval myriapod (a kind of millipede). When newly hatched these have only three pairs of walking legs and short abdomens like insects (see Figure 13.8). The vertebrates, the most successful of all animals from the point of view of their colonization of land, sea and air, as well as their general level of organization, are thought to have arisen from larvae of the invertebrate ascidians, the humble sea squirts (Figure 13.9). Many authorities believe that retardation of somatic development in both myriapod and ascidian larvae allowed gonadal maturation to occur before metamorphosis to adult forms, with the creation of new breeding groups. Such individuals would not only be relatively unspecialized and therefore capable of divergent specialization, but would also be

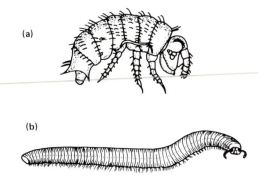

Figure 13.8. Larval and adult millipedes (Myriapoda).
(a) A newly emerged larva of the millipede *Strongylosoma*. It has a short, segmented body with only three pairs of legs, closely resembling the insects. (b) An adult millipede, *Iulus terrestris*. The adult has very many segments. (Not to scale. Redrawn from Parker and Haswell, 1949.)

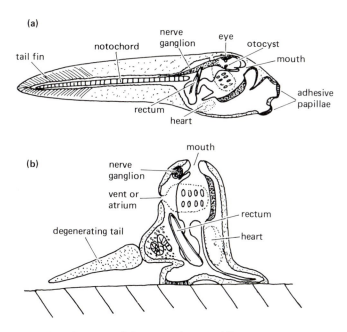

Figure 13.9. Larval and young adult sea squirts (Ascidia).
The larva when newly emerged (a) resembles an amphibian tadpole, with a long, finned tail containing a notochord and a dorsal nerve cord that expands anteriorly into a sensory vesicle carrying an eye and a balancing organ (otocyst). It also has a gut and a heart. A few hours later it begins metamorphosis into the adult (b) following adhesion to a solid substratum. Metamorphosis involves loss or modification of many of these structures so that the adult bears little resemblance to a vertebrate. (After Parker and Haswell, 1949.)

equipped with masses of redundant genetic information to provide abundant raw material for new evolutionary advance.

The origin of *Homo sapiens*

A strong argument has been put forward in support of the idea that among the primates man is a paedomorphic species. Human adults show striking features of resemblance to young or foetal apes. Among the most obvious are our lack of body hair, brow ridges and cranial crests, our large brain cases with late-closing sutures, our flat faces, the relative proportions of our limbs and trunk, the structure of our feet and the forward orientation of the vagina. Furthermore, our babies take 21 months from conception to reach the average state of maturity of the great apes at birth, gestation in the great apes being on average three weeks shorter than in man. In fact, compared with the apes the whole of human development is slowed down. Our milk teeth erupt 6–24 months after birth, whereas those of apes appear at 3–13. Our permanent teeth begin to emerge at six years, those of apes appear at half that age. Our growth period is 20 years, in apes it is 11, and whereas we can look forward to a lifespan of 70 years, an ape can expect only 30–35 (see Figure 13.10).

This retardation of physical development has a profound bearing on our physical and psycho-social development. During the first year after birth we are essentially extra-uterine foetuses, entirely dependent on maternal care, and during this time the foundations of some of our most basic attributes are laid. The skeletal ossification that takes place during the latter part of pregnancy in apes, when the foetus is curled up inside its mother's womb, occurs after birth in

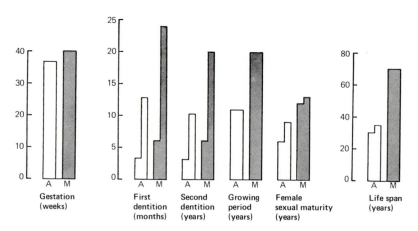

Figure 13.10. Comparative rates of development of men and apes.
Average values for the orangutan, chimpanzee and gorilla are represented (A) beside those of man (M). All these measures except period of gestation are considerably extended in humans compared to the great apes. (Data from Gould, 1977.)

human babies who can stretch out freely in their cots. The result is that our spines, instead of being bowed, are recurved and our pelvises become constructed with the acetabula directed caudally rather than ventrally, allowing an upright stance. During this first year we also receive sensory and mental stimulation crucial to our later development. Babies deprived of human contact during their first year fail to develop simple social skills and remain for ever mentally backward. The slowing down of somatic maturation also means that parental influence is very prolonged compared with that of all other species, so we have many years of childhood at our disposal during which we can acquire the multitude of capabilities of human adults.

Gonadal maturation is also retarded in humans, but not to such an extent as that of the rest of the body. In apes, female fertility is acquired at about six to nine years, in humans at 12 to 13.

The reason for the difference in maturity at birth of man compared with that of his nearest relatives is related to the size of the human brain. In man the onset signal for brain growth occurs earlier relative to general body growth, so the brain becomes disproportionately large. Despite the compensatory evolutionary broadening of women's hips, after 40 weeks' gestation the baby's head is so big it will only just pass through the hoop of its mother's hip girdle. So the baby is born at 40 weeks even though, by primate standards, it is still very immature. Retardation of general somatic maturation, relative to the onset signal for brain growth, therefore seems to have been the original trigger which set off the evolution of *Homo sapiens*, the species we fondly consider the crown of creation.

Natural selection and late-acting genes

From the point of view of the transmission of hereditary traits, the forces of natural selection apply only to those characters that appear before reproduction takes place. Regression of sexual maturity to earlier developmental stages or retardation of somatic maturation relative to that of the gonads puts genes that are expressed in older adults beyond the pale of selective control. Thus in our own species the allele that causes the extremely disabling nervous disorder called Huntington's chorea is not naturally eliminated from the population, even though it is inherited as a Mendelian dominant, because it acts at around 40 years of age, by which time many carriers have passed the gene to their children.

Allometric growth

During ontogenetic development the shapes and relative sizes of organs and body parts gradually become changed as the animal increases in overall size. During the evolution of species heterochronic changes in growth patterns also bring about changes in the shapes and relative sizes of body parts. The type of analysis, known as allometric analysis, rests on the observation that the definition of one shape can

be mathematically transformed into that of a different but topologically equivalent shape, by distortion of certain co-ordinates. This permits comparison of the
shapes of an organ at different stages of ontogeny, or of homologous organs born
by different individuals or species. In many cases the relationship between the
growth of a part and that of the whole approximates to the simple expression:

$$y = bx^a$$

in which x is the weight or volume of the whole organism and y that of some part
of it at any chosen stage; b and a are constants. A plot of y against x usually
produces a curve, but this can be adjusted to a straight-line relationship if we
convert the values into logarithms:

$$\log y = \log b + a \log x$$

The values of log b and a can then be obtained as the intercept and slope of the
plot (see Figure 13.12).

When the organ grows faster then the body as a whole it gets disproportionately
larger and larger. It is said to show positive allometry, a is greater than unity and
the slope of the logarithmic plot is greater than $45°$. If as growth proceeds the
organ becomes progressively smaller compared to the whole animal it shows
negative allometry, a is less than unity and the slope of the line is less than $45°$.
Growth that takes place according to this rule is called allometric growth. A slope
of $45°$ represents isometric growth and a = 1.

During human development the legs become progressively longer in proportion
to total body length, while although the head continues to grow into adulthood,
it grows more slowly than the rest of the body and becomes proportionately
smaller (Figure 13.11). In ontogenetic terms head growth therefore shows

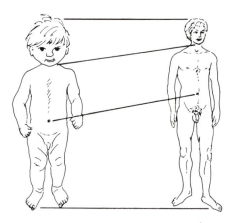

Figure 13.11. Allometric growth in man.
During human post-natal development the head shows negative allometry while the legs
are positively allometric.

negative allometry, but from an evolutionary point of view human head size is positively allometric in relation to that of other vertebrates. As we saw above, the evolutionary increase in human head size seems to be due to recession of the onset signal for brain growth relative to those for general body growth.

Most examples of allometric evolutionary change involve adjustments to growth rates, but some depend on variations in the timing of offset signals. A notable example in the latter category is the antlers of the Irish Elk (*Megaloceras giganteus*). This animal, the largest of the cervine deer, is now extinct, but in its day stood nearly 2 m tall at the shoulders with truly magnificent antlers each over 2 m long. Such enormous adornments must have been very unweildy and expensive in terms of body resources and may well have contributed to the animal's eventual demise. However, in allometric terms, taking into account the relative growth of antlers compared with that of the rest of the body, they were the 'correct' size. They were so large because the animal as a whole grew to a very great

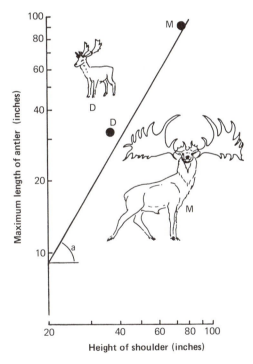

Figure 13.12. Allometric growth of antlers in cervine deer.
The graph reveals the relationship between maximum antler length and shoulder height at maturity. The Irish Elk, *Megaloceras giganteus* (M), lies on the line plotted for members of the group as a whole, despite its enormous antlers. The smaller fallow deer, *Dama dama* (D), has antlers that are actually disproportionately large for its size although they appear well proportioned. The value of a is shown.

size. Those of the smaller fallow deer (*Dama dama*) are of a more resonable proportion, being not more than 60 cm long in relation to an average shoulder height of about 90 cm. When viewed in relation to the cervine deer as a whole, however, *Dama's* antlers show positive allometry, or in other words, *Dama's* antlers are actually unusually large (see Figure 13.12).

The vertebrate limb provides some interesting examples of allometric variation. In limb development the fate of mesenchyme cells is partially determined by their exposure to the inductive influence of the apical ectodermal ridge (AER) during their time in the progress zone (see Chapter 6). This experience not only instructs the cells which part of the limb they should form, but also dictates how many mitotic events they must perform in doing it. It will be appreciated that any allele which alters the position or size of the AER, or a heterochronic change in the timing of the inductive experience, could produce major modifications to the form of the limb which later develops. In pythons, limb buds occur only in the pelvic position and their AER is reduced to a vestige, the inductive influence of which terminates early. The hind limbs therefore develop only as rudiments corresponding to just the thigh and these have taken over the new function of sexual stimulation during mating. In other snakes the limb buds do not appear at all. In *Anguis* (the slow worm) the limb buds are transient and lack an AER. Two other extreme limb forms are the wings of bats and the front legs of moles (Figure 13.13). We might expect that the grossly elongated bones of the former and truncated elements of the latter also owe their anatomy to differences in inductive exposure of progress zone cells to their respective apical ectodermal ridges.

One of the classic examples of evolutionary change is the reduction of digit numbers in the horse. The major feature in the evolution of a single-toed hoof from a five-toed foot seems to have been the relatively faster growth of the central digit compared to the lateral ones. Positive allometry of the central digit continued over a prolonged period as the animal grew to a very much greater size, allowed the central digit to become relatively massive and to take over support of the entire body weight (Figure 13.14).

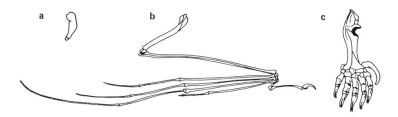

Figure 13.13. Skeletal structure of vertebrate limbs.
(a) Hind limb of a boid snake, *Trachyboa*. Only the femur is represented in this species. (Redrawn from Bellairs, 1969.) (b) The wing of a flying fox, *Pteropus*. (Redrawn from Parker and Haswell, 1954.) (c) The left fore limb of the mole, *Talpa*. (Redrawn from Young, 1958.)

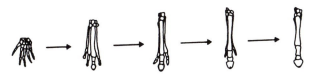

Figure 13.14. Evolution of the forelimb of the horse showing progressive loss of side digits and progressive increase in size of the central digit.
(Redrawn from Stormer, 1978; not to scale.)

13.6. Limits to growth

One of the puzzling features about growth is why it should stop when the animal eventually reaches adult size. How does the genome appreciate that its activities can be slowed down because the major task of constructing the phenotype has at last been achieved? In fact not all organisms are limited with respect to size. Some species, such as fish, have genetically indeterminate growth and will continue growing until they reach a maximum supportable size in relation for example to available food.

Body size is to a large extent governed by the length of the long bones of the skeleton. In these a race takes place between accretion of cartilage and its ossification. The growing regions, or epiphyses, eventually become overtaken by bone formation, and growth ceases as adulthood is approached. This occurs in man and many other mammals with determinate growth patterns, but in rats the epiphyses remain open and can be restimulated into mitotic activity by somatotrophin.

One can postulate a number of additional growth-limiting mechanisms. The cells may be pre-programmed to undergo only a limited number of cell divisions, the processes of construction may become balanced by those of decay, a growth-promoting substance may cease to be produced, or growth-limiting substances (chalones) may build up to critical levels. All of these mechanisms and others contribute to the limitation of size increase.

13.7. Summary and conclusions

Increase in size is one of the normal features of development of embryos and, especially during the early stages, is frequently associated with changes in body shape. Morphogenetic changes can also occur however without appreciable increase in body size. Most morphogenetic processes originate at the cellular level in directed cell movement, differential patterns of mitosis, or orientated mitosis and programmed cell death. Cells become distributed in accordance with rules of contact guidance, contact inhibition of movement and density-dependent inhibition of mitosis. In these and in many other aspects of cell behaviour and organization the surface of the cell is of particular importance. Specific interactions between cell surfaces are probably critical for the generation of correct linkages in the nervous system, although much reorganization of neuronal contacts and cell destruction also occurs.

Some of the most important changes in body form occur due to hydrostatic pressure, the swelling of glycosaminoglycan gels and contraction of cell extensions.

Changes in the timing of developmental events are important in evolution, especially adjustments to the relative rates of somatic and gonadal maturation. Heterochronic change can be based on modifications of growth rates, or changes in the timing of onset or offset signals. Paedomorphosis due to relative acceleration of sexual maturity may have accounted for the creation of many major new life-forms, including mankind. The original trigger for the creation of man from a pre-ape-like ancestor may well have been a general slowing down of somatic development relative to the onset signal for brain growth.

Evolutionary and ontogenetic changes in the proportional sizes of body parts frequently occur in accordance with the rules of allometry, and although some species continue growing until they reach the limits of their external resources, in others growth is limited by internal genetically controlled factors.

Bibliography

Aiello, L. C., The allometry of primate body proportions. In *Vertebrate Locomotion Symposium of the Society of Zoology, London, No. 48*, edited by M. H. Day. Academic Press, London (1981).

Ambrose, E. J. and Easty, D. M., *Cell Biology*, 2nd edn. Thomas Nelson and Sons, Sunbury on Thames (1977).

Bard, J. B. L., The cellular origins of tissue organization in animals. In *Developmental Control in Animals and Plants*, edited by C. F. Graham and P. F. Wareing. Blackwell, Oxford, pp. 265–289 (1984).

de Beer, G., *Embryos and Ancestors*. Clarendon Press, Oxford (1962).

de Beer, G., *Some General Biological Principles Illustrated by the Evolution of Man (Oxford Biology Readers)*, edited by J. J. Head and O. E. Lowenstein (1971).

Bellairs, A., *The Life of Reptiles*, Vol. 1. Weidenfeld and Nicolson, London (1969).

Bennett, D., Embryological effects of lethal alleles in the t-region. *Science*, **144**: 263 (1964).

Bowen, I. D. and Lockshin, R. A. (eds), *Cell Death in Biology and Pathology*. Chapman and Hall, London (1981).

Carter, N., *Development, Growth and Ageing. Biology in Medicine Series*. Croom Helm, London (1980).

Comfort, A., *Biology of Senescence*, 3rd edn. Churchill Livingstone, Edinburgh (1979).

Dormer, K. J., *Fundamental Tissue Geometry for Biologists*. Cambridge University Press, Cambridge, (1980).

Ede, D. A., *An Introduction to Developmental Biology*. Blackie, Glasgow (1978).

Edelman, G. M., Cell adhesion molecules. *Science*, **219**: 450 (1983).

French, V., Pattern formation in animal development. *Developmental Control in Animals and Plants*, edited by C. F. Graham and P. F. Wareing. Blackwell, Oxford, pp. 242–264 (1984).

Garcia-Bellido, A., Lawrence, P. A. and Morata, G., Compartments in animal development *Sci. Am.*, **241**: 102 (1979).

Gould, S. J., Geometric similarity in allometric growth: a contribution to the problem of scaling in the evolution of size. *Am. Nat.*, **105**: 113 (1971).

Gould, S. J., *Ontogeny and Phylogeny*. The Belknap Press of Harvard University Press, Cambridge, Mass. (1977).

Gustafson, T. and Wolpert, L., Cellular movement and contact in sea urchin morphogenesis. *Biol. Rev.*, **44**: 442 (1967).

Hopkins, W. G. and Brown, M. C., *Development of Nerve Cells and their Connections*. Cambridge University Press, Cambridge (1984).

Huxley, J., *Problems of Relative Growth*. Methuen, London (1932).

Matsuda, R., The evolutionary process in talitrid amphipods and salamanders in changing environments, with a discussion of "genetic assimilation" and some other evolutionary concepts. *Can. J. Zool.*, **60**: 733 (1982).

Murray, J. D., On pattern formation mechanisms for lepidopteran wing patterns and mammalian coat markings. *Philos. Trans. R. Soc. Lond. [Biol. Sci.]*, **295**: 473 (1981).

Newth, D. R., *Animal Growth and Development*. Edward Arnold, London (1970).

Oota, K., Makinodan, T., Iriki, M. and Baker, L. S., *Aging Phenomena, Relationships among Different Levels of Organization*. Plenum, New York (1980).

Oster, G. F., Murray, J. D. and Harris, A. K., Mechanical aspects of mesenchyme morphogenesis. *J. Embryol. Exp. Morphol.*, **78**: 83 (1983).

Parker, T. J. and Haswell, W. A., *A Textbook of Zoology*, Vols I & II, 6th edn. Macmillan, London (1949).

Reinert, J. and Holtzer H., *Cell Cycle and Cell Differentiation*. Springer-Verlag, Berlin (1975).

Sinclair, D., *Human Growth after Birth*, 2nd edn. Oxford University Press, Oxford (1973).

Steinberg, M. S., Does differential adhesion govern self-assembly processes in histogenesis? Equilibrium configurations and the emergence of a hierarchy among populations of embryonic cells. *Exp. Zool.*, **173**: 395 (1970).

Stormer, L., Fossil record. *Encyclopaedia Britannica. Macropaedia*, 15th edn. Vol. 7, pp. 555–557 (1978).

Thompson, D'Arcy W. *Growth and Form*. 2nd edn. Cambridge University Press, London (1949).

Townes, P. and Holtfreter, J., Directed movements and selective adhesion of embryonic amphibian cells. *J. Exp. Zool.*, **128**: 53 (1965).

Wolpert, L., Constancy and change in the development and evolution of pattern. In *Development and Evolution*, edited by B. C. Goodman, N. Holder and C. C. Wylie. Cambridge (Mass.), Cambridge University Press, pp. 49–57 (1983).

Waddington, C. H., *Principles of Development and Differentiation*. Macmillan, New York/Collier-Macmillan, London (1967).

Wourms, J. P., Reaggregation of completely dispersed amoeboid blastomeres during the normal development of annual fishes. *Am. Zool.*, **6**: 549 (1966).

Wyllie, A. H., Kerr, J. F. R. and Currie, A. R., Cell death: the significance of apoptosis. *Int. Rev. Cyt.*, **68**: 251 (1980).

Young, J. Z., *The Life of Vertebrates*. Clarendon Press, Oxford (1958).

Chapter 14 The principles of animal development

During their lifetimes, organisms tend to pass through a standard sequence of major stages and events, as outlined in the introduction to Chapter 2. Development is canalized along certain pathways, tissues acquire competence, become induced to adopt new phenotypes and become determined. These events occur by a variety of means, but if we analyse the basis of development in many systems we recognize the application of several general principles that are applied in various ways in all or most embryos. These relate to the chemical composition of living matter, the interrelations between different levels of organizational order, gradients and biochemical differentials, internal equilibrium, the adoption of definable developmental 'strategies' and the application of natural selection within the body. In this chapter we will discuss these issues and also draw together some of the previously described ideas on the structure and control of eukaryote genes, which are probably also common to most eukaryotes.

The first question we can ask is: why do animals develop at all? Why do they not stay as simple single cells? There seems to be a certain amount of inevitability about development at the current stage of evolution: once development starts, the only thing it can do is to proceed along previously defined courses to ever larger and more complex outcomes. But the complicated systems which result would not exist now if evolutionary forces had not favoured their development in the past. So one wonders whether increase in size and complexity are just incidental features of evolutionary progress, or whether large size and complexity are themselves evolutionarily advantageous.

14.1. Developmental principles

Increase in size and complexity

Observation indicates that complex organisms are capable of existence in less strictly defined environmental conditions than simple ones. Therefore, they are able to spread more widely, to colonize ecological niches that were previously not occupied, to exploit food resources that were formerly unavailable, and generally to be more successful than their competitors in spreading their genes far and wide.

286

Mastery of the environment necessitates complexity. For example, among the vertebrates the amphibians first colonized the land surface, however, they still need to return to water to reproduce and they cannot survive in dry atmospheres since a moist skin is required for respiration. It was the reptiles which really conquered dry land, and in order to do this they evolved more efficient lungs, an impermeable skin and the capacity to lay eggs that provide an aqueous and nourishing environment for their embryos even in the driest localities. However, the reptiles were ever dominated by environmental temperature. They failed to colonize cold regions and are still forced to find shelter in the cool of night when their metabolism slows down and they cannot defend themselves so readily. Their descendants, the birds and mammals, solved this problem by creating heat within their bodies and insulating themselves from heat loss by modifying their dermal scales into feathers and fur. They took care of the young a step further and in mammals the requirement for a nest was obviated by acquiring the ability to raise embryos within their own bodies, nutrients being transmitted directly from the maternal bloodstream instead of being stored within an egg. The dominant mammal, man, has taken giant steps beyond this stage in modifying his environment to suit himself, rather than the other way around, and in so doing is creating an increasingly hostile world for his competitors. We are therefore currently witnessing one of the great transitions between geological periods, similar to those of which we see records in the rocks beneath us, during which many very ancient life-forms will be extinguished and new ones created.

As suggested above, there are two aspects to the issue of complexity. On the one hand, organisms that are free of environmental strictures must necessarily be complex. On the other hand, complex polygenic systems and networks of biochemical pathways probably have a capacity for self-buffering (canalization; see Chapter 2) which resists the effects of external influences that would upset simpler systems (see Chapter 8). Both aspects contribute to the body's internal equilibrium; the attainment of this is one of the basic conditions around which life is organized (see below).

A large body is relevant to evolutionary success in more than one way. Self-sufficiency demands that an organism develops organs for nutrition, reproduction, respiration, for the appreciation of and response to its surroundings, and so on, and these just cannot all be packed into a very small body. A large body mass is also another stabilizing feature that helps the animal resist environmental fluctuations in humidity, temperature or food supply and when competition between members of the same species involves physical conflict, large size is again often advantageous. There is thus frequently a general tendency for small animals to evolve into larger ones.

Composition

The most basic developmental principle relates to the composition of living matter. All species select particular atoms and molecules from their environments for incorporation into their bodies. All life is based on the carbon atom because it is abundant, it has only four electrons in its outer orbit and its dioxide is a water-soluble gas. This makes it tetravalent and uniquely capable of contributing to an enormous variety of complex chemical structures widespread on land, in water and in the air. Carbon is the only element with this combination of properties and all organisms seek out and sequester carbon. Oxygen, nitrogen and water, the other major components of our atmosphere, are also essential constituents of all living matter. As Joseph Joubert so poetically phrased it, "life is a tissue woven of wind". The other major essential constituents are phosphorus and sulphur.

Among the simple molecules we encounter we reject the L isomers of sugars but incorporate the D forms. We ignore the D forms of amino acids but avidly seek the L isomers. In all organisms the most important chemical processes are catalysed by proteins. Almost all organisms use DNA as their hereditary material. Some viruses use RNA and only one known species, that which causes scrapie disease in sheep, uses anything else: we have so far failed to discover what.

Molecular structure and chemical reactivity are related closely to energy supply, since chemical bonds represent stored forms of energy. It is only those chemical reactions which result in net increase in disorder that can take place spontaneously, that is without input of energy. For reactions to occur which involve net increase in order, they must be coupled to others in which energy is released. In the majority of cases this involves, directly or indirectly, the hydrolysis of ATP, as the energy in ATP is stored in its phosphate bonds. The generation of ATP is thus critical to all life processes and this is accompanied by utilization of the light and heat originally given out by the sun. Green plants are the major converters of the sun's energy into ATP and other complex molecules produced as a result of its hydrolysis. These molecules enter the bodies of animals in the form of vegetable food and although we ourselves can synthesize ATP by other processes, the energy in its bonds is normally considered to derive ultimately from sunlight. Recently there have been reports of organic molecules being generated abiotically around thermal vents in the ocean floor. It seems likely therefore that some of the energy bound up in living systems possibly originated instead from the earth's internal heat.

Elements of order

The third principle concerns the interrelations between levels of order. Organizational features of high order generally derive from lower-order features. For example, the three-dimensional structure of each protein molecule is a function of its amino acid sequence. Phospholipid molecules will assemble spontaneously into membranes at interfaces between aqueous and lipid fluids, because one end of the molecule is hydrophilic and the other hydrophobic, the interface providing conditions that permit this to take place. When phospholipid membranes form *de novo* their assembly probably requires similar permissive conditions.

In the early chicken embryo, haemopoietic cells form into aggregates called blood islands. At first all the cells in these aggregates are apparently equivalent, but before long those on the outside become flattened and link up as an endothelium, the blood vessel, while the inner ones become stem cells that eventually give rise to true blood cells. Differentiation of the two cell types arises as a direct result of the aggregation of their progenitors and the types of cells produced are specified by their position in the aggregate. Differences in the accumulation of cellular waste products, and the availability of oxygen to the cells, are probably important in this differentiation.

Since most biological systems and structures arise by the aggregation of subunits or the integration of simpler systems, the conventional investigative approach is to split biological units into smaller and smaller subunits, on the assumption that eventually sufficient information will accrue for the investigator theoretically to reconstruct and understand the whole system (Figure 14.1(a)). However, there are situations in which the converse approach may be more valid, since the properties of 'the parts' are in some cases defined by 'the whole'. To understand animal development then we need occasionally to think in these terms also. A physical illustration of this principle is the shapes adopted by soap bubbles crammed inside a closed space, such as a bottle, the shapes of the bubbles being defined ultimately by the shape of the bottle. The iris of the eye provides a biological example. It is round because it represents a section through the spherical eyeball. In this case its essentially two-dimensional form is defined by the essentially three-dimensional shape of the eyeball (see Figure 14.1(b)).

Another illustration is provided by consideration of what are known as embryonic fields. A field is a part of the body that has spatial unity in the sense that

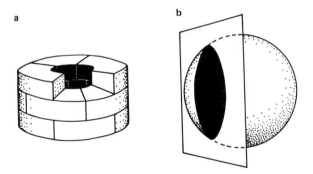

Figure 14.1. Levels of order in the derivation of form.
(a) A complex tubular structure derived by the assembly of blocks of simpler structure. The conventional investigative approach followed by biologists is to subdivide complex systems in order to understand how they are derived. The concept implicit in this approach is illustrated by the diagram. (b) A disc derived as a section through a sphere. In this case a simple structure is produced from a more complex one. This illustrates a less widely appreciated means by which morphologies are derived. This specific example illustrates the derivation of the shape of the iris of the eye.

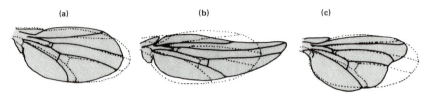

Figure 14.2. The insect wing as a morphogenetic field. The internal vein patterns of wings of *Drosophila* mutants are related to the overall form of the wings, as the whole wing structure is integrated.
The mutants are: (a) *approximated*; (b) *Blade*; (c) *dumpy*. The structure of the normal wing is shown in dotted outline. (Redrawn from Waddington, 1940.)

it must be considered as an organized whole, not merely as the combination of its constituent parts. Some fields exist as such because their components are organized within limiting boundaries. An example which is sometimes quoted is the wing of the fruit fly *Drosophila*, in which mutant genes such as *dumpy* and *Blade* cause distortion of the outline, co-ordinated with the whole internal pattern of venation (see below and Figure 14.2).

According to strict definition, if you cut an embryonic field in two the halves will develop as if each is a whole, since the tendencies of the individual cells or cell groups are subordinated to an overall influence which causes them, even after subdivision, to form a single organ. This is the case with the lens field in the head ectoderm of an early vertebrate embryo. The retinal field is another example. At the early neural plate stage there is a single retinal field which subdivides into two lateral fields under the inductive influence of the underlying archenteron roof (see Chapter 3). In this case the self-organizing property of the field is exploited by the embryonic system to produce bilaterally symmetrical organs. If this induction is prevented a single central eye results.

An intriguing example of an embryonic field unique to the primates is the pattern of dermal ridges that we see on the palms of our hands and the soles of our feet. These are related to the number of digits by a mathematical expression known as Penrose's dermatoglyphic formula. This has the form:

$$T + 1 = L + D$$

where T is the number of triradii, D is the number of digits and L the number of loops, a whorl counting as two loops (see Figure 14.3). The validity of the formula has been demonstrated not only with normal limbs, but also on malformed hands and feet with from one to six digits.

Characteristic dermatoglyphic patterns develop due to close packing of regiments of fibroblasts within the confines of the embryonic palm. Their movement is limited by a boundary, which in the normal hand is broken by six 'gates' opposite the wrist and the areas of necrosis between the digits (see Figure 13.6). The orientation and movement of fibroblasts is limited by contact inhibition of

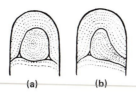

Figure 14.3. Dermatoglyphic characters of the human hand.
(a) A whorl with two triradii. (b) A loop with one triradius.

movement and contact guidance (see Chapter 13), which cause them to merge in closely packed arrays parallel to the boundary except where they spill out, producing ridges perpendicular to the edge of the hand. Away from the boundary the fibroblasts are thrown into inter-twining streams. As far as the field concept is concerned, the important point is that Penrose's formula can apply only if each palm or sole acts as an isolated integral area and the properties of the whole, in terms of numbers of digits, define its parts, as represented by relative numbers of triradii, loops and whorls.

The field concept is important in theoretical biology in that it postulates a situation which can be considered as a microcosm of the complete organism, perturbations in one part of the body frequently being reflected by disturbances in another. This idea of the healthy organism as a unified whole is one of the tenets of the homeopathic approach to medicine.

Multicellularity and cytodifferentiation

Apart from the protozoans like *Amoeba*, *Euglena* and *Paramecium*, all eukaryote animals are built up of a multitude of cells, in man about 100 000 000. A typical cell contains a single nucleus and a variety of other structures within a phospholipid plasma membrane. Cell types become differentiated from one another and cells of broadly similar type become grouped together as tissues. Characteristic assemblages of tissues of different types constitute organs.

No one cell type expresses all the genes that are expressed in that individual *in toto*, cytodifferentiation involves the selection and stabilized expression of characteristic groups of genes chosen from the entire complement. As suggested in Chapter 7, it is very likely that in many cases selection of a tissue-specific pattern of gene expression begins as a change in cytoplasmic enzyme activities, brought about by external influences, and only later becomes consolidated by nuclear changes. This would be particularly important where regulative as distinct from mosaic aspects of development are concerned (see Chapters 2 and 15), but how it might operate is as yet unknown. Other cases of cytodifferentiation, such as those that occur during development of insect imaginal discs, are almost certainly initiated by primary changes at the genic level (see Chapter 7).

Unusual features of cytodifferentiation in specialized systems include rearrangement of transcription rate modulator sequences, as with respect to the α-amylase gene in mice, gene rearrangements of the type that occur in the immune system, and transient gene amplification as in the case of the α-actin gene during muscle differentiation in chickens (see Chapter 12). Whether or not these specialized features occur in other systems is as yet uncertain. The Leader Cell Hypothesis (Chapter 7) suggests that cytodifferentiation may be pioneered in some tissues by unusual cells that are particularly responsive to embryonic induction and which then influence their less responsive neighbours to follow suit. This could apply in numerous situations in which an embryonic tissue is competent to differentiate in several different ways under the influence of appropriate inducers, but as yet evidence in support of this idea is scanty.

Gradients and differentials

A fifth principle involves the creation of and response to biochemical differentials and gradients. Biochemical differentials become established by separation of materials as for example under gravity, by localized differences in metabolic activity and by the erection of semi-permeable barriers. These include phospholipid membranes and extra-cellular collagenous basement membranes, such as that of the kidney glomerulus. Membranes are important for isolation of the contents of nucleus, mitochondria and other organelles, for the structure of the endoplasmic reticulum and for the separation of tissues and body segments.

Gradients exist within cells and around them. Cells adhering to substrata tend to ascend gradients of adhesivity. When slime mould amoebae aggregate they migrate towards a chemical attractant, which in *Dictyostelium discoideum* is cyclic AMP (see Chapter 4). Gradients of diffusible molecules may be used for defining positional information within insect eggs, in early echinoderm embryos (Chapter 2), in the limbs of vertebrates (Chapter 6) and in many other situations.

Natural selection within the body

A sixth principle of development is the application of natural selection within the body. Since the publication of Darwin's theory of evolution, the concept of natural selection of favourable variants has been widely accepted at the level of whole organisms and species, but the same principle also applies within the body. For example, in the vertebrate immune system each of a multitude of antibody-producing cells actively expresses a different combination of immunoglobulin genes (see Chapter 11). Antigens foreign to the body stimulate these cells to proliferate and this stimulation is specific, so that clones of cells are produced, each synthesizing antibody directed specifically against the stimulatory antigen. The antigen (an external environmental factor) therefore selectively directs the proliferation of a subset of the vast population of potential antibody-producing cells.

The immune system itself also exerts selective effects on the rest of the body, identifying and eliminating potential cancer cells at the rate of some tens of thousands per day in each one of us. Occasionally this internal scrutiny goes awry and the animal starts to destroy its own tissues, producing what is known as 'autoimmune disease'.

Specificity among connections in the nervous system is also achieved partly by natural selection. Although many nerve connections are probably made correctly at the outset, many others are not and the final pattern of specific connections is the result of reselection and destruction of incorrectly synapsed neurones (see Chapter 13).

Another illustration is provided by bone, the internal structure of which depends on the selective retention of small elements called trabeculae that are aligned appropriately to resist stress, while the rest are destroyed or realigned (see Chapter 15).

Internal equilibration

A seventh developmental principle is the trend shown by all forms of life towards stabilization of their internal metabolism (see above). Animals avoid extremes of climate and tend to stabilize conditions within them. The evolution of stable systems has created enormously complex regulatory mechanisms and at the biochemical level this very complexity probably also contributes to stability, since most metabolites can be produced or depleted in more than one way. Large body size introduces an element of metabolic inertia, since very many molecules must be eliminated before an appreciable change in composition is produced (see above). The possible role of enzymes as sequestrators of metabolites may also be important in this connection (see Chapter 8).

Physiological and morphological stability are sometimes achieved by establishment of balance between opposing influences. The forward and reverse performance of any chemical reaction is governed by the Law of Mass Action, which depends on balance derived from the relative concentrations of reactants and products. Transcription of 5S rRNA is regulated by competition for binding of transcription initiation factor between the RNA transcript itself and the DNA of the gene which codes for it (see Chapter 12). Within salamander eyes, metaplasia of iris pigment epithelium into lens cells probably depends on disruption of the normal balance between inhibitors emitted by the natural lens and stimulatory factors from the retina (see Chapter 7).

The competitive principle applies also to physical phenomena. For example, expansion of the eyeball is assisted by swelling of the vitreous body within it, which inflates the eye like a balloon, the spherical form adopted resulting from the balance between internal pressure and tension in the eye wall (see Chapter 3). As we saw in the last chapter, morphogenesis of *Drosophila* wings proceeds through a sequence of balanced fluctuations in relative expansion and contraction of the wing rudiment, until the organ's final form eventually becomes fixed by desiccation.

Strategy and tactics

It is not in the nature of embryos or their parts to strive toward goals or to pursue 'intentions' in the way that a human being would comprehend the idea. Embryogenesis results from the operation of genetic programmes that are historical records of ancestral success, and living systems build on what is already established, taking their directions from their foundations. From a human viewpoint, however, it is easy to imagine that events occur according to purposive strategy, and this terminology has been used by some authors. In military terms, developmental programmes involve many common 'strategies' such as growth, gastrulation, cytodifferentiation, and so on, but the 'tactics' of development can vary considerably. Determination of sex is a good example.

Sexual differentiation is a feature of very many species, but it is brought about by a host of different means and as with most phenomena, with evolutionary advance there has been a tendency to replace environmentally caused or environmentally modified controls by stable genetic regulation.

The marine annelid *Ophryotrocha* starts life as a male, then becomes female as it grows larger. If surgically amputated it reverts to male again, since sex is determined by body size. In another annelid species, *Dinophilus*, the female lays small eggs that develop into males and large ones, 27 times the size, that become females, sex being specified by cytoplasmic volume. Larvae of another marine worm, *Bonellia*, develop as females if they settle in isolation, but if they happen to alight on the proboscis of a female they become males parasitic on their mates.

In teleost fish sex is partly determined by the genome, but is also influenced very much by hormones in the diet and other factors. One per cent of platyfish develop into the sex opposite to that determined by the genome. In many reptiles sex is determined genetically, but in others it depends on the temperature of incubation of the egg. Alligator eggs incubated at $30\,^{\circ}$C develop as females whereas males form from similar eggs maintained at $36\,^{\circ}$C.

Even genetic determination of sex occurs by a variety of means. In honeybees females are diploid with 32 chromosomes and males usually haploid with only 16. The queen lays fertilized eggs in wax cells built to accommodate worker females and unfertilized but fertile eggs in the larger drone cells. This is achieved by control of the sphincter of her sperm storage receptacle in relation to the size of the wax cell. However, it is not haploidy and diploidy that actually determine sex, but heterozygosity, as compared to homozygosity or hemizygosity, at a particular multi-allelic locus. A heterozygous egg develops as a diploid female, unfertilized hemizygous eggs develop as normal males, while homozygotes begin development as diploid males, but are destroyed by the workers soon after hatching.

The fruit fly, *Drosophila melanogaster*, has sex chromosomes like those of mammals, an X and a Y in males and two X chromosomes in females. But whereas in mammals the autosomes carry feminizing factors while the Y chromosome probably carries factors that specify maleness, in *Drosophila* a male-determining factor is on the third chromosome and a feminizing factor on the X.

To further confuse the picture, among birds males are the homogametic (ZZ) sex, while females are heterogametic (ZW).

   There are then eight basic principles of development that we can discern, which seem to be applied within the bodies of all or most developing animals. In summary they are:

   1. The acquisition and incorporation of a defined selection of atoms and molecules.
   2. The derivation of organizational elements of high order from those of lower order, and the establishement of embryonic fields as organizational entities.
   3. Multicellularity and the development of different cell types.
   4. The creation of biochemical differentials and gradients and their exploitation as organizing features.
   5. Selective destruction or retention of cell types, neuronal connections, structural features, etc.
   6. General stabilization of internal conditions.
   7. Pursuit of common 'objectives' by a variety of means.
   8. General increase in size and complexity.

14.2. A molecular interpretation of embryological terms

Throughout this book attempts have been made to interpret the somewhat mystic concepts of early embryology in terms of meaning to the modern molecular biologist. Few of these interpretations have actually been demonstrated, but in this section these ideas are collected together for better appraisal of their value. Embryological descriptions of their meanings are given in Chapters 1, 2, 3, and 7. To the embryologist these concepts also imply broad principles by which embryogenesis is achieved. At the molecular level each is probably based on a variety of more basic principles.

Competence

The inductive competence of a cell develops cumulatively and may involve cytoplasmic as well as nuclear states. These could include:

   (a) presence of hydrophilic hormone receptor protein and other associated proteins in the plasma membrane (see Chapter 4);
   (b) presence of hydrophobic hormone receptor protein in the cytoplasm (see Chapter 4);
   (c) presence of an appropriate density of mitochondria (see Chapter 4);
   (d) presence of appropriate species of masked mRNA in the cytoplasm (see Chapter 9);

(e) low-level transcription, processing and export to the cytoplasm of appropriate species of mRNA (see Chapters 7 and 9);

(f) suitable patterns of relaxation of the chromatin solenoid (see Chapter 10);

(g) the correct DNA conformation (B or Z form) at appropriate regions of the chromosomes (see Chapter 12);

(h) chemically modified (for example acetylated) histones and absence of histone H1 from the chromatin, on a gene- or domain-specific basis (see Chapter 10);

(i) appropriate patterns of association of specific NHC proteins with the DNA (see Chapter 10);

(j) appropriate gene-specific patterns of cytosine methylation and demethylation (see Chapter 12);

(k) at the tissue level, the presence of appropriate 'leader cells' (see Chapter 7).

Induction

Embryonic induction of tissues could occur by the following means:

(a) synthesis by the inducing tissue of extra-cellular matrix material and response by competent target tissue to contact with it (see Chapter 4);

(b) physical pressure or tension caused by swelling of glycosaminoglycans or packing of tissues, causing modification of cytoplasmic structure (see Chapter 4);

(c) modification of cytoplasmic enzyme activities by transfer of ions (including H^+ and OH^-), small metabolites, co-enzymes or enzyme co-factors between cells (see Chapter 8). Transfer could occur through gap junctions (Chapter 4) or directly through the plasma membrane;

(d) uptake of Ca^{2+} ions and modification of the properties of cell proteins through activation of calmodulin (see Chapter 4);

(e) unmasking of masked mRNA (possibly derived from the mother as a constituent of the egg cytoplasm) (see Chapter 9);

(f) modulation of gene expression at selection of mRNA for translation through inter-cellular transfer of translation initiation factors (see Chapter 9);

(g) activation of cyclic AMP-mediated or other second messenger systems by extra-cellular factors (see Chapter 4);

(h) prolongation of mRNA survival and recruitment of mRNA species into the cytoplasmic high abundance class (see chapter 9);

(i) transfer of transcription initiation factors such as Mn^{2+} or Mg^{2+} ions (see Chapter 12);

(j) stimulation of transcription of specific gene sequences by hydrophobic hormones (see Chapter 4);

(k) control of synthesis or activity of nuclear-processing RNA species and related proteins (see Chapter 9).

Determination

Possible bases for determination are listed at the beginning of Chapter 7. At the cellular level determination may involve:

(a) fixation by means as yet unknown, of patterns of gene expression originally initiated as physiological modulations (see Chapter 7);

(b) establishment of biochemical feedback loop systems and build-up of metabolite pools (see Chapter 8);

(c) stabilization of cell internal structure and metabolism by synthesis of extra-cellular matrices (see Chapter 4);

(d) establishment of defined patterns of cytosine methylation and demethylation (see Chapter 12);

(e) establishment of defined patterns of DNA in B and Z conformations (see Chapter 12);

(f) loss of mitochondria (see Chapter 4);

(g) establishment of defined patterns of association of NHC proteins and histones with the DNA (see Chapter 10);

(h) establishment of defined patterns of histone modification (see Chapter 10);

(i) rearrangement of nuclear DNA sequences (see Chapter 11);

(j) destruction of nuclear DNA (see Chapter 11);

(k) closure of the inter-cellular channels in gap junctions (see Chapter 4).

14.3. *Control sequences of structural genes*

It is still too soon to be able to state with confidence which of the nucleotide sequences within or beside structural genes are of importance in the control of their expression, but we can draw attention to some which may have that role. The following description applies to genes that code for non-histone polypeptides and are transcribed by RNA polymerase type II. In accordance with convention the deoxyribonucleotide sequence of the non-codogenic DNA strand is described in the direction 5′ to 3′. This corresponds directly to the sequence of ribonucleotides in the RNA transcript in those regions which are transcribed.

There are probably control sites both within the gene itself and up to many kilobases on either side within a large domain that could include other genes of related or unrelated function. Some of these sites may be repetitive and may include the sequence CCGG, which in vertebrates (but apparently not in arthropods) is subject to variable methylation at the second cytosine, the presence of methylcytosine being related to adoption of the Z conformation by the DNA (see Chapter 12).

A few hundred bases upstream (5′) of the cAp site (which corresponds to the beginning of the RNA transcript) is a region rich in A and T, then at nucleotide positions -70 to -80 relative in the cAp site is the 'CAAT box' containing the seaquence GC^C_TCAATCT. This whole region is probably involved in binding the RNA polymerase and may regulate the rate at which the gene is transcribed.

At − 31 relative to the cAp site is the 'Hogness box' containing the sequence TATA^A_TA^A_T, which seems to assist in defining the cAp site three turns of the double helix further down. Between the two is TTGCT in a variable position and of unknown function.

Somewhere in the upstream region there are probably also gene-specific control sequences, possibly inverted repeats, that are recognizable by gene-specific NHC proteins responsible for controlling transcription. One might expect these to be upstream of the sites that bind and locate RNA polymerase.

The cAp site is an A surrounded by pyrimidines (C or T), in the general sequence Py‒‒‒Py A Py Py Py Py Py. Downstream of this, but before the translation-start signal ATG, is the region which corresponds to the mRNA 'leader'. This contains a sequence such as CTTC‒G, complementary to part of the 18S ribosomal subunit which, it is thought, is essential for the formation of translation initiation complexes (see Chapter 8).

Between the ATG translation start signal and the transcription terminator are the coding exons, interrupted by non-coding introns. The downstream 'donor' end of each exon is signposted by a sequence such as AGGTAAGT and the upstream 'acceptor' end of the next exon by a partially identical sequence such as ^C_TAGG. The variability of such sequences and the significance of this variation are still uncertain.

The total length of the introns is on average about twice that of the exons and within the last exon is the translation stop codon, which in the DNA is TAA, TAG or TGA. Beyond this is the region that corresponds to the messenger 'trailer'. Within this, the sequence AATAA may precede the polyadenylation site GC by some 16 to 25 bases. Following the polyadenylation site there are frequently up to five or more adjacent T residues, which may be involved in the termination of transcription. This region is very variable, however, and we do not yet know which sequences are important in this respect.

Downstream of the gene there are probably more repeat sequences concerned with accessibility of the domain, which may include CCGG, as on the upstream side.

14.4. A typical example of gene expression

The current literature contains no complete description of the expression of a single gene in any eukaryote. However, there are numerous incomplete descriptions and by combining these it is possible to build up a picture of the probable sequence of events that takes place during the expression of a typical eukaryotic gene. It should be stressed however that much of this description is speculative. In this example the gene is considered to be activated by several non-histone chromosomal (NHC) proteins that are either produced within the cell or derived from the egg cytoplasm.

It is likely that three classes of NHC protein are important: one that recognizes and opens up the chromatin domain around a specific structural gene and which is specific to that domain; a non-specific set that effectively blocks the histone mask in the opened-up domain; and a gene-specific class that identifies individual gene sequences once they have been unmasked and prepares them for transcription.

The NHC proteins are synthesized or activated in the cytoplasm, transported to the nucleus, through the nuclear membrane and eventually into contact with the DNA. During interphase each chromosome is in the form of an extended coiled coil, a very long solenoid. In man the total length of DNA per cell is about 2 m, distributed between 23 pairs of chromosomes and wound around histone beads linked by histone H1. Much of it is in the untranscribable left-handed spiral or Z form.

The domain- and gene-specific NHC proteins carry charge distributions which probably confer upon them specific affinities for certain sequences of nucleotides. It is suggested that in the neighbourhood of the gene in question, and for some kilobases on either side, are short sequences possibly containing inverted repeats and possibly present in several or many copies, which act as attachment sites for the domain-specific class of NHC protein. Their binding to the DNA may interfere with methylation of cytosine residues in CCGG or CG sequences within the domain, or by some other means modify the conformation of the DNA in that region, to produce an extended strand of the B-form right-handed helix. This change probably allows entry of enzymes that acetylate, phosphorylate or otherwise modify the histones, causing them to become less highly charged, for histone H1 to be released, and for the chromatin to relax out of the solenoid form and into the open beaded-string structure. The end-result of binding the domain-specific NHC proteins is therefore a relaxation of the chromatin within the domain and the opening up of the genes it contains to other controlling proteins.

It is suggested that the latter include the non-specific class of NHC proteins, the high mobility group proteins, HMG14, HMG17 and ubiquitin, which possibly intercalate themselves between DNA and nucleosomes, binding to the histones and freeing the DNA from its close contact with the nucleosome cores. This in turn may provide access for a gene-specific NHC protein that can bind specifically to gene-specific control sequences upstream of the structural gene, so providing the all-important signal beacon for RNA polymerase. It is suggested that this protein enables the RNA polymerase to identify this region of the chromatin, to bind to neighbouring sequences and to commence transcription on a gene-specific basis.

The RNA polymerase then slides 'downstream', taking its cue from the position of the TATA box. Three turns of the helix further down it comes to the cAp site and here commences synthesis of an RNA sequence complementary to the codogenic strand, that is similar to that of the non-codogenic strand, but with each T replaced by a U. The first-formed RNA transcript consists of a leader sequence, introns, exons and trailer. When the polymerase reaches the transcription terminator it falls off the DNA and releases its transcript.

The heterogeneous nuclear RNA molecule so produced is then processed into messenger RNA by enzymes and ribonucleoprotein molecules present in the nucleoplasm. These add a cap of methylated guanosine triphosphate in reverse polarity at the 5′ end and methylate the first one or two residues in the leader. They may trim back the trailer if it extends beyond the GC sequence and then add a poly-A tail. Introns are excised and the cut ends of the exons spliced together by ribonucleoprotein molecules containing 'processing RNA', which scan the molecule for the key sequences at the donor and acceptor ends. The resultant mRNA is then released from the nucleus and discharged into the cytoplasm.

Translation of the messenger requires initiation factors, some of which are messenger specific, which facilitate attachment of a small ribosomal subunit to the leader sequence. This then moves down the RNA until it reaches the AUG translation start signal. A large ribosomal subunit then joins the small one and the completed ribosome moves down the messenger, translating the RNA into polypeptide. When it reaches one of the stop signals, UGG, UGA or UAA, the translation complex falls apart and the polypeptide and messenger are released. The messenger may be translated many more times before it is eventually destroyed by ribonucleases.

Before it can function as an enzyme or structural protein, the polypeptide usually needs to be modified. The N-terminal methionine produced by translation of the AUG start signal is cleaved off and sometimes the penultimate two or three N-terminal amino acids are acetylated. Cleavage may occur elsewhere in the polypeptide and other kinds of chemical modification of the amino acid side-chains may take place. Such modified polypeptides then often combine with a co-factor, such as a metal ion, plus one or more other polypeptides. Its actual activity as an enzyme then depends on the supply of co-enzymes, such as NAD, and substrate molecules, the rate of evacuation of its product and a variety of other conditions.

From this point on, the interpretation of genotype into phenotype is best described at cellular and supra-cellular levels. The modification of enzyme activity that comes about due to transcription and translation of the gene may cause changes in cell behaviour and in the cell's synthetic activities. Some of its products may be secreted through the plasma membrane, or be transferred via gap junctions to neighbouring cells. The integration of these activities causes developmental changes, perhaps bringing previously distant tissues into close proximity and allowing inductive interactions to occur between them. These initiate new rounds of gene activity which may in turn modify expression of the original gene.

Eventually expression of the gene is discontinued, perhaps because either the gene-specific or the domain-specific NHC protein is no longer available, or because the mRNA is out-competed for translational facilities. Surplus mRNA is degraded and irreversible changes may take place at the chromatin level, preventing further transcription. However, control is frequently exerted on these processes also.

This then is a possible outline of the typical sequence of events that may take place during expression of a single structural gene in a eukaryote cell. There are many variations on this scheme and it should be remembered that no gene is expressed in isolation. In any one cell 10 000 or more genes may all be in various stages of expression at any one time. As many as 180 may be involved in the synthesis of something so apparently straightforward as the chorion of a moth egg. It is interesting to compare this extremely complex situation that exists in eukaryote cells with the exceptionally elegant and simple systems devised by the bacteria and described in Chapter 1.

14.5. Summary and conclusions

Comparative analysis of ontogenetic mechanisms in many species reveals that development takes place in accordance with certain principles of composition, organization and operation. All organisms are built up of a similar combination of atoms and molecules and there is a general tendency among the eukaryotes towards evolutionary increase in size and complexity. Most eukaryotes are multicellular, cells of diverse types each expressing a different fraction of the genetic information.

Structural and biochemical elements of high order are generally derived from those of lower order, but the reverse can also be true, as exemplified by the existence of embryonic fields. Many organizational features are most easily explained in terms of the response of cells to biochemical differentials and gradients, and there is a general tendency in all organisms towards the establishment of internal physiological and physical equilibria. It is suggested that the embryological concepts of competence, induction and determination are each derived from a variety of molecular and cellular conditions.

It is suggested that the opening up of a gene for transcription by RNA polymerase II occurs by a three-step process, in which a domain many kilobases in extent, and surrounding the gene, is first identified and prepared by a NHC protein specific to that domain. Next, the histone mask on the DNA within the domain is rendered ineffective by incorporation of non-specific NHC proteins. Gene-specific NHC proteins then identify sites beside genes within the domain and provide markers that indicate sites for attachment of RNA polymerase. Initiation and termination of transcription are signalled by sequences at each end of the gene and these are probably similar in most genes transcribed by RNA polymerase II. The RNA transcript is then processed by capping, excision-splicing and polyadenylation, before transfer to the cytoplasm.

The mRNA is then translated into a polypeptide, which may be processed in a variety of ways to produce a functional enzyme. Subsequent expression of the gene is best understood in terms of changed activity at the cellular level.

Bibliography

Bellairs, R., Curtis, A. and Dunn, G. (eds), *Cell Behaviour. A Tribute to Michael Abercrombie*. Cambridge University Press, Cambridge (1982).

Bonner, J. T. (ed.), *Evolution and Development*. Springer-Verlag, Berlin (1982).

Cale, G. H. Jr and Rothenbuhler, W. C., Genetics and breeding of the honey bee. In *The Hive and the Honey Bee*. Dadant and Sons, Hamilton, Illinois, pp. 157–184 (1975).

Crick, F., *Life Itself. Its Origin and Nature*. MacDonald, London (1982).

Cummins, H. and Midlo, C., *Finger Prints, Palms and Soles*. Dover Publications, New York (1963).

Elsdale, T. R. and Wasoff, F., Fibroblast cultures and dermatoglyphics: the topology of two planar patterns. *Wilhelm Roux Archiv.*, **180**: 121 (1976).

Gerisch, G., Chemotaxis in *Dictyostelium*. *Ann. Rev. Physiol.*, **44**: 535 (1982).

German, T. L. and Marsh, R. F., The scrapie agent: a unique self-replicating pathogen. In *Progress in Molecular and Subcellular Biology*, Vol. 8, edited by F. E. Hahn, D. J. Kopecko and W. E. G. Müller. Springer-Verlag, Berlin, pp. 111–121 (1983).

Grant, P., *Biology of Developing Systems*. Holt, Rinehart and Winston, New York (1978)

Holt, S. B. and Penrose, L. S., *The Genetics of Dermal Ridges*. Charles C. Thomas, Springfield, Illinois (1968).

Kerr, W. E., Advances in cytology and genetics of bees. *Ann. Rev. Entomol.*, **19**: 253 (1974).

Lima-de-Faria, A., *Molecular Evolution and Organisation of the Chromosome*. Elsevier, Amsterdam (1983).

Macmahon, T. A. and Bonner, J. T., *On Size and Life*. Freeman, Oxford (1984).

O'Connor, R. F., *Chemical Principles and Their Biological Implications*. Hamilton, Santa Barbara (1974).

Pitts, J. D. and Finbow, M. E., *The Functional Integration of Cells in Animal Tissues*. Cambridge University Press, Cambridge (1983).

Riedl, R., *Order in Living Organisms. A Systems Analysis of Evolution*. Wiley, Chichester (1978).

Roberts, D. F., Population variation in dermatoglyphics: field theory. In *Progress in Dermatoglyphic Research*, edited by C. G. Bartsocas. Liss, New York, pp. 79–91 (1982).

Suzuki, D. T., Griffiths, A. J. F. and Lewontin, R. C., *An Introduction to Genetic Analysis*, 2nd edn. W. H. Freeman, New York (1981).

Waddington, C. H., *Organisers and Genes*. Cambridge University Press, Cambridge (1940).

Waddington, C. H., *Principles of Development and Differentiation*. Macmillan, New York/Collier-Macmillan, London (1967).

Weiss, P., The nature of biological organisation. In *Biological Organisation. Cellular and Sub-Cellular*, edited by C. H. Waddington. Pergamon Press, Oxford, pp. 1–21 (1959).

Chapter 15 An epigenetic theory of evolution

The general plan throughout this book has been to work down through the various levels of control, from external factors and gross phenotype, through protein and RNA to the nuclear DNA, a course which broadly follows the historical development of our subject. It now appears that this could also be the order in which controls upon gene expression become assimilated during evolution, external controls gradually being replaced by internal ones, cytoplasmic factors by nuclear ones. This chapter will consider the inadequacy of conventional theory in explaining how evolution takes place. By investigating certain peculiarities of animal development it becomes possible to construct an essentially new theory of evolution, with the emphasis on developmental rather than genetic concepts, which offers solutions to some of the outstanding problems of evolution. From the new standpoint then attained we will be in a position to look back over the contents of the previous chapters and appreciate their relevance in quite a different way.

15.1. Development in space and time

An organism is the tangible expression of the genes it carries, interpreted in the context of appropriate environmental conditions. Under variant conditions that same combination of genes might produce other very different forms, many of which would be grossly abnormal and die early deaths. To produce phenotypes which evolutionary forces favoured in the ancestors of that individual, all the many systems which control gene expression must inter-play in a regular fashion and follow a standard sequence. In other words, for normal development gene expression must be co-ordinated both temporally and spatially, and in this the external environment and the genome both play crucial roles.

Three time-scales are particularly important: enzyme reaction time, measureable in milliseconds to hours; ontogenetic time, referring to events that occur during the life of the individual and covering the range from hours to decades; and evolutionary time, extending through several generations or geological periods measured in millennia. Much common misunderstanding about biological phenomena arises from lack of recognition of these very different

303

time-scales over which biological events take place. Evolutionary rearrangements of genetic material and changes of allele frequencies within populations occur very much more rapidly than the acquisition of new mutations and can be just as significant in the evolution of new life forms (see Chapter 4).

Where 'space' is concerned we can distinguish the internal dimensions of an organism from the 'external space' which surrounds it. We can consider both intra-cellular and extra-cellular controls. From the point of view of the nucleus there are intra-nuclear and extra-nuclear features, while at the level of the gene there are controlling elements located within or beside structural genes, as well as extra-genetic controls based elsewhere.

Morphogenetic phenomena occur with reference to local and distant parts of the body, as well as to elements of the external world. The environment not only provides selective filters that ensure preferential survival of the fittest, but, by imposing challenges, it promotes adaptive changes that fuel evolutionary advance. It has therefore been suggested that evolution can be viewed as the control of development by ecology. There is however a general tendency during evolution to replace reliance upon external cues by recognition of others arising within the body.

Every organism or part of an organism develops within this imaginary 'reference frame', constructed between the axes of space and time, but they shift their positions within it. Within the compartments of the space-time reference frame organisms demand defined conditions and upon these impose their own kinds of organization. Systems gradually move from one compartment to another and the states of organization contained within them gradually evolve in complexity.

15.2. *Ontogeny and phylogeny*

It is relatively easy for the student of evolution to account for the evolutionary loss of organs that are no longer of value, such as the eyes in cave-dwelling and burrowing animals and the limbs of snake-like vertebrates, in that there are few selective forces to ensure their retention. For example, deletion or destructive mutation of an allele essential for eye development would be unlikely to affect survival through deprivation of vision if the individual lived in a location where there was no light by which to see. That deleterious mutation could therefore be retained in the breeding population. There are also no particular difficulties in accounting for very gradual quantitative changes, such as the transition from five toes to three, to one during the evolution of the horse. It is much more difficult to account for rapid adaptive evolutionary change, for the creation of entirely novel structures or tissues, or for differential rates of evolution.

It would perhaps be useful at this stage to summarize the most widely held ideas about how evolution takes place. In essence, Charles Darwin's explanation of a century ago was that individual members of any species vary in their phenotype and that features of this phenotype are heritable. Those individuals that are best suited for survival ('the fittest') tend to leave more offspring than the less fit, so their features become more widespread in the population. When this process is continued over a long period of time it can lead to modification or subdivision of the original species into new species, especially when populations are geographically isolated from one another. The term 'neo-Darwinism' is applied to the updated versions of this general theory, incorporating the ideas of Mendelian genetics. Darwin knew nothing about genes as we understand them, but in neo-Darwinist thinking modification of the genome is seen as the primary source of variability between individuals. This includes the genetic recombination involved in reproduction, nucleotide substitutions, insertions, deletions, duplications, chromosome rearrangements and the capturing of genetic material from other organisms through the agency of viruses. All evolutionary change cannot be accepted as the product of natural selection however. Many shifts in gene frequency are probably the result of random events (genetic drift), or because an unselected allele just happens to be in close genetic linkage to another that is under strong positive or negative selection and 'hitch-hikes' along with it. Some phenotypic features are gained or lost because they are pleiotropic expressions of alleles maintained at high or low frequencies on the basis of other distinctly different properties. However, although very many aspects of evolution are accounted for by neo-Darwinist or 'neutralist' explanations, many others remain a puzzle.

Deficiencies of current evolutionary theories

One of the principal deficiencies of current evolutionary theory is its failure to account for the acquisition of the observed number of favourable phenotype modifications in the time available. This is because it is usually assumed that new phenotypes can arise only from new genetic mutations and new mutations just do not occur that frequently (see Chapter 4). Moreover, the vast majority of new mutants are less, not more, fit for survival than the normal wild-type forms. The rate of arrival of new mutations is not actually the limiting factor however. Phenotypes are to a large extent the result of adaptation by the individual to the environment, while developmental phenomena may direct the recognition and retention of polymorphic alleles that are already present within the population.

Several leading evolutionary theorists have held the opinion that the creation of new species sometimes occurs by 'saltation', by great evolutionary leaps, rather than as the sum total of many small changes, as would be expected if due to the slow accumulation of minor mutations. This is because the selective advantage conferred by any new modification would be unlikely to be significant until several contributory mutations have been accumulated and considerable adaptation has been accomplished. The essence of the idea is that new major changes in phenotype, such as those that accompany the establishment of new species, are produced by catastrophic mutations. The arthropod homeotic mutants, in which appendages develop quite inappropriately, would fall into this category (see

Chapter 6). Goldschmidt called such weird forms 'hopeful monsters' and suggested that in suitable circumstances they could survive to found new species. There are no strictly analogous mutants among the vertebrates however, even though they show similar evidence of saltation. In this group therefore the explanation seems to lie in another direction.

Neo-Darwinist theory dictates that phenotypic features arise from new combinations of alleles or from new genetic mutations. This means that all members of a species which share a common phenotypic feature are descended from a small population of common ancestors in which that combination of alleles or those mutations first occurred. This implies that every species must be the product of a great deal of inbreeding, although the degree of genetic polymorphism in modern species would not seem to support this view.

Another striking and unexplained feature of evolution is the contrast between the great conservatism of some traits, particularly the most ancient, and the remarkable evolutionary lability of others. For example, the pentadactyl limb has evolved into a multitude of variant forms (Figure 13.13), but its basic pattern and its situations beside pectoral and pelvic girdles have been conserved since the outset some 400 000 000 years ago. The pentadactyl limb emerged with the bony and cartilaginous fishes in the Silurian period (see Figure 15.1). Before that time and in other fishes there was a variety of different distributions and structures of fins, but in the bony fishes, which later gave rise to the terrestrial vertebrates (including ourselves), two fore and two hind limbs, with associated girdles became adopted as the norm. The adoption of this basic pattern presumably involved the acquisition of a genetic programme for development of an ectodermal ridge as a source of inducer, on the apex of each of the fore- and hind-limb buds, plus a zone of polarizing activity at the rear margin (see Chapter 6). Further evolution of the limb then involved tinkering with this basic system rather than radical reorganization. Clearly the basic limb pattern is conserved by much stronger restraints than are the relative sizes and shapes of its bones and the exact numbers of digits. There is currently no theory that adequately accounts for such differences in evolutionary stability, suggesting why very ancient features should have different status in this respect compared with those more recently acquired.

One of the biggest problems is how to account for situations in which precisely the right genes are mobilized *in the right parts of the body, at the right times in development*, and the 'right times' often means *before* exposure to the sort of *environmental* stimuli that might be expected to invoke their expression. For example, how does it come about that the genes for epidermal keratinization, which normally come into play only in regions of the skin subjected repeatedly to pressure, express themselves on the soles of a baby's feet long before the baby will begin to walk? What causes creases to appear at the finger joints in babies with fused joints that will not bend? Before trying to provide explanations of such phenomena we will first examine some of the clues to the evolutionary puzzle that are offered by ontogeny, most of which in themselves constitute separate additional puzzles.

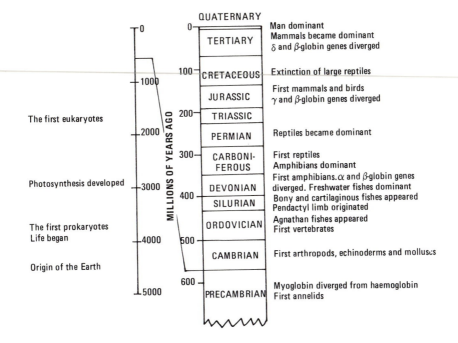

Figure 15.1. Major events in the evolution of life.
Some of the most important steps in evolution are indicated in relation to geological periods and time before the present. (Data from Dobzhansky, 1977, Zuckerkandl, 1965, and Tribe *et al.* 1981.)

15.3. Clues to the evolutionary puzzle

Recapitulation and atavism

During development every animal recapitulates in outline some of the more important steps of its own evolution. This recapitulation is summed up in the phrase "ontogeny repeats phylogeny", and by the all-too-memorable idea that "during his development each one of us climbs his own evolutionary tree". For example, at an early stage in our own ontogeny we have gill pouches reminiscent of those of our fishy ancestors which we share with very distant relatives such as the birds and reptiles. The exact relevance of this kind of observation to evolution was at one time the most hotly disputed topic in biology. The debate cooled with the recognition that such features represent *embryonic* characters of ancestors, rather then features of mature ancestral adults. In other words, *speciation is now seen to have proceeded by divergent maturation of juveniles*, not as some claimed

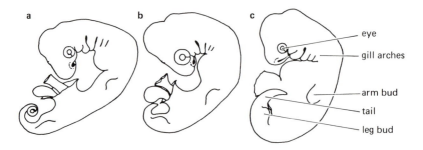

Figure 15.2. Gross appearance of embryos from widely separated vertebrate groups. (a) Reptile. (b) Bird. (c) Human. The embryos are very similar in appearance at this stage. All have gill arches and pouches which are absent from the adult, as well as rudimentary tails.

by enhanced maturation of adults. During our development we repeat some of the embryonic stages followed by our ancestors, which are also shared by our relatives. Thus present-day species from separate taxonomic phyla are strikingly similar in appearance during early embryonic stages, but gradually diversify as development proceeds (see Figure 15.2).

There are a few curious instances in which ancestral relict features are expressed erroneously in individuals that do not normally express them. For example, guinea-pigs normally have only three digits on their forefeet and four on their hind. Guinea-pigs are descended from five-toed ancestors and several mutations have been found that re-establish the ancestral conditions. Polydactylous phenocopies can also be produced by a variety of environmental agents, including temperature and the age and nutritional status of the mother. Polydactylous horses are also well known to modern science (Figure 15.3), and Julius Caesar is recorded as having owned and ridden a remarkable horse with feet cleft like a human foot. Human babies are sometimes born with a tail or with body hair. The recurrence in descendants of a character which had been expressed in an ancestor, after an interval of several or many generations, is called atavism. Similar conditions can arise from other causes and should not be confused with true atavisms. Polydactyly, for example, can be caused by disruption in pattern formation in the limb bud in a way which does not represent a return to the ancestral situation (see Chapter 6).

One of the most astonishing illustrations of the retention of relict genes was provided by an experiment in which epithelium from chicken embryo gill arches was cultured in contact with mesenchyme taken from the molar tooth region of a mouse embryo. The chicken epithelium induced dentine synthesis in the mouse mesenchyme, while the mesenchyme induced synthesis of tooth enamel in the epithelium, even though birds do not normally synthesize enamel anywhere in their bodies. In some cultures complete teeth were formed, demonstrating the

a   b

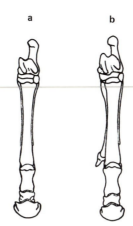

Figure 15.3. Atavism in horses.
The limb of a polydactylous horse (b) is shown beside that of a normal horse (a). This should be compared with the evolution of the horse limb shown in Fig. 13.14.

reactivation of avian genes that have probably remained unexpressed since the days of their last toothed ancestor, *Hesperornis*, some 80 000 000 years ago. Atavisms demonstrate that *the evolutionary loss of a feature does not necessarily involve loss of the genes which code for it.* A change in the developmental conditions that brought those genes into expression can produce the same effect.

Functional adaptation

The second step in piecing together the evolutionary puzzle is the realization that all animals individually adapt physiologically and anatomically to the way of life they are forced to adopt, although according to Schmalhausen, species which develop as predominantly mosaic embryos, such as annelids and ascidians, generally show less capacity for functional adaptation than those derived predominantly by regulative processes (see Chapter 2).

Human beings living at high altitudes produce red blood cells prolifically in response to the low partial pressure of oxygen in their blood. If blood vessels around the heart are clamped shut in early embryos, alternative circulatory systems develop that are equally as good as normal, while the heart also forms in a modified shape. Mammals can even develop new kidney glomeruli when the concentration of metabolic products in their blood is raised to high levels. Exercise strengthens our muscles causing them to increase in thickness, while elastic fibres and tendons develop in the direction of tension. Already formed bones are remoulded by the co-ordinated action of osteoblast cells which build them up and osteoclasts which break them down. Bone develops an internal structure appropriate to the magnitude and direction of applied stresses. This is due to selective

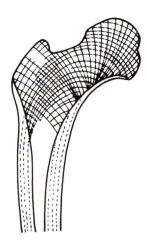

Figure 15.4. The internal structure of the head of the human femur.
Thin plates of bone called trabeculae become aligned in relation to the pressure and tension developed in the bone with use. A genetic factor is also involved however, since bone grown in culture develops a somewhat similar internal structure. (Redrawn from Sinclair, 1973.)

destruction or reorientation of developing bone trabeculae orientated at oblique angles to applied stresses, as these are subjected to very high shear forces. The result is that strong struts develop that are perfectly adapted to resist physical forces with the lightest possible structure (Figure 15.4). In short, functional adaptation is so effective it can cause immeasurably greater changes within one individual than can the most intensive selection of hereditary variation carried out over hundreds of generations. The remarkably overdeveloped physiques of male 'body builders' are eloquent testimony to this fact.

Furthermore, adaptive changes in one tissue, or one part of the body, stimulate modification in other tissues or parts so that, for example, when we undergo

(a)   (b)

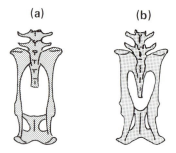

Figure 15.5. Modification of pelvic structure in relation to bipedalism.
(a) The pelvis of a normal quadrupedal goat. (b) The pelvis of a goat born without forelimbs. This animal was forced to adopt a form of locomotion similar to that of a kangaroo and its pelvis became modified accordingly. (After Slijper, 1942.)

athletic training we develop not only our muscles, but also our tendons, our skeleton, our circulatory, nervous and digestive systems, while at the same time decreasing our deposits of accumulated fat. *The important points about functional adaptation are that it arises due to performance of the function itself and that individuals vary in their capacity to adapt.*

One of the most striking examples of adaptation of body form is Slijper's goat. This animal was born with undeveloped forelimbs so that it was forced to adopt a kangaroo-like means of locomotion. As a result its skeleton and musculature became modified, with a pelvis like that of a biped instead of a quadruped (see Figure 15.5).

Genetic assimilation

The third key piece of information is provided by observations on the genetic assimilation of phenocopies (see Chapter 4).

When cabbage white butterflies of the genus *Pieris* undergo pupation the majority turn into black and white pupae, but a proportion, around 20% in *P. napi*, fail to synthesize pigment and appear green due to their haemolymph showing through. When raised under orange glass the proportion of green pupae increased to over 90%. The offspring of the latter (green) group, raised under white light, included an increased proportion (30%) of green pupae, and this proportion was doubled among the offspring of butterflies emerging from green pupae following exposure to orange light for two generations. The probable explanation is that deprivation of the blue end of the spectrum revealed an underlying genetic predisposition towards formation of green pupae, and by breeding from this group the alleles which confer susceptibility to blue deprivation were caused to increase in frequency. In other words, the selected breeding group had started to assimilate alleles that favour the green phenotype into its communal genome.

In our own species, the existence of different degrees of skin pigmentation in the different races is due to fixation (homozygosity) of the appropriate alleles at three or four loci in relation to natural sunlight intensity. This has probably come about by genetic assimilation of those alleles that favour a suitable degree of sun tanning, permitting sufficient synthesis of vitamin D (which is promoted by sunlight), while preventing toxic vitamin overdose, sunburn and skin cancer.

Analysis of *Drosophila* stocks that have genetically assimilated initially environmentally induced features such as a broken wing vein (see Chapter 4) suggests that the effects are related to increase in the frequency of suitable alleles of genes of both major and minor effect. However, this interpretation was challenged recently in the case of induction of the four-winged *bithorax* phenocopy of *Drosophila* by ether treatment of the egg (see Chapter 5). In this experiment no selection was applied to the stocks, but the phenocopy increased in frequency nevertheless. (In Waddington's original experiments frequency of the phenocopy increased and decreased respectively in the lines selected for positive and negative expression.)

Furthermore, whereas a cross between control males and females from the ether-treated line produced F_1 and F_2 offspring that displayed the bithorax phenotype in high frequency without ether treatment, the reciprocal cross did not. This suggests very strongly that the bithorax effect can be caused by the cumulative effects of ether on the maternal system, transmitted through the ovum, rather than by selection of modifying alleles, which would also be transmissable by males. This example may be atypical, however, as an explanation of 'genetic' assimilation, since it concerns overall body patterning which could be related to the *distribution* of maternal components in the egg cytoplasm, rather than their specific nature (see Chapter 2).

Whether or not the allele selection theory is correct in other cases, it is an entirely plausible means by which genetic predisposition to a certain response *could* become enforced in later generations in a population under selection. At the population level genetic assimilation would be represented by a swing in the frequency of favourable modifying alleles. In summary, the observations on genetic assimilation suggest that *selective breeding from individuals which demonstrate strong expression of an environmentally induced character can result in expression of that character in descendants, even if they develop away from the appropriate environmental stimulus.*

Specificity and timing of embryonic inductions

The fourth clue is that very many, if not most, morphologies and physiologies of complex eukaryotes are the direct or indirect products of embryonic inductions (see Chapters 2, 3 and 4). As we have seen, induction is caused in some cases by hormones, but more usually by close proximity of dissimilar tissues. Induction is essential for normal cytodifferentiation to occur and such events are related in a complex way to the physical contortions of morphogenesis. It is probably of the utmost significance that a diversity of substances and conditions seem to be able to perform any one induction (see Chapter 3). This is because *specificity of response depends on the state of competence of the target tissue rather than on the nature of the inducer.*

Investigations into the induction of the vertebrate lens suggest that three separate inductive influences from foregut endoderm, presumptive heart mesoderm and optic vesicle produce similar effects in competent ectoderm (see Figure 3.4), lens differentiation being the final expression of the cumulated inductions. It will be appreciated that in situations like this an increase in duration or intensity of an early inductive exposure, or replacement by a more powerful inducer, could render subsequent inductions redundant. In *Rana esculenta*, the hybrid edible frog, a lens will develop in the correct place even if the last and usually most important inducer, the optic vesicle, has been removed. This is because in this species another tissue, the chordamesoderm, has taken over as inducer. In *Rana fusca* the source of inducer depends on temperature. At 25 °C the optic vesicle exerts the main influence, but at 12 °C it is the chordamesoderm.

The idea of the replacement of a later induction by an earlier one is of considerable importance in evolutionary theory, since *heterochronic change of this sort permits recession of developmental events back to earlier stages*. It is probably by means such as this that adaptive features appropriate to the adult are expressed during embryogenesis, before exposure to the environmental stresses that would otherwise produce similar effects in later life.

The other side of the coin concerns the timing of inductive competence. Competence, it seems, results from previous inductions, so a changing regime of induction affects the temporal acquisition of competence. The cornea is normally induced in the overlying head ectoderm by the lens, but if ectoderm of the stage when it is competent to form cornea is challenged with a young optic cup, the latter will induce cornea instead of its normal inductive product, the lens. This gives us a clue to the possible evolutionary origin of the lens. The cornea, or an equivalent external 'window', almost certainly arose before the lens, possibly under the inductive influence of the primitive retina. Evolution of the lens would require and probably followed the acceleration of competence of the head ectoderm to respond to optic cup induction, in association with the recognition of earlier supplementary inductive stimuli arising from other embryonic tissues.

It is interesting that the conditions required for some kinds of inductive response include mechanical as well as chemical stimuli. The two are not necessarily unrelated since mechanical forces can be applied intracellularly by microtubules and microfilaments, which are under the control of biochemical influences (see Chapter 4).

Modulatory tissues and the physiological control of cytodifferentiation

A concept of crucial importance to the development of the theory outlined below is that *some aspects of cytodifferentiation are regulated through cytoplasmic physiology*. Embryonic cells pass through modulatory states in which their future fates become decided by exposure to inductive influences, some of which certainly exert their effects physiologically. For example, chondrogenesis in limb bud mesenchyme is stimulated by NAD (see Chapter 4). The work on transdifferentiation of neural retina (Chapter 7) confirms that cytodifferentiation can, in some cases at least, be controlled by cell physiology. In Chapter 7 it was suggested that phenotype modulation in cultured pigment epithelial cells is possibly based on competition for limited translational facilities between species of mRNA.

There are other tissue instabilities that could fall into the modulatory category, which operate on an inter-cellular, rather than an intra-cellular, basis. The conversion of keratin-synthesizing epithelium into mucus-secreting on exposure to vitamin A operates by activation of mucus secretion in a previously unspecialized cell type (see Chapter 4). The converse change was perhaps relevant to the evolution of reptilian skin from that of the amphibian.

Chondroid bone, osteodentine and the fossil record

The fossil record provides little direct information on the evolution of soft tissues, but where bones and teeth are concerned we are more fortunate. Bone, cartilage and dentine as we know them today were already present in the earliest vertebrates, the jawless Agnathan fish of the Ordovician period (Figure 5.1). These animals also contained a tissue intermediate between bone and cartilage called chondroid bone, and another tissue between bone and dentine called osteodentine. The existence of these intermediate tissues indicates that bone, cartilage and dentine must be very closely related to one another.

The intermediate tissues disappeared in the early Silurian period, as the shark-like fish, with cartilaginous skeletons and dermal denticles, diverged from those with skeletons of bone. The line which gave rise to the bony fishes also provided the ancestors of the amphibians, reptiles, birds and mammals, and in their embryos we still see relics of the earlier intermediate skeletal materials. Thus chondroid bone is formed in chicken embryos if they are immobilized, since normal development requires mechanical stimulus.

The co-existence of chondroid bone and osteodentine with true cartilage, bone and dentine in the Agnatha indicates that at that early stage the intermediate tissues were, even in the adult, probably still in a state capable of modulation by external stimuli. By the time the more advanced fish had arrived, development of bone, cartilage and dentine was apparently completed in the embryo, since the intermediate tissues had by then been lost from adults. It has been suggested that the evolutionary changes from the Agnathan situation to that in bony and carti-laginous fishes several million years later could have occurred by recognition of internal stimuli and their appropriation as embryonic inductions, so producing the same differentiative effects, but earlier in ontogeny.

In summary then, this argument suggests that non-modulatory tissues could have evolved from more ancient tissues which were modulatory in the adult, *by the replacement of regulatory factors external to the body by internal inductive stimuli acting during the late stages of ontogeny.* In later descendants other induc-tive stimuli may have taken over acting at even earlier ontogenetic stages.

These then are six clues to the way in which adaptive evolution could come about. In summary they are: the evidence of recapitulation and atavism; the enormous capacity of organisms for functional adaptation; the experimental demonstration of genetic assimilation of phenocopies; the potential of heterochronic change in embryonic inductions; the fact that cytodifferentiation can be controlled by extra-cellular influence on cell physiology and the fossil evidence which suggests that intermediate tissues of ancestral adults gave rise to current differentiated tissue types through ontogenetically earlier resolution of modulatory states.

15.4. The evolutionary assimilation of adaptive phenotypes

The idea that different kinds of organisms can be transformed one into the other has existed since the dawn of human culture. The overwhelming evidence of our senses however is that although individuals can adapt, species are essentially permanent. Jean Baptiste de Lamarck was the first modern naturalist to discard the concept of fixed species and to state explicitly that complex organisms have evolved from simpler ones. His theory of evolution, published in 1809, is best known for the concept that the functional adaptations acquired by organisms during their own lifetimes can be inherited directly by their offspring. A familiar interpretation of this idea is that the children of blacksmiths should have strong arms because they would inherit the massive muscles developed by their fathers. This theory is apparently so easily disproved that it has become conventional to scoff at any evolutionary theory that smacks of 'Lamarckism'. Nevertheless, even if we include natural selection as a major directive factor, there is no currently accepted explanation of how adaptive evolution can possibly proceed as rapidly and as effectively as it evidently does (see Chapter 4).

One example of the direct inheritance of an acquired character which is sometimes quoted is the transmission of modified cortical patterns in *Paramecium* (see Chapter 4). This can perhaps be dismissed as a feature of growth rather than inheritance. We have also already considered assimilation of the bithorax phenocopy in a stock of *Drosophila* flies treated with ether at the egg stage. This appears to be due to a cumulative effect on the flies' physiology which is transmitted maternally. However, with our current understanding of animal development, there are at least two other means by which acquired characters may credibly be introduced as new aspects of inheritance. One of these is distinctly 'Lamarckist', acting in the short term, the other is a much more complex theoretical construct, in which it is suggested that acquired characters could, over a long period of time, be brought under genetic control. These theories should be considered as supplementary to the more conventional neo-Darwinist and other ideas outlined at the beginning of the chapter, rather than alternatives to them. It is interesting that even Charles Darwin accepted Lamarck's suggestion that acquired characters can be inherited, although he felt this process to be subsidiary to evolution by natural selection.

Modification of maternal influences on development: short-term adaptation

The short-term adaptation theory relates to the accumulation of maternal RNA and protein products within the cytoplasm of the egg and is based on a suggestion by Gell (1984; see also Chapters 2 and 9). It will be recognized that a healthy, well-fed and well-adapted female is more likely to lay eggs containing a good balance of nutrients than would a starved unhealthy female in an alien environment, and that embryos developing from the former eggs would have the better chance of survival. However, the argument can be taken further if it is postulated that maternal way of life or well-being may exert a qualitative or differential quantitative effect on the mRNA or protein species she contributes to the cytoplasm of her eggs. Accumulation of a greater variety of molecular species or failure to accumulate the normal molecular spectrum would affect enzyme activities in the early embryo and could bring about profound effects on subsequent patterns of gene expression in the offspring. If the altered phenotype produced should lead to a similar imbalance in yolk content of the eggs laid by the offspring, then a self-perpetuating series of acquired phenotypes could become established.

In locusts a situation exists in which something of this sort may occur. As was mentioned in Chapter 4, locusts (*Locusta migratoria* and *Schistocerca gregaria*) can exist in two extreme phases, solitary and gregarious, which differ in behaviour, structure and colouration. All intermediate stages are also found. Locusts which develop in isolation tend to adopt the solitary phase and solitary females lay eggs which also tend to develop as solitary-phase insects. However, both the pattern of development and the type of egg laid by the new female is modified if conditions become crowded. Under crowded conditions individuals tend to develop into the gregarious phase and eggs laid under these conditions also tend to produce gregarious-phase individuals, or intermediate individuals if conditions later become less crowded. Thus it takes two or more generations for formation of fully gregarious or fully solitary-phase individuals.

Development of the solitary form seems to be related to excessive secretion of juvenile hormone. The simplest explanation for the maternal effect is that the composition of the locust egg cytoplasm depends on the social environment experienced by the mother and that this controls hormone secretion in the offspring. This situation, which allows alternative phenotypes to develop from the same genome, has adaptive value since the highly active, gregarious form develops wings that allow it to migrate in search of food, while the sluggish solitary form is less conspicuous to predators and has a much higher rate of egg production. However, there seems to be no theoretical reason why a new phenotype acquired by a species in this way, through an accidental or environmental effect on maternal physiology, should not become a self-perpetuating permanent feature if it were of permanent value to the species.

The Adaptive-Inductive-Genetic Theory of evolution

The second hypothesis accommodates many of the observations on genetics and development, described above and in earlier chapters, within a new theory of progressive genetic assimilation of adaptive phenotypes which links functional adaptation, regulative and mosaic aspects of development in a long-term evolutionary continuum.

As described in Chapter 2, embryologists working at the beginning of this century suggested that animal life develops according to two basic plans, mosaic and regulative. The parts of 'mosaic embryos', it was believed, develop independently, without interaction with other parts, whereas in 'regulative embryos' tissue interaction is a vital feature. We now know this distinction to be an over-simplification, both developmental modes seem to apply in varying degrees to all or most organisms. However, we should perhaps not yet abandon the concept that expression of individual genes, rather than development overall, may be controlled in accordance with comparable modes. We can conceive the situation in which important controls upon gene action are exerted both by factors intrinsic

to the cell which is expressing those genes, and by others which originate externally to the cell. Intra-cellular control could involve the switching of genes by
maternal mRNA translation products or maternal proteins present in the cytoplasm, or could entail cascades of gene activity in which the product of one gene
triggers others and so on (see Chapter 12). Extra-cellular controls could include
the effects of inductive molecules and hormones arising from other cells (internal
extra-cellular controls), as well as the influence of conditions outside the body
altogether (external extra-cellular controls). It is suggested that external adaptive
control features tend to become replaced by internal extra-cellular ones and then
by intra-cellular controls, as favourable alleles become assimilated by the genome
and new mutations arise. This, it is suggested, is a process complementary to
evolution directed by natural selection acting on genetic mutations which in terms
of the DNA are produced at random.

As a first step in developing the theory we need to consider the relationship
between functional adaptation to external conditions and embryonic induction.
Functional adaptation involves changes in cell physiology, behaviour and structure
which have far-reaching effects on the capacity of an organism to survive. It occurs
by a host of mechanisms, all of which relate ultimately to external conditions (see
above). Embryonic induction can be considered as the provision of, and cellular
adaptation to, defined *internal* environments, and in this light it represents a
relatively straightforward extension of cellular adaptation to the external
environment.

It is suggested that evolutionary progress involves initially the replacement of
external cues for adaptive response, by internal embryonic inductions acting late
in development. This could occur by genetic assimilation of favourable alleles
already present in the population, as proposed by Waddington (see above and
Chapter 4), selection always favouring predisposition towards the appropriate
adaptation. Eventually, it is suggested, appropriate adaptation could be achieved
without or before exposure to the external stimulus which invoked it in the
ancestors. For example, growth of feathers can be stimulated in competent avian
epidermis by pressure from within (see Chapter 4), although one of the most
important original stimuli was probably external chilling. Chilling still promotes
hair growth in mammals.

It is suggested that in descendants late-acting inductions are supplemented
or replaced by others acting progressively earlier in ontogeny, promoting
acquisition of inductive competence and bringing about heterochronic acceleration in the development of the adaptive phenotype. Thus the ontogeny of new
aspects of phenotype would become initiated at progressively earlier developmental stages as evolution proceeds. The genetic basis of this change could be
further genetic assimilation, possibly following the admixture of a new set of
genetic polymorphisms, due to the merging of the original population with
another from which it was previously isolated, or by a mutational event and subsequent dissemination of the mutant gene through the population.

Such ontogenetic progression would be most unlikely to occur if the specificity of an inductive response were defined by the inducer. As we know, it is governed by the state of competence of the responsive tissue, so that many different conditions may induce the same response in a competent target tissue (see Chapter 2). It is suggested that the regulative aspects of development operate within this range of extra-cellular controls and the situation is labile, in an evolutionary sense, provided that a sufficient degree of polymorphism is maintained among the relevant controlling alleles.

It will be appreciated that in this scheme the most ancient traits are specified at several levels in the control hierarchy, due to a temporal series of inductions followed by environmental modelling, while more recent ones depend on controls arising from the highest degree of interaction with the surrounding ecosystem. This could have several consequences which are in fact borne out by observation. For example, development of any feature would be canalized towards a defined phenotypic outcome and cell types would show progressive determination of phenotype (see Chapter 2). Ancestral features would appear progressively earlier in ontogeny and, being controlled at many levels, some very distant from environmental influence, would be well buffered against mutational or other disturbance. Recently acquired characters would be much more susceptible to change and since the most recently acquired are maintained only by genes that predispose towards favourable functional adaptations, reversed genetic assimilation could occur. The trait would then be lost, evolution could proceed in other directions and this would be seen as divergent maturation of the juvenile.

Mosaic development indicates an even more basic type of regulation, in which the gene-controlling elements are intrinsic to individual cells and tissues. These could include cytoplasmic organelles and other maternal contributions to the ovum, the incorporation of morphogens into the cell interior (see Chapter 2), and the development of self-contained biochemical systems that function relatively independently of extra-cellular events. Such features could be distantly ancestral in origin, being derived by further progression of the sequence described above or they could arise rapidly *de novo*, since the egg cytoplasm is really only a superficial phenotypic feature of the mother, albeit a very sophisticated one. In the latter situation cytoplasmic morphogens could be subject to the same evolving kinds of control as other aspects of phenotype, or be modified as described in the first model above.

Discussion of the theory

This model ties together a lot of the worrying inconsistencies of previous evolutionary theory and integrates adaptation, gene action and the regulative and mosaic aspects of ontogeny in a continuing sequence. It is stressed that in this scheme, apart from functional adaptation by animals individually responding to environmental stress, no change is innovative at the phenotypic level and in this it represents a radical departure from conventional neo-Darwinist thinking in which new phenotypes are considered to arise as a result of genetic change. Each step merely involves facilitation, stabilization, reinforcement or acceleration of development of a feature originally developed as a simple adaptation by an individual to the external world, plus selection of appropriately responsive individuals. This would allow evolution to proceed very rapidly in the 'right' direction, driven by the necessity for each part of every individual to adapt optimally for its role in relation to its surroundings.

When considered alongside current neo-Darwinist theory, the Adaptive-Inductive-Genetic model provides an explanation for the apparent co-existence of two different rates of evolution. It was suggested long ago by Goldschmidt that the gradual accumulation of mutations of minor importance (microevolution) proceeds in quite a different way from the major saltational leaps (macroevolution) that characterize the transition from one species to another. The theory outlined above accounts for saltation in that major phenotypic changes are ascribed to functional adaptation acting in conjunction with genetic assimilation, both of which we know can produce new phenotypes with rapidity (see Chapter 4). Paedo-morphosis is another means by which major evolutionary steps come about (see Chapter 13). It is suggested that microevolution proceeds in the conventional neo-Darwinist fashion, by gene mutation and selection.

Another outcome of adaptation-led evolution of this sort is that different species which colonize the same environment and therein adopt similar ways of life should evolve towards similar body forms. In these circumstances, convergent evolution is precisely what does happen. In Chapter 4, when comparing the bodies of three large aquatic carnivores — the ichthyosaur, the dolphin and the shark — the point was made that the environment itself seems to have moulded them into similar forms, even though their ancestors came from widely separated taxonomic groups. It now begins to look as if that is actually the case and that such environmental sculpturing proceeds from functional adaptation to assimilation of polymorphic alleles which favour that adaptation. Next internal inductions arise which produce the same phenotype in the absence of adaptation, and new mutations are acquired that lay its foundations at progressively earlier stages. Eventually in some species, appropriate morphogenetic gradients are laid down in the early embryo, and the cells acquire specialized cytoplasm which allows that development and differentiation to proceed undisturbed by extra-cellular influences. As with other aspects of neo-Darwinist theory, natural selection is a key feature, but in the model presented here the ecosystem exerts a directive as well as a selective influence.

Predictions of the theory

If the main theory outlined above is essentially valid, several consequences can be predicted which should be amenable to verification.

1. There should be a relationship between the level of control of genes which specify a character and the length of time that character has been in existence. We

would expect few very ancient features to be controlled solely by environmental factors or late-acting inductions. In contrast, external or superficial controls would be expected to be important in the development of recently evolved characteristics such as those which typify the mammals. At the cellular level, recently acquired aspects of cytodifferentiation should be controlled predominantly by extra-cellular influences, whereas very ancient ones should be regulated predominantly by intrinsic intra-cellular controls.

2. During the history of a taxonomic group we would expect the rate of incorporation of new features to be appreciably greater in its early compared with its later stages. This has been demonstrated from fossil evidence in the lungfishes.

3. Creatures which develop as predominantly 'mosaic' embryos, such as annelids, molluscs and ascidians, should currently evolve at maximal speeds related to the DNA mutation rate, whereas the evolution of those which develop as predominantly 'regulative' embryos, such as the vertebrates, should be virtually unrelated to mutation rate and should proceed much more rapidly.

4. Because phenotype modification is considered to be led by the demands of the environment and supported by a variety of favourable genotypes, adaptive phenotypes of similar type could occur independently, in unrelated individuals. According to conventional theory, variants of similar type are all descended from common ancestors among which a particular important mutation or combination of alleles first occurred. This has far-reaching implications in relation to the depression of viability that normally occurs with inbreeding.

If these predictions are substantiable this would lend considerable support to the Adaptive-Inductive-Genetic Theory as described above.

15.5. *The evolution of new cell types*

So far we have dealt implicitly with the evolutionary reorganization of parts or systems of the body that have already become established, the creation of new tissues and organs are altogether a different matter. During development the major distinctions between the different organs is initiated at the levels of differences between cells. Cytodifferentiation is thus the thin end of the wedge of organ differentiation. In evolutionary terms it is suggested that new cell phenotypes could have arisen by the evolution of determinative strictures applied to one or the other alternative phenotypes adoptable by a modulatory tissue. The example of bone, cartilage and dentine deriving from chondroid bone and osteodentine is described above. Another possibility is that a cell type might manifest a new type of modulation. This could occur by changes in the tissue environment, or a change in the timing of competence, or because of mutational changes arising in the DNA.

As discussed in Chapter 7, in some cases determination seems to *follow* overt cytodifferentiation, in others it precedes it. In the former case cytodifferentiation is apparently governed by extra-cellular physiological influences. It is suggested that this is the more recent situation, which in ancient systems has been replaced by intra-cellular controls intrinsic to that cell type. In the latter, determination precedes cytodifferentiation, as for example in insect imaginal discs.

The Leader Cell Hypothesis (see Chapter 7) provides an interesting new theory in this area, since it allows the possibility that a new aspect of tissue competence could evolve through admixture of only a small proportion of potential leader cells with a predisposition to differentiate in a new way among a large population of undetermined followers. The acquisition of the new option this would provide need have no overt effect on tissue differentiation unless and until it became exposed to the appropriate stimulus. This would therefore represent a semi-conservative advance, rather than the radical departure from the old situation that would be expected to occur if new cell types arise by simple mutation of key enzymes in already established cell types.

15.6. The evolution of biochemical pathways

Evolution can be considered also from a wholly biochemical viewpoint. An interesting question is whether biochemical pathways have evolved forwards or backwards. The answer seems to be both.

A convincing argument can be advanced that the first 'organisms' were complete heterotrophs which engulfed abiotically generated 'organic' molecules that already existed in the environment around them and incorporated these into their own structures. Such molecules are still being created on this planet around hydrothermal vents in the ocean floor. As these molecules became scarce in the primitive ocean the early life-forms derived from them found means to elaborate related molecules into those which were previously accepted and so exploited new resources. When the supply of these in turn became depleted, other molecules were used and so on. If we can accept that such simple 'organisms' could acquire or produce catalysts capable of facilitating the necessary molecular interconversions, we can envision whole pathways evolving stepwise in reverse order to that in which they now appear in our biochemical reaction schemes.

However, if we look more carefully we find that the final steps of several very basic pathways depend on aerobic metabolism, whereas the earlier ones are entirely anaerobic. The prime example is glycolysis, in which the TCA cycle and electron-transport chain responsible for efficient production of ATP are tagged on to the end of the anaerobic Embden–Meyerhoff pathway. The same occurs in the synthesis of sterols and carotenoids and in fatty acid synthesis. We believe that oxygen was unavailable when life was created and that aerobic metabolism became possible and indeed essential only after the Cyanobacteria had arisen and poisoned the atmosphere with oxygen derived as a byproduct of photosynthesis. The later steps in the above pathways must therefore post-date the acquisition of aerobic metabolism, which followed assimilation of mitochondrial endosymbionts into eukaryote cells (see Chapter 1).

Common sense also persuades us that the biochemical back-up needed for development of advanced systems must of necessity have involved elaboration of simpler ones. It seems obvious that the haemoglobin tetramer must be derived from monomeric evolutionary precursors. One would also expect that the biochemical pathways involved in the synthesis of feathers and fur would involve forward or sideways, but not backward extension of those that operate in the production of reptilian scales. It seems likely, therefore, that biochemical pathways can evolve in both directions, backwards to exploit a new food source, or forwards to elaborate a new addition to our molecular make-up.

At the same time the enzymes which catalyse these reactions are themselves evolving in terms of amino acid sequence, shape and catalytic properties. Such changes arise at the DNA level mainly as point mutations, but deletions, insertions and sequence rearrangements are also important.

15.7. Summary and conclusions

Current evolutionary theory is complex, but derives largely from 'neo-Darwinist' principles. These propose that the primary basis of phenotypic variation is gene mutation or genetic recombination and that new favourable mutations increase in frequency through natural selection. Neither these theories, nor those based on the assumption that some alleles are of neutral value, offer a convincing explanation for the great speed at which adaptive evolution can undoubtedly take place, nor for its apparent variations in pace. Moreover, ancient features are in evolutionary terms more stable than those more recently acquired, but no theory explains this inconsistency.

A new concept of evolution is outlined, which offers explanations for these paradoxes and which draws its inspiration from several unusual features of animal development. These are:

1. The observation that much of the postulated evolutionary history of an organism is repeated in an abbreviated form in early ontogeny and that evolutionary loss of function can be based on a change of gene expression, rather than loss of genetic information.

2. The recognition that phenotype modification by adaptation of the individual is very much more effective in producing well-adapted animals than straightforward selection of hereditary variants without individual adaptation.

3. Waddington's experiments on genetic assimilation, which show that 'phenocopies' can be produced *without* exposure to environmental stress by accumulation of alleles that favour phenocopy development.

4. The observation that the result of embryonic induction is defined by the

state of competence of the target tissue, rather than the nature of the inducer, and the deduction that a late induction could be strengthened or rendered redundant by an earlier one.

5. The experimental findings that cytodifferentiation can be directed by extracellular control of cell physiology.

6. The evidence of the fossil record that bone, cartilage and dentine replaced the intermediate modulatory tissues, chondroid bone and osteodentine.

The new evolutionary theory is based on three possibilities. The first of these is that the early development of embryos could be modified by environmentally induced variations in the spectrum of informative macromolecules which the mother contributes to the egg cytoplasm, and that this could lead to the foundation of new cytoplasmically inherited variations.

The second concept extends Waddington's idea that phenotypic and genotypic evolution can be achieved by accumulation of alleles already present in the population and which facilitate favourable adaptation to a specific environment. It is suggested that this represents the first step in the progressive genetic underpinning of phenotypic adaptation, pre-adapted states and, later, states of inductive competence, arising by the recognition and appropriation of progressively earlier internal stimuli for exploitation as internal inductions. It is envisaged that with evolutionary advance, late-acting inductions are reinforced or replaced by earlier ones, so that development of ancient traits is initiated progressively earlier in ontogeny and becomes buffered (canalized) by a sequence of controls. In contrast, recently acquired traits arise late in ontogeny and would be more accessible to forces capable of causing modification. This series of events should cause evolution to proceed very rapidly in an adaptive direction. Eventually the most basic aspects of phenotype might become defined by morphogens and other gene-controlling molecules laid down within the cells of the early embryo, strongly protected from disruptive external influences. This would represent the acquisition of 'mosaic' as compared to 'regulative' control of phenotype. Irreversible steps in this sequence would correspond to the acquisition and fixation of new favourable alleles and depletion from the population of genetic polymorphisms that could cause divergent development.

The third aspect of the theory deals with the evolution of new cell types. It is suggested that cells of new types might arise by the acquisition of a new modulatory alternative for a tissue, or the stabilization of one or the other alternative modulatory forms it can already adopt. This could occur through the mobilization, by inductive or mutational means, of genes not normally expressed in that tissue. Alternatively, a new phenotypic option could arise through the introduction of a new type of 'leader cell' into a population of competent followers. It is suggested that biochemical pathways may evolve both forwards and backwards.

Several predictions of the main theory are outlined which offer a test of its validity.

Bibliography

Bonner, J. T. (ed), *Evolution and Development*. Springer-Verlag, Berlin (1982).

Cohen, J., Maternal constraints on development. In *Maternal Effects in Development*, edited by D. R. Newth and M. Balls. Cambridge University Press, Cambridge, pp. 1–28 (1979).

Darwin, C., *On the Origin of Species by Means of Natural Selection*, 6th edn (1972). Murray, London (1859).

Day, W., *Genesis on Planet Earth*. Shiva Publishing, Nantwich (1981).

De Beer, G., *Embryos and Ancestors*. Clarendon Press, Oxford (1962).

Dillon, L. S., *The Genetic Mechanism and the Origin of Life*. Plenum, New York (1978).

Dobzhansky, T., Azala, F., Stebbins, G. L. and Valentine, J. W., *Evolution*. Freeman, San Francisco (1977).

Eigen, M., Gardiner W., Schuster, P. and Winkler-Oswatitsch, R., The origin of genetic information. In *Evolution Now, A Century after Darwin*, edited by J. Maynard Smith. Nature Publications, pp. 11–27 (1982).

Eldredge, N. and Gould, S. J., Punctuated equilibria: an alternative to phyletic gradualism. In *Models in Palaeobiology*, edited by T. J. M. Schopf. Freeman Cooper, San Francisco, pp. 82–115 (1972).

Flickinger, R. A., Sequential gene action, protein synthesis and cellular differentiation. *Int. Rev. Cyt.*, **13**: 75 (1962).

Flickinger, R. A., Relation of an evolutionary mechanism to differentiation. *Differentiation*, **3**: 155 (1975).

Frazetta, T. H., *Complex Adaptations in Evolving Populations*. Sinauer Associates, Sunderland, Mass. (1975).

Gell, P. G. H., Destiny and the genes: genetic pathology and the individual. In *The Encyclopedia of Medical Ignorance*, edited by R. Duncan and M. Weston-Smith. Pergamon Press, Oxford (1984).

Goldschmidt, R. B., Ecotype, ecospecies and macro-evolution. *Experientia*, **4**: 465 (1948).

Gould, S. J., *Ontogeny and Phylogeny*. The Belknap Press of Harvard University Press, Cambridge, Mass. (1977).

Gould, S. J., *Hen's Teeth and Horse's Toes*. Penguin, Harmondsworth, Middlesex (1983).

Gutfreund, H. (ed.), *Biochemical Evolution*. Cambridge University Press, Cambridge (1981).

Hall, B. K., Epigenetic control in development and evolution. In *Development and Evolution*, edited by B. C. Goodwin, N. J. Holder and C. G. Wylie. Cambridge University Press, Cambridge, pp. 353–379 (1983).

Hall, B. K., Developmental mechanisms underlying the formation of atavisms. *Biol. Rev.*, **59**: 89 (1984).

Halvorson, H. O. and Van Holde, D. E., *The Origins of Life and Evolution*. Alan R Liss, New York (1981).

Harrison, J. W. H., Induced pigmentation in the pupae of the butterfly *Pieris napi* L. and their inheritance. *Proc. R. Soc. Lond.* [*Biol.*], **102**: 347 (1928).

Ho, M.-W. and Saunders, P. T., *Beyond Neo-Darwinism*. Academic Press, London (1984).

Holder, N., The vertebrate limb: patterns and constraints in development and evolution. In *Development and Evolution*, edited by B. C. Goodwin, N. J. Holder and C. G. Wylie. Cambridge University Press, Cambridge, pp. 399–425 (1983).

Horder, T. J., Embryological bases of evolution. In *Development and Evolution*, edited by B. C. Goodwin, N. J. Holder and C. G. Wylie. Cambridge University Press, Cambridge, pp. 315–352 (1983).

Horowitz, N. H., On the evolution of biochemical synthesis. *Proc. Natl. Acad. Sci. USA*, **31**: 153 (1945).

King, M.-C. and Wilson, A. C., Evolution at two levels in humans and chimpanzees. *Science*, **188**: 107 (1975).

Kollar, E. J. and Fisher, C., Tooth induction in chick epithelium: expression of quiescent genes for enamel synthesis. *Science*, **207**: 993 (1980).

Lamarck, J. B., *Zoological Philosophy*, translated by H. Elliot (reprinted 1963). Hafner, New York (1809).

Lewontin R. C., *The Genetic Basis of Evolutionary Change*. Columbia University Press, New York (1974).

Ninio, J., *Molecular Approaches to Evolution*. Pitman, London (1982).

Riedl, R., *Order in Living Organisms, A Systems Analysis of Evolution*. Wiley, Chichester (1978).

Roberts, D. F., *Paradoxes of Evolution in Recent Man. Twentieth Raymond Dart Lecture*. Witwatersrand University Press, Johannesburg, pp. 1–16 (1983).

Romer, B. S., *The Vertebrate Body*, 2nd edn. W. B. Saunders, Philadelphia (1955).

Schmalhausen, I. I., *Factors of Evolution*, translated by I. Dordick, edited by T. Dobzhansky. Blakiston Company, Philadelphia (1949).

Schopf, J. W., Evolution of the earliest cells. *Sci. Am.*, **239** (3): 84 (1978).

Schwemmler, W., *Reconstruction of Cell Evolution: A Periodic System*. C. R. C. Press Inc., Boca Raton, Florida (1984).

Sinclair, D., *Human Growth After Birth*. 2nd edn. Oxford University Press, London.

Slijper, E. J., Biologic-anatomical investigations on the bipedal gait and upright posture in mammals, with special reference to a little goat, born without forelegs. II. *Proc. Ned. Akada. Wetensch. Amsterdam*, **45**: 407 (1942).

Smith, J. M., *The Theory of Evolution*. Penguin, Harmondsworth, Middlesex (1975).

Stebbins, G. L. and Ayala, F. J., Is a new evolutionary synthesis necessary? *Science*, **213**: 967 (1981).

Torrey, T. W. and Feduccia, A., *Morphogenesis of the Vertebrates*, 4th edn. Wiley, Chichester (1979).

Tribe, M. A., Morgan, A. J. and Wittaker, P. A., *The Evolution of Eukaryote Cells*. Edward Arnold, London (1981).

Van Brink, J. M. and Vorontsor, N. N., *Animal Genetics and Evolution*. Dr W. Junk, The Hague (1980).

Waddington, C. H., *The Evolution of an Evolutionist*. Edinburgh University Press, Edinburgh (1975).

Westoll, T. S., On the evolution of the Dipnoi. In *Genetics, Palaeontology and Evolution*, edited by G. L. Jepsen, E. Mayr and G. G. Simpson. Princeton University Press, Princeton, pp. 121–184 (1949).

Wigglesworth, V. B., *The Life of Insects*. Weidenfeld and Nicolson, London (1964).

Wilson, A. C., Carlson, S. S. and White, T. J., Biochemical evolution. *Ann. Rev. Biochem.*, **46**: 573 (1977).

Young, J. Z., *The Life of Vertebrates*. Clarendon Press, Oxford (1950).

Zuckerkandl, E., The evolution of hemoglobin. *Sci. Am.*, **212** (5): 110 (1965).

Glossary

Abundance	The abundance of a species of mRNA is the average number of molecules of that species per cell.
Acceptor splicing site	The boundary between the right (3′) end of an intron and the left (5′) end of an exon.
Acrasin	A chemotactic substance produced by certain slime moulds which causes the aggregation of their cells.
Actin	A protein occurring in muscle where it combines with myosin to form actomyosin. Also found in flagella and in microfilaments in the cytoplasm of many cell types.
Actinomycin D	An antibiotic produced by *Streptomyces chrysomallus* that inhibits transcription by eukaryote RNA polymerase, especially transcription of rRNA by Pol I.
Activation	The stimulation of an unfertilized ovum by some means other than fertilization by a spermatozoon, to undergo cleavage.
Adaptation	Some feature of an individual's phenotype, whether inherited or not, that improves its chances of survival and reproduction in a specific environment.
Adenine	A purine base that pairs with thymine in the DNA double helix.
Adenosine	The nucleotide containing adenine as its base.
Adenylate cyclase	A membrane-bound enzyme that converts ATP to cyclic AMP on stimulation by a hydrophilic hormone acting through a membrane-bound receptor protein and an intermediary called G protein.
Adrenalin(e)	*See* 'Epinephrine'.
Adrenocorticotrop(h)ic hormone (ACTH)	A polypeptide hormone secreted by the adenohypophysis which controls the growth and activity of the adrenal cortex.
AER	Apical ectodermal ridge.
Aerobic metabolism	The utilization of elementary oxygen for the oxidation of organic compounds, with the formation of water and carbon dioxide, and the liberation of energy. Aerobic metabolism utilizes the enzymes of the Krebs tricarboxylic acid (TCA) cycle and the cytochrome system, which are carried on the mitochondria. (cf. 'Anaerobic metabolism').
Agouti	The grizzled colour of the fur of many wild mammals resulting from alternate light and dark bands on the individual hairs.
Allele	One of several alternative forms of a gene occupying a given locus.
Allelic exclusion	The expression in a clone of immune cells of only one allele of an immunoglobulin gene for which the organism is heterozygotic.
Allometric growth	Growth in size which correlates with change of shape. It often approximates to the expresson, $y = bx^a$ where y = the size of a body part, x = whole body size and a and b are constants. If a > 1

the organ shows positive allometry; if a < 1 it shows negative allometry.

Allometry — Change of shape correlated with increase or decrease in size.

Allosteric protein — A protein that has the capacity to modify its interaction with another molecule due to changes in shape following reversible interaction between another site on that protein and a third molecule.

Alpha-amanitin — An octapeptide derived from the poisonous toadstool *Amanita phalloides* that inhibits transcription by eukaryotic RNA polymerases, especially transcription of mRNA precursors by Pol II.

Alpha helix — A right-handed helix which forms the most stable structure in most polypeptide chains (*cf.* 'Beta pleated sheet').

Alternation of generations — The occurrence in the typical life-cycle of a species of two or more different forms produced in a different manner, generally an alternation of a sexually produced with an asexually produced form. This occurs for example in the hydroids, jelly fish and tapeworms.

Alu-family — A set of dispersed and similar sequences each about 300 base pairs long in the human genome.

Amino acid — The basic building block of polypeptides and proteins, carrying both an amino ($-NH_3$) and a carboxyl ($-COOH$) group.

AMP — Adenosine monophosphate, a nucleotide composed of adenosine plus one phosphate group.

Anaerobic metabolism — Fermentation, or liberation of energy by breakdown of organic substances without utilization of elemental oxygen as electron acceptor (*cf.* 'Aerobic metabolism').

Analogy — A likeness in certain respects, a similarity between organisms or their parts due to independent evolution of a similar feature, for example, the wings of bats and butterflies (*cf.* 'Homology').

Animal pole — The point on the surface of an egg nearest to its graded distribution of cytoplasmic components. Yolk is usually at lowest concentration at the animal pole (*cf.* 'Vegetal pole').

Antero-posterior axis — The axis from front to rear. With respect to a human hand the axis that passes from thumb to little finger.

Antibody — A protein (immunoglobulin) molecule produced by the immune system that recognizes a particular foreign antigen and binds to it.

Anticoding strand — *See* 'Non-codogenic strand'.

Anticodon — A triplet of nucleotides in a constant position in the structure of transfer RNA that is complementary to the codon(s) in messenger RNA to which the transfer RNA responds.

Antigen — A molecule which triggers the production of antibodies that will specifically bind to it.

Apoenzyme — The protein part of an enzyme which requires a co-enzyme to become a holo-enzyme.

Apical ectodermal ridge (AER) — A ridge of epithelium of ectodermal origin situated at the tip of the vertebrate limb bud.

Apoinducer — In positively controlled prokaryote operons a protein that binds to the DNA and switches on transcription by RNA polymerase.

Archenteron — The central cavity within the gastrula of many embryos, which later becomes the gut cavity.

Atavism — Recurrence in descendents of a character which had been possessed by an ancestor, after an interval of several or many generations.

ATP — Adenosine triphosphate, a nucleotide cofactor made up of adenosine and three phosphate groups. Two of the phosphates are readily lost to form either adenosine diphosphate (ADP) or adenosine monophosphate (AMP) with the release of energy. ATP acts as a form of 'energy currency' in all living systems.

Autogamy	A sexual process found in some Protozoa and Diatoms in which the nucleus of an individual divides into two parts which reunite.
Auto-immune disease	A disease state in which the animal mounts an immune response against a constituent of its own tissues.
Autoneuralization	The process of formation of a neural tube by excised gastrula ectoderm in the absence of stimuli from other tissues.
Autoradiography	A means of detecting radioactively labelled molecules by their effect in creating an image on photographic film.
Autosome	In species which have sex chromosomes, any chromosome that is not a sex chromosome.
5-Azacytidine	An analogue of cytidine with the fifth carbon atom replaced by nitrogen.
Barr body	In female mammals the condensed X chromosome which appears as a dense body in somatic cell nuclei.
Basal lamina	The collagenous basement membrane which underlies most epithelia.
Base pair	A pair of hydrogen-bonded bases, one purine and one pyrimidine, that when linked beside other base pairs constitute the DNA double helix.
Beta pleated sheets	A regular folding of polypeptide chains found in some proteins (*cf.* 'Alpha helix').
B-form DNA	The normal right-handed double helical conformation of DNA (*cf.* 'Z-form DNA').
Bilateral symmetry	Capable of being halved in one and only one plane in such a way that the two halves resemble mirror images of one another. Bilateral symmetry is characteristic of freely moving animals.
Binary fission	Reproduction by splitting into two equal parts.
Bithorax complex (BX-C)	A series of closely linked alleles in *Drosophila melanogaster* that define segment identity. According to the Lewis model they are activated sequentially, segment by segment, in a progressive fashion down the length of the body.
Blastocoele	The cavity which in many species appears within the mass of cells formed by cleavage of the fertilized egg, marking the blastula stage.
Blastoderm	Superficial sheet of cells formed as a result of cleavage of a yolky egg, such as a bird's or insect's egg, corresponding to the blastula stage of development.
Blastodisc	The disc of material on the surface of the yolk of a bird's egg which gives rise to the embryo.
Blastomere	Any of the embryonic cells produced during the first few cleavages of the zygote.
Blastopore	A transitory opening on the surface of an embryo at the gastrula stage by which the archenteron communicates with the exterior. In many species it later becomes the anus.
Blastula	The stage of embryonic development at or near the end of cleavage and immediately preceding gastrulation. In most species it consists of a hollow ball of cells.
CAAT box	Part of a conserved sequence at about 75 bases upstream from the cAp site of eukaryote genes.

Calcium gate · A specialized pore in the plasma membrane which opens when the membrane becomes depolarized, or following binding of a ligand, allowing Ca^{2+} ions to enter the cell where they function as a second messenger.

Calmodulin · A highly conserved allosteric calcium-binding protein, present in all animal and plant cells that have been examined, which plays an important part in processes regulated by Ca^{2+} ions.

Canalization of development · The concept, illustrated by the epigenetic landscape, that as development proceeds the number of alternative differentiative pathways open to an embryonic cell becomes progressively restricted.

cAp site · The transcription initiator site. An adenine residue surrounded by pyrimidines at which transcription commences, defining the site which in mRNA carries the cap.

Capping · The addition to the 5′ end of eukaryotic mRNA of a guanine nucleotide in reverse polarity, coupled with methylation of this additional guanine plus the first two or three adjacent nucleotides.

Casein · The principle protein constituent of milk.

Caste · A type of structurally and functionally specialized individual among a colony of social insects. For example, in honey bee colonies there are three castes: queens, workers and drones.

cDNA · Complementary DNA or copy DNA. A single-stranded DNA complementary to an RNA molecule and synthesized from it by reverse transcription *in vitro*.

Cell hybrid · *See* 'Hybrid Cell'.

Cell lineage · The succession of cells from the zygote onwards, culminating in a particular cell population. A pedigree of cells related through asexual division. Developmental history in terms of descent by cell division of later cells from earlier cells.

Centriole · A self-reproducing organelle of characteristic structure that functions during the separation of chromosomes at cell division, being situated at the end of the mitotic spindle.

Centromere · The part of the chromosome located at the point lying on the equator of the spindle at metaphase and dividing at anaphase, controlling the separation of the chromosomes. The point of attachment of the spindle fibre, the kinetochore.

Cephalization · Specialization of the anterior end of a bilaterally symmetrical animal as the site of the mouth, the principal sense organs and the major ganglia of the central nervous system. Evolutionary formation of the head.

Chalone · Internal secretion believed to inhibit or depress cell proliferation, typically in the organ which produced it. Some chalones have been isolated and characterized as glycoproteins.

Character · An attribute of the individuals within a species for which heritable differences can be defined.

Charge relay network · A system which exists in enzymes that enables the catalytic activity of the active group to be enhanced by exchange of protons with amino acids elsewhere in the molecule.

Chiasma (plural chiasmata) · The cytological manifestation of crossing over.

Chondroid bone · An ancestral tissue intermediate in character between bone and cartilage.

Chorion · The superficial envelope of an insect egg. The embryonic membrane of an amniote vertebrate consisting of outer ectodermal epithelium plus mesoderm. In mammals, the superficial layer enclosing all the embryonic structures.

Chromatid   One of two replica chromosomes produced by chromosomal replication before their separation into daughter nuclei.

Chromatin   The complex of DNA, RNA and protein in the nuclei of eukaryote cells, originally recognized by its reaction with stains specific for DNA.

Chromocentre   A dark-staining structure formed by the co-aggregation of heterochromatin from different chromosomes.

Chromosome   A discrete unit of the eukaryote genome carrying many genes and consisting of a very long molecule of DNA complexed with proteins and RNA. Chromosomes are visible as morphological entities only during cell division.

Chromosome puff   An expansion of a band of a polytene chromosome associated with active transcription in that region.

Cis configuration   In coupling, on the same chromosome as distinct from its homologue (*cf.* '*Trans* configuration').

Cisternae   Flattened fluid-filled vesicles of the Golgi body or endoplasmic reticulum.

Cistron   Equivalent to a structural gene in comprising a unit of DNA representing a protein. The term is derived from the *cis/trans* test. In a double heterozygote, if two mutations are in the same cistron wild-type gene product will result when they are in *cis* configuration, but not when in *trans*.

Citric acid cycle   *See* 'Tricarboxylic acid cycle'.

Cleavage   Repeated subdivision of the zygote cytoplasm accompanying nuclear mitosis, with formation of smaller cells. Scision of a polypeptide, as for example in the conversion of pro-insulin into insulin.

Clone   A large number of cells, molecules or whole organisms produced from a single ancestor with which they are genotypically identical.

Coding strand   *See* 'Codogenic strand'.

Codogenic strand   Sense strand. The strand of duplex DNA which acts as template for RNA synthesis. It is identified by its ability to hybridize specifically with its complementary mRNA..

Codon   *See* 'Triplet codon'.

Co-enzyme   A non-protein accessory substance necessary for the proteinaceous part of an enzyme (apoenzyme) to function and which is only weakly bound to the protein (cf. co-factor or prosthetic group).

Collagen   Literally 'maker of glue'. A class of glycoprotein of highly defined structure which is the major contributor to the extra-cellular matrix. It contains an unusual amino acid, hydroxyproline, that occurs in very few other proteins.

Collagen helix   A special kind of helical secondary structure found in collagen which differs from the more usual alpha helix of other proteins.

Commitment   Usually considered synonymous with 'Determination' (q.v.).

Compartment (subcellular)   An anatomically, biochemically, or temporally isolated aspect of cell metabolism.

Competence   Capacity of a tissue to respond appropriately to an inductive stimulus. Tissues acquire competence as a result of their past histories of inductive experience.

Conjugation   The process by which sexual reproduction is effected in most ciliates. Two individuals partially fuse, their macronuclei disintegrate and each micronucleus undergoes meiosis to produce two gamete nuclei. These are then exchanged to produce a zygote nucleus containing genetic material from both partners in each individual. The organisms then separate and their nuclei undergo further divisions, each ultimately giving rise to a new macronucleus and micronucleus.

Contact guidance The direction of cell movement by physical features of the substratum.

Contact inhibition of growth *See* 'density dependent inhibition of growth'.

Contact inhibition of movement Cessation of cell motility caused by contact with another cell.

Controlling element *See* 'Transposable element'.

Core particle The DNAase digestion product of the nucleosome that retains the histone octamer, together with 146 base pairs of DNA.

Co-repressor A small molecule that triggers repression of transcription by binding to a regulator protein.

Cortex The plasma membrane of an egg, together with the immediately underlying layer of semi-solid cytoplasm.

Cortisol Hydrocortisone; a glucocorticoid very similar to cortisone. (Cortisone is produced by the adrenal body and exerts many complex effects including healing, reduction of inflammation and promotion of carbohydrate formation from fat and protein.)

C_0t curve A graph of the fraction of DNA reannealed relative to normalized time (C_0t), used for studying the renaturation of denatured DNA.

C_0t value The product of initial DNA concentration and elapsed time of incubation in a DNA reassociation reaction. Cot is the value required to proceed to half completion and is directly proportional to the unique length of the reassociating DNA.

Covalent bond The union of a pair of atoms by the sharing of a pair of electrons.

Crossing-over The reciprocal exchange of material between homologous chromosomes that occurs during meiosis and is responsible for genetic recombination of linked genes.

Crystallins A class of proteins of diverse structure found in the vertebrate eye lens. Their close packing contributes to its optical properties.

C terminus The end of a polypeptide at which the carboxyl ($-COOH$) group which could participate in a peptide bond is free.

C value The haploid DNA content of a eukaryotic cell.

Cyclic AMP (cAMP) A form of AMP in which the phosphate group is bonded to both the $3'$ and $5'$ positions of the ribose, forming a ring-like molecule. Cyclic AMP is concerned with regulation of metabolism and gene expression, constituting the 'second messenger' in the action of some hydrophilic hormones.

Cytodifferentiation The creation of phenotypic differences between cells within the body. Cells are said to be differentiated when they manifest forms or physiologies that distinguish them from one another, or which represent the properties of mature cells.

Cytidine The nucleoside containing cytosine as its base.

Cytoplasm All the protoplasm of a cell apart from the nucleus. It usually includes organelles such as mitochrondria and the Golgi apparatus, plus smaller bodies such as ribosomes, in a transparent, slightly viscous fluid.

Cytosine A pyrimidine base that pairs with guanine in double stranded DNA.

Cytoskeleton	The complex network of protein filaments that confer form, organization and capacity for movement upon the cell cytoplasm.
Denaturation	Conversion of DNA from the double-stranded to the single-stranded state; 'melting' of DNA.
Density-dependent inhibition of growth	Inhibition of mitosis brought about by cell contact.
Desmosome	A specialized structure of the cell surface that is involved in binding cells to one another or to a basement membrane. There are three main types: belt desmosomes, spot desmosomes and hemidesmosomes.
Determinate growth	Growth which terminates at pre-set limits.
Determination	The process by which the differentiation potential of a cell or nucleus becomes restricted during ontogeny.
Determined	A term applied to embryonic cells and tissues, meaning that their differentiative fate is fixed so that thereafter they can form only one particular type of adult tissue or organ.
Development	Ontogeny. The enactment of a series of events which bring about the conversion of a fertilized ovum into a mature individual. The elaboration of a simple type of organization into a more complex one.
Dictyostelium	A genus of slime moulds, a group which in some respects represents the transition between protozoa and fungi. They have the remarkable property of being able to exist as amoeboid unicellular individuals, as well as aggregates that develop a defined structure and behave as integrated multi-cellular creatures.
Differentiation	*See* 'Cytodifferentiation'.
Diploid	Having two complete sets of chromosomes; an individual with two complete sets of chromosomes.
Diptera	The large order of two-winged flies.
Directed somatic mutation theory	An early theory of cytodifferentiation which suggested that phenotypic distinctions between cell types arise due to the progressive acquisition during ontogeny of irreversible modifications to the nuclear DNA. In its original form the theory was rejected, but some of its elements have been incorporated into later more acceptable theories.
DNA	Deoxyribonucleic acid. The genetic material of most organisms. A very long molecule consisting of numerous nucleotides linked side by side into two chains hydrogen bonded together and coiled into a helix. Each nucleotide contains the sugar deoxyribose, plus a base, either guanine, adenine, cytosine or thymine. The sequence of these bases provides the molecular basis for inherited information.
DNAase	An enzyme that digests DNA.
Domain	(1) A discrete continuous part of the amino acid sequence of a protein that can be equated with a particular function. (2) An extensive region of chromosome, including an active gene, which has heightened sensitivity to degradation by the enzyme DNAase-1. (3) A region of a chromosome within which supercoiling is independent of that in other regions.
Dominant	The allele in diploid organisms which determines the phenotype when present in heterozygous combination with another allele, which is therefore recessive to it.
Donor splicing site	The boundary between the right (3′) end of an exon and the left (5′) end of an intron in hnRNA.

DOPA (L-DOPA)
: 3,4-Dihydroxyphenylalanine, produced by hydroxylation of tyrosine by tyrosinase.

Dorso-ventral axis
: The axis from the upper to the lower surface; with respect to a human hand, the axis from the back of the hand to the palm.

Downstream
: Sequences located in the direction followed by RNA polymerase during gene expression. The coding region is downstream from the initiating codon.

Drosophila
: Fruit fly or banana fly. A genus of the order Diptera. Much of our knowledge of genetics and development is derived from studies on *Drosophila melanogaster* introduced as a laboratory species by T. H. Morgan and colleagues.

Ecdysone
: Insect moulting hormone. A steroid hormone produced by the prothoracic gland in insects and the Y-organ in Crustacea, which stimulates growth and moulting.

Eco RI
: A commonly used restriction endonuclease that recognizes and cleaves the sequence GAATTC.

Ectoderm
: The superficial germinal layer of the gastrula. It gives rise to the epidermis, the nervous system and the nephridia.

Effector
: The end-product of a metabolic pathway which has the property of controlling its own synthesis by feedback inhibition of an enzyme earlier in the pathway.

Elastin
: An elastic protein, the principle component of the elastic fibres of connective tissue.

Elongation
: The second phase of translation of a mRNA molecule by a ribosome, during which amino acids are added to the polypeptide chain.

Embryo
: An animal in the process of development, before hatching or birth.

Embryonic field
: *See* 'Field'.

Embryonic induction
: *See* 'Induction'.

Endoderm
: One of the germinal layers of a triploblastic gastrula composed of cells that have moved into the embryo's interior. It gives rise to the alimentary canal and digestive glands.

Endoplasmic reticulum
: A meshwork of double membranes in the cytoplasm which are continuous with the plasma and nuclear membranes. When lined with ribosomes it is described as 'rough'.

Enhancer
: A nucleotide sequence that increases the utilization of some eukaryote 'promoters' in *cis* configuration, but which can function both upstream and downstream of the promoter.

Enzyme
: A protein with the capacity to catalyse a chemical reaction.

Epigenetic landscape
: A visual model proposed by Waddington to illustrate the progressive determination and differentiation of cells during development. The usual representation is of a ball poised at the top of an eroded slope. The ball represents a cell's cytoplasmic state while the diverging pattern of valleys represents the influence of the genetic system. As the ball rolls down the slope at first there are many pathways open to it, but the number becomes increasingly restricted as development proceeds.

Epigenetics
: The study of development in relation to inheritance, or of the interpretation of genotype into phenotype; the causal analysis of development. Derived from 'epigenesis', the word used by Harvey to distinguish his concept of gradually emerging complexity from the doctrine of preformation. The word was coined by C. H. Waddington to link this approach with the science of genetics.

Epimorphosis	Regeneration by outgrowth of new tissue from a wound surface (*cf.* 'Morphallaxis').
Epinephrine	Also known as 'Adrenaline', a hormone secreted by the adrenal medulla and by nerve endings of the sympathetic nervous system in response to stress. Its action is to increase blood pressure by stimulation of cardiac action and constriction of the blood vessels.
Epiphysis	The separately ossified ends of a growing mammalian limb bone or vertebra. Growth occurs by encroachment of new bone from the main body of the bone into the epiphysis and formation of new cartilage on the opposite side.
Erythropoietin	A glycoprotein released from the kidney which stimulates bone marrow cells to produce erythrocytes.
Euchromatin	The chromatin which makes up and shows the staining behaviour of the bulk of the chromosomes, (*cf.* 'Heterochromatin'). It contains the active genes, is decondensed during interphase, condensed during cell division.
Eukaryote (eucaryote)	Literally 'true nucleus'. Having the genetic material in the form of chromosomes contained within a nucleus separated from the cytoplasm by a nuclear membrane (*cf.* 'Prokaryote').
Evolution	Descent with change. The cumulative modification of characteristics of a population of organisms related by descent, occurring during the course of successive generations.
Excision	The removal of introns from hnRNA.
Exon (or extron)	A coding segment of an interrupted eukaryote structural gene. Exons are represented in the mature messenger RNA product.
Expressivity	The strength of phenotypic manifestation of a given mutant allele in terms of deviation from the normal phenotype.
Extra-cellular matrix	The material in which animal cells are embedded.
Facultative heterochromatin	A transcriptionally inert state of DNA sequences which can also exist as active copies (*see* 'Heterochromatin').
Feedback inhibition (end-product inhibition)	The inhibition of the synthesis of a group of enzymes in a metabolic pathway by its end-product, which combines with and inhibits an enzyme earlier in the pathway.
Fertilization	The union of male and female gametes to form a zygote.
Fibronectin	A fibre-forming, extra-cellular glycoprotein that promotes cell adhesion.
Field (embryonic)	A part of the body which during development has spatial unity with respect to the organization of its constituent parts and in which the fate of these parts is determined by the whole system. For example, the lens field if subdivided will form a perfect lens in each of its two halves, as the fate of the individual lens cells is controlled by an overall influence governing the cell community.
Fitness	The relative probability of survival and reproduction of an individual or genotype.
Foetus	A mammalian embryo after recognizable appearance of the main features of the fully developed animal, in man after about two months' gestation.
Foldback DNA	DNA which renatures immediately after melting, due to self-complementation with formation of hairpin loops. This occurs because it contains inverted repeat sequences.
Follicle-stimulating hormone (FSH)	A glycoprotein gonadotrophic hormone secreted by the anterior pituitary which stimulates growth of Graafian follicles, oestrogen secretion and spermatogenesis.

Gap junction
: A type of porous linkage between animal cells which allows physiological and electrical communication between them. Also known as a 'junctional complex'. At gap junctions the interacting plasma membranes are separated by a 'gap' of 2–4 mm (*cf.* 'Tight junction').

Gastrula
: The stage of embryonic development following the blastula, when the movements of gastrulation takes place.

Gastrulation
: The complex of cell movements which occurs in almost all animals following cleavage, giving rise to the germinal layers. As a result of gastrulation most of those cells whose descendants will form the future internal organs are carried to their definitive positions inside the embryo.

Gene
: The hypothetical entity which defines a particular character. At one time it was assumed that the hereditary units defined by mutation, recombination and function were the same. With the abandonment of that belief the term 'gene' became restricted to the unit of function, i.e., the 'Cistron'. It is now considered to be a segment of DNA which occupies a (usually) fixed chromosomal locus, which through transcription has a specific effect on phenotype and which can mutate to various allelic forms. It includes regions preceding and following the coding region as well as intervening sequences (introns) between the individual coding segment or exons. Alternatively, the information encoded by that segment of DNA.

Gene amplification
: The production of additional copies of a gene sequence during ontogeny. The extra copies may be incorporated into the chromosome or be extrachromosomal, as is the case with amplified rRNA sequences in the nucleolus.

Gene cloning
: The production of multiple copies of a selected gene sequence, usually by insertion into the genetic material carried by a microorganism, which is then encouraged to proliferate.

Genetic assimilation
: The process by which a modified phenotype initially formed as a response to unusual environmental conditions comes under the control of the genotype in descendants selected for expression of the modification. It probably usually occurs through the collection together in selected individuals of previously hidden alleles already present in the population which favour expression of the modified phenotype.

Genetic background
: The genetic 'context' within which a specific gene operates; the remainder of the genome.

Genetic code
: The set of correspondences between nucleotide triplets in mRNA and amino acids in the polypeptides.

Genetic drift
: Random fluctuations of gene frequencies unrelated to selection.

Genetic mosaic
: An individual composed of cells of different genotypes. Mosaics can be formed as a result of somatic mutation occurring during ontogeny, through transplantation of cells between individuals or fusion of individuals other than monozygotic twins.

Genetics
: The science of heredity and variation.

Genome
: A term used in the strict sense to refer to all the genes carried by a single gamete. Less strictly it is used to describe the total genetic endowment of an individual, inclusive of non-coding DNA sequences.

Genotype
: The genetic constitution of an organism, as distinguished from its physical characteristics or 'phenotype'.

Germ cells	Gametes and their precursor cells.
Germplasm (poleplasm)	A specialized type of cytoplasm in the ovum, which becomes restricted to the future germ cells.
Germinal layers	The main layers or groups of cells which can be distinguished in an embryo during and after gastrulation. In triploblastic animals they are the ectoderm, the mesoderm and the endoderm.
Globin	The protein constituent of haemoglobin.
Glucagon	A polypeptide hormone secreted by the α-cells of the islets of Langerhans in the pancreas, which stimulates glycogenolysis in the liver.
Glycolysis	The anaerobic breakdown of glucose via pyruvic acid, with production of acetyl CoA (which enters the TCA cycle), lactic acid or ethanol.
Glycosaminoglycan	Long unbranched polysaccharide chains composed of repeating disaccharide units, one of the two sugars in the unit being an amino sugar. (Formerly known as 'Mucopolysaccharide'). A general term which includes hyaluronic acid, chondroitin sulphate, dermatan sulphate, heparan sulphate, heparin and keratan sulphate.
Golgi apparatus	A cytoplasmic body concerned with the glycosylation and other modification of secreted proteins and other molecules. Its function is not completely understood.
Grey crescent	The pale crescent-shaped belt around the equator of an amphibian zygote or cleavage stage embryo, opposite the point of entry of the sperm. The grey crescent gives rise to the 'Primary organizer' (q.v.).
Growth	Increase in size of an organism or of its parts due to synthesis of body components (*see* 'Allometric growth').
Growth factor	A substance that stimulates mitosis, often on a tissue-specific basis.
Growth hormone	Any of several growth-promoting endocrine secretions, especially somatotropin.
G-protein (GTP-binding protein)	A membrane-bound protein that binds GTP and acts as an intermediary between hydrophilic hormone receptor protein and adenylate cyclase.
Guanine	A purine base that pairs with cytosine in double stranded DNA.
Guanophore cell	A cell type which contains guanine, found in the epidermis of some animal species.
Guanosine	The nucleoside having guanine as its base.
Gyrase	A topoisomerase with the ability to introduce negative supercoils into DNA.
Haem	An iron-containing porphyrin molecule that forms the oxygen-binding portion of haemoglobin.
Haemoglobin (Hb)	The respiratory blood pigment in the vertebrates. There are several types of haemoglobin related to developmental stage. Adult haemoglobin consists of a pair of α-globin and a pair of β-globin protein chains, plus four haem groups coiled together into an almost spherical molecule.
Haltere	A drumstick-like appendage which represents a modified hind wing in dipteran flies. Its function is to assist with maintenance of equilibrium during flight.
Haploid	Having only one set of chromosomes. Unfertilized eggs and sperms are haploid.

Heat-shock proteins	A characteristic small set of proteins produced by many organisms in response to elevated temperature.
Heat-shock response	A physiological response to raised temperature shown by most organisms from bacteria to man. Synthesis of most proteins ceases or slows down and is replaced by that of a characteristic set of heat-shock proteins.
Hemimetabolous	Referring to insects which acquire the adult form gradually, by a series of moults, e.g. cockroaches (*cf.* 'Holometabolous').
Helix	A spiral.
Hensen's node	The primary organizer in birds and mammals equivalent to the dorsal lip of the blastopore in amphibians.
Heterochromatin	Highly condensed chromatin which shows maximum staining in interphase (*cf.* 'Euchromatin'). Heterochromatin is of two types, 'constitutive', which is uniform throughout the body, and 'facultative' for example, sex chromatin, which shows variation in different parts of the body. It contains few active genes.
Heterochrony	Evolutionary change in the timing of developmental events.
Heterogametic sex	The sex that has heteromorphic sex chromosomes, for example, males (XY) in mammals, females (ZW) in birds.
Heterogeneous nuclear RNA (hnRNA)	A diverse assortment of RNA types found in the nuclei of eukaryote cells and transcribed by RNA polymerase II. It includes the precursor of mRNA.
Heterokaryon	A cell that contains nuclei from genetically different origins. The stage following fusion of two cells of different origin, before mitosis brings about mingling of their chromosomes (*cf.* 'Hybrid cell').
Heterotroph	An organism that requires complex organic molecules as nutrients from which to obtain energy and build body components.
Heterozygote	An individual with different alleles at a particular locus on homologous chromosomes.
Histones	Basic nucleoproteins characteristic of eukaryote chromosomes. There are five main species, H1, H2A, H2B, H3 and H4, plus a less frequent type, H5.
HMG proteins	'High mobility group' proteins. A class of small highly charged non-histone chromosomal proteins that move rapidly during electrophoresis. HMG 14 and HMG 17 are associated with the nucleosomes within actively transcribed genes.
hnRNA	*See* 'Heterogeneous nuclear RNA'.
Hogness box	The TATA box.
Holoenzyme	An active enzyme consisting of an apoenzyme and a co-enzyme neither of which is active by itself.
Holometabolous	Referring to insects that undergo complete metamorphosis in their ontogeny from larva to pupa to imago, the latter differing strongly from the larva, for example, butterflies (*cf.* 'Hemi-metabolous').
Hom(o)eopathy	A system of therapeutics founded by S. C. F. Hahnemann, based upon the observation that certain drugs when given in large doses to a healthy individual will produce conditions similar to those relieved by the same drug in small doses when occurring as symptoms of a disease.
Hom(o)eotic mutation	A class of mutations in arthropods in which entire organs, such as legs, wings or halteres, develop at inappropriate parts of the body, replacing the organ which would normally be found

at that position. For example in *Drosophila melanogastar* the *ophthalmoptera* mutation causes the eyes to develop as wings.

Homogametic sex	The sex with homologous sex chromosomes, for example, females (XX) in mammals, males (ZZ) in birds.
Homology	A similarity between two organisms or their parts due to inheritance of the same feature from a common ancestor, for example, the wing of a bird and the arm of a man are homologous (*cf.* 'Analogy'). Chromosomes are homologous if they carry the same genetic loci, a diploid cell having two copies of each chromosome homologue, one from each parent.
Homozygote	An individual with the same allele at corresponding loci on homologous chromosomes.
Housekeeping enzymes	Enzymes present in most cell types that are concerned with functions of general maintenance and growth, as distinct from 'luxury enzymes' concerned with the specialized aspects of cytodifferentiation.
Huntington's chorea	A human disorder inherited as an autosomal dominant and characterized by involuntary movements of the face and limbs together with progressive mental deterioration. The average age of onset is around forty years.
Hyaluronic acid	A polysaccharide found as an integral part of the extra-cellular matrix of animal connective tissue.
Hybrid cell	A cell containing hereditary material from genetically different origins, typically within a single nucleus (*cf.* 'Heterokaryon'). They can be produced artificially by several means including cell fusion.
Hybridization	(1) The pairing of complementary RNA and DNA strands to give an RNA–DNA *hybrid*. (2) The fusion of two cells of dissimilar origin. (3) The creation of a plant or animal by crossing parents that are genetically dissimilar, usually of different species.
Hydrogen bond	A weak bond involving the sharing of an electron with a hydrogen atom.
Hydrophilic	Water soluble.
Imaginal disc	In holometabolous insects, larval structures composed of thickened epidermis, together with underlying mesenchyme, which at metamorphosis give rise to adult organs.
Imago	The sexually mature adult form of a holometabolous insect.
Immunoglobulin	Any globulin protein capable of activity as an antibody.
Indeterminate growth	Growth which continues without pre-set limits.
Induction	(1) Switching on of transcription as a result of interaction of an inducer with a regulator protein. (2) Communication between cells required for their cell-type-specific differentiation, morphogenesis and maintenance.
Informosome	An inactive polysome found in unfertilized ova, translation being blocked by protein.
Initiation	The first phase of translation of a mRNA molecule by a ribosome, during which the ribosome assembles at its 5′ end.
Initiation complex	The complex formed at the commencement of translation by a small ribosomal subunit, methionyl tRNA carrying methionine, and a variety of initiation factors together with a strand of mRNA.
Initiator	The transcription initiation site, or 'cAp site'.

In situ hybridization	A technique for locating a DNA sequence within the chromosome by incubating a chromosome preparation with radioactive copies of specific RNA transcripts of the gene and detecting its radioactivity after the RNA and DNA have hybridized.
Instar	A stage in the development of an insect between two moults.
Insulin	A protein hormone secreted by the pancreas in vertebrates which controls the amount of glucose in the blood.
Interferon	A small protein which inhibits viral action and is produced by host cells in response to infection.
Intron	A non-coding DNA sequence within a eukaryote structural gene. Introns are transcribed but excised from the RNA transcript before translation (*cf.* 'Exon').
Inverted repeat	A nucleotide sequence repeated in opposite orientation on the same molecule of double stranded DNA. Adjacent inverted repeats constitute a palindrome, reading the same on one strand from left to right as on the other strand from right to left.
Isometric growth	Growth without change of proportion.
Isotope	One of several forms of an atom having the same atomic number but different atomic masses.
Isozyme (isoenzyme)	One of two or more proteins with the same enzymic specificity, but differing in other properties such as pH optimum or isoelectric point.
Junctional complex	*See* 'Gap junction'.
Juvenile hormone (JH)	One of several hormones produced by the corpus allatum in insects, which maintains larval characteristics in the young insect and is inhibited at the final moult.
Keratin	A tough fibrous protein that occurs in the epidermis of vertebrates forming its resistant outermost layer plus hair, feathers, scales, nails, claws, hooves and the outer surface of horns.
Keratinization	Transformation of epidermal cells into keratin.
Kinase	An enzyme which transfers phosphate groups. A substance which transforms zymogens into enzymes.
Kinety	Structure on the outer surface of ciliated protozoans consisting of a cilium and a basal body or kinetosome used for locomotion.
Klinefelter syndrome	An abnormal human male phenotype characterized by an extra X chromosome (XXY). Patients are sterile feminized males, frequently mentally retarded.
K_m value	The substrate concentration at which a particular enzyme operates at half its maximum activity.
Krebs cycle	*See* 'Tricarboxylic acid cycle'.
Lac operon	An operon involved in lactose metabolism of *Escherichia coli* controlling synthesis of permease and β-galactosidase.
Lactate dehydrogenase (LDH)	The enzyme which catalyses the interconversion of lactate and pyruvate. It is composed of four subunits each of which may be of type M or type H. Five different isozymes are thus possible, and in extracts of most tissues these normally exist in binomial proportions, depending on the relative abundance of the M and H subunits in that tissue.
Laminin	A large glycoprotein found in epithelial basement membranes.
Lampbrush chromosome	The form taken by chromosomes in the oocytes of vertebrates. They are usually up to 1 mm long and covered by a profusion of hair-like projections that are actually loops of DNA undergoing transcription.

Larva	The pre-adult form in which some animals hatch from the egg. Larvae typically possess specialized organs necessary for their existence which are absent from the adult.
Leader	The non-translated sequence at the 5′ end of mRNA that precedes the 'start' codon.
Leader cell	A cell which it is thought directs its neighbours to adopt a state or type of differentiation similar to its own.
Lentoid bodies	Rounded structures that appear in cell cultures of lens epithelium, retinal or iris pigment epithelium and neural retina of several species. They contain cells which by morphological and biochemical criteria are equivalent to lens fibres and are considered to indicate that differentiation of lens-specific cells has taken place.
Lewis Model	A theory which explains the determination of segment identity in *Drosophila* in terms of sequential activation of members of the *bithorax* gene complex (q.v.).
Limb bud	An outgrowth of loose mesenchyme cells surrounded by an epithelial sheet of ectoderm, which develops into the limb in vertebrates.
Linkage	The tendency of alleles of different genes to be inherited together, as a result of their location on the same chromosome. Linked genes do not obey Mendel's law of independent assortment.
Lipophilic	Fat-soluble, hydrophobic.
Locus	The position on a chromosome at which the gene for a specific trait resides. A locus may be occupied by any one of the alleles of a gene.
Luteinizing hormone (LH)	A gonadotrophic pituitary hormone which stimulates theca-lutein cell formation in the ovary and the interstitial cells of the testis.
Lysosome	Membrane-bounded body containing hydrolytic enzymes (lysozymes) present in large numbers in the cytoplasm of animal cells. They liberate their enzymes in injured cells and assist in the digestion of dead cells. Believed to be important agents of 'Programmed cell death' (q.v.).
Macroevolution	The study of evolutionary events and processes that require times of the order of geological periods for their occurrence or operation.
Maintenance methylase	A hypothetical enzyme believed to maintain patterns of methylation in newly synthesized DNA throughout mitosis.
Malpighian tubules	Tubular glands concerned with excretion, which open into the interior part of the hind gut of insects, arachnids and myriapods.
Maternal effect	An influence of the mother's physiology on the phenotype of the offspring.
Maternal inheritance	The transmission of a phenotypic character from female parent (only) to offspring, usually by means which do not involve the normal mediation of nuclear genes.
Meiosis	A type of cell division that occurs in the formation of germ cells. Two successive divisions reduce the chromosome number from 2n to 1n in each of the four product cells.
Meiotic drive	Any mechanism which may affect the genetic composition of a population through inequalities in chromosomal transmission at gametogenesis or fertilization.
Melanin	An animal pigment usually appearing black, brown or yellow,

depending on concentration, distribution and chemical modification. Eumelanin is the black form, phaeomelanin a yellow modification. Both are produced from the amino acid tyrosine.

Melanocyte
A cell containing melanin; a pigment cell.

Melanosome
A cytoplasmic body in melanocytes that carries the pigment melanin.

Melting
Denaturation of DNA, conversion from the double- to the single-stranded state, most often accompanied by heating.

Mendel's laws
The two laws of inheritance defined by Gregor Mendel can be described in modern terms as follows: (1) The law of segregation: The two members of each pair of alleles possessed by a diploid parental organism become separated during gamete formation and transferred to different offspring. (2) The law of independent assortment: Members of different pairs of alleles are distributed independently of one another among the gametes (provided they reside on different chromosomes) and the subsequent pairing of male and female gametes is also random (i.e., irrespective of the alleles they carry).

Mesenchyme
Embryonic connective tissue of mesodermal origin, consisting of widely scattered irregularly branching cells in a jelly-like matrix. The condensation of mesenchyme around epithelia is especially important in the formation of organ primordia.

Mesoderm
One of the germinal layers of a triploblastic gastrula composed of cells that have moved from the embryo surface into its interior. It gives rise to tissues situated between gut and ectoderm, to the muscles, the blood vascular system, connective tisssue and sex organs.

Mesothorax
The middle of the three segments of the insect thorax. It bears a pair of walking legs and in flying insects a pair of wings.

Messenger RNA
A short-lived transient form of RNA transcribed from the genes, which serves to carry genetic information from them to the ribosomes for translation into proteins.

Metabolism
The sum total of all the physical and chemical processes by which living tissues are produced and maintained and by which energy is made available for the cells of the organism.

Metabolite
A substance which takes a part in the processes of metabolism, whether of external or internal origin.

Metamorphosis
The period of rapid transformation of an animal from larval to adult form.

Metaplasia
Transformation of one sort of tissue into another. Strictly it refers to alteration to the normal pathway of differentiation of undifferentiated stem cells, *cf.* 'Transdifferentiation' which describes the change from one differentiated cell type to another.

Metathorax
The most posterior of the three segments of the insect thorax. It bears a pair of walking legs and in many species a pair of wings, or in dipteran flies a pair of halteres.

Methylation
Replacement of a hydrogen atom by a methyl ($-CH_3$) group.

Micro-evolution
An evolutionary pattern usually viewed over a short period of time, such as changes in gene frequency within a population over a few generations.

Microfilament
Filament of actin found in the cytoplasm of eukaryotic cells.

Micropyle
A pore in the chorion of an insect egg through which the sperm reaches and fertilizes the ovum.

Microtubule	A hollow filament composed of the protein tubulin found in cilia, flagella and the cytoplasm of eukaryote cells. Microtubules link the centrioles and chromosome centromeres in the mitotic spindle.
Mitochondria	Self-reproducing organelles that occur in the cytoplasm of practically all eukaryote cells. They are the sites of the reactions of oxidative phosphorylation which result in the formation of ATP.
Mitosis	The natural division of a eukaryotic somatic cell.
Mitotic spindle	A bipolar fibrous structure largely composed of microtubules radiating out from centrioles at two foci, which plays an integral part in the separation of the chromosomes at mitosis.
Modulation	Fluctuation of cell phenotype within the already established range of its determination, but with retention of capacity to revert to the original phenotype.
Modulator	An enhancer-type sequence.
Molecular biology	A branch of biology in which phenomena and processes are studied both from a phenomenological and a physicochemical or biochemical approach. It aims to account for biological events in terms of the established principles of physics and chemistry.
Morphallaxis	Regeneration by reorganization of the remaining portions of the body to produce the missing structures (*cf.* 'Epimorphosis').
Morphogen	A biochemical considered to be present in the cytoplasm of the egg or embryo which directs the adoption of some aspect of body form.
Morphogenesis	Development of form or structure during ontogeny or regeneration.
Mosaic	An individual composed of cells of different genotypes.
Mosaic theory	An outdated theory that the early embryo is constructed as a mosaic, each cell fitting into its predetermined location in the larva and differentiating independently of influence from other regions. The theory postulated that the membranes or cytoplasm of the egg contain precise instructions for development which become packaged around individual nuclei and dictate the entire future course of differentiation. The idea arose due to lack of appreciation of the importance of patterns and timing of cleavage. Many insect species for example show extreme mosaicism, but still undergo regulation.
mRNA	*See* 'Messenger RNA'.
Mucopolysaccharide	Any of a group of polysaccharides containing an amino sugar and uronic acid; now known as 'glycosaminoglycan'.
Mucus	A slimy solution or gel composed of protein–polysaccharide complexes secreted by animal cells.
Multi-gene family	A group of genes of related function located side by side on the chromosome.
Multipotency	*See* 'Pluripotency'.
Mutation	Any change in the sequence of nucleotides genomic DNA.
Myoblast	The cell type that gives rise to muscle.
Myoglobin	A monomeric haem protein of muscle concerned with oxygen transport and storage.
NAD	Nicotinamide adenine dinucleotide. A co-enzyme involved in many metabolic pathways as a hydrogen acceptor, forming

NADH₂; formerly called DPN and co-enzyme 1.

Negative control
Prevention of biological activity by presence of a specific controlling molecule. Usually used in relation to transcription of a pro-karyote operon, in which diffusible repressor protein acts at an operator site on the DNA, preventing transcription. In inducible systems a diffusible inducer prevents repressor function, in repressible systems a diffusible co-repressor stimulates repressor function (*cf.* 'Positive control').

Neo-Darwinism
Essentially the post-Darwinian concept that species evolve by the natural selection of adaptive phenotypes caused by mutant genes. The central idea is that the prime cause of natural variation is reassortment and mutation of the genes.

Neoteny
Paedomorphosis due to retardation of somatic development.

Neural crest
Embryonic material of vertebrates initially at both sides of the neural plate as this rolls up to form the neural tube. The neural crest gives rise to a variety of very different cell types: melanocytes, nerve ganglia, Schwann cells, adrenalin-secreting cells, cartilage and bone.

Neural plate
A region of ectoderm on the dorsal surface of the embryos of vertebrates at the neurula stage which gives rise to the central nervous system.

Neural tube
Longitudinal tube of neural tissue in vertebrate embryos formed by rolling up the neural plate. It gives rise to the brain and spinal cord.

Neurectoderm (neural ectoderm)
The ectoderm on the dorsal surface of the embryos of vertebrates at the gastrula stage, which gives rise to the central nervous system.

Neurula
The stage in the development of a vertebrate embryo after most of the gastrulation movements have ceased, during which the neural tube forms from the neural plate. An embryo which has a neural plate.

Neurulation
The process of formation, rolling and closure of the neural plate to form the neural tube.

NHC protein
Non-histone chromosomal protein. Any member of a group of acidic proteins that exists in chromatin.

Notochord
A skeletal rod of large vacuolated cells of mesodermal origin packed within a firm sheath that lies lengthwise between the central nerve cord and the gut at some stage in the development of all chordates. In the vertebrates remnants of it persist in the adult in the inter-vertebral discs.

N terminus
The end of a polypeptide at which the amino (–NH₂) group which could participate in a peptide bond is free.

Nucleolar organizer
The chromosomal region which carries the repetitive sequences coding for 28S, 18S, and 5.8S rRNA.

Nucleolus
A dense frequently spherical body within the nucleus of a eukaryote cell during interphase containing amplified copies of the ribosomal DNA sequences and specialized for the synthesis of rRNA. A typical cell contains one or several nucleoli, but amphibian oocytes may have over 1000.

Nucleosome
The basic structural subunit of chromatin. It consists of about 200 base pairs of DNA wound around an octamer of histone proteins.

Nucleotide	A compound of a purine or pyrimidine base linked to ribose or deoxyribose and phosphoric acid.
Oestrogen	Any of several steroid hormones, such as oestradiol, oestriol and oestrone, produced mainly in the ovaries, which promote oestrus and the development of female secondary sexual characters.
Offset signal	A signal assumed to direct the termination of a stage in development or growth.
Onset signal	A signal assumed to direct the commencement of a stage in development or growth.
Ontogeny	The whole course of development during an individual's life history.
Oocyte	A cell type which undergoes meiosis to form an ovum.
Operator	The site on the DNA of a prokaryote operon at which a repressor protein binds, preventing transcription becoming initiated at the adjacent promoter.
Operon	A unit consisting of adjacent cistrons that function co-ordinately through production of a polycistronic messenger, under the control of an operator gene.
Order	(1) Non-randomness. (2) One of the kinds of groups used in classifying organisms. Orders are subgroups of Classes and are themselves subdivided into Families.
Organ	A part of an organism which forms a structural and functional unit composed of cells of more than one differentiated type.
Organelle	A persistent structure of characteristic form and specialized function forming part of a cell. The term is most frequently used to denote self-reproducing bodies such as mitochondria, centrioles and plastids.
Organism	A living being.
Organizer	Any part of an embryo that performs an inductive or organizing influence on another part, especially the dorsal lip of the blastopore or 'Primary organizer' (q.v.).
Osteoblast	A cell type responsible for formation of the calcified inter-cellular substance of bone.
Osteoclast	A type of cell which remodels bone shape by breaking down the calcified inter-cellular material of bone.
Osteodentine	An ancestral tissue intermediate between bone and dentine.
Ovalbumin	The major protein component of egg white.
Ovum	An unfertilized egg cell containing a haploid nucleus.
Oxytocin	A polypeptide secreted by the posterior pituitary which in mammals induces contraction of uterine and other smooth muscle.
Paedomorphosis	The retention of ancestral juvenile characters in later ontogenetic stages of descendants.
Palindrome	Inverted repeat.
Parthenogenesis	Development of an ovum into a new individual without fertilization.
Penetrance	The proportion of individuals with a specific genotype who manifest that genotype at the phenotypic level. Penetrance can vary due to environmental factors or the presence of other specific alleles in the genetic background.
Pentadactyl limb	A limb having, or derived from one with, five digits. It originally evolved as an adaptation to life on land and occurs in amphibians, reptiles, birds and mammals.

Peptide bond

A covalent bond joining two amino acids, with the amino ($-NH_2$) group of one bound to the carboxyl ($-COOH$) of the other, with the elimination of water.

Phenocopy

A non-inherited departure from the normal phenotype, induced by an environmental influence usually acting on an organism during its early development, that mimics the effect of a known mutant allele.

Phenotype

The observable properties of an organism, resulting from the interaction of its genetic constitution, or genotype, with the environment.

Pheromone

A chemical substance released by an animal which influences the behaviour or development of other individuals of the same species.

Philadelphia chromosome

An unusual small chromosome found in the bone marrow of most patients with chronic myelogenous leukaemia. It is a deleted copy of chromosome 22, its distal segment being translocated elsewhere, usually to the long arm of chromosome 9.

Phylogeny

The evolutionary history of a lineage.

Plasma membrane

The thin membrane which surrounds body cells, consisting basically of a phospholipid bilayer in which proteins and glycoproteins are embedded.

Plasmid

An autonomous self-replicating extra-chromosomal loop of DNA characteristically found in bacteria.

Pluripotency

The capacity of a cell or nucleus to give rise to several differentiated cell types.

Polar Co-ordinate Model

An explanation originally proposed by French, Bryant and Bryant to explain bizarre aspects of limb regeneration in cockroaches and amphibians. The model conceives a three-dimensional display of positional information within the limb, such that the position of any part is defined with respect to a set of radial co-ordinates numbered as on a clock face, plus another set defining position along the proximo-distal axis. Regeneration occurs as if to intercalate missing numbers by the shortest possible route around the circumference, and to replace those parts distal to the cut.

Poleplasm

See 'Germplasm'.

Poly-A tail

The sequence of polyadenylic acid which is attached to the 3' end of eukaryotic mRNA.

Polypeptide

A condensation product of many amino acids linked together by peptide bonds. Polypeptides are the first products of translation of mRNA, which are then processed into proteins.

Polyploid

Having more than two complete sets of chromosomes.

Polysaccharide

A biological polymer composed of sugar subunits.

Polysome (polyribosome)

A strand of messenger RNA associated with a series of ribosomes engaged in translation.

Polytene chromosome

A giant chromosome which is multi-stranded, being much broader than a normal chromosome and visible throughout interphase. Polyteny results from repeated DNA replication without cell division. Such chromosomes are characteristic of certain tissues, such as those in the salivary glands of larval dipteran flies.

Porphyropsin

The retinal visual pigment of freshwater vertebrates derived from the aldehyde of Vitamin A_2.

Position effect

A change in the expression of a gene brought about by its translocation to a new site in the genome.

Position effect variegation	Variegation in phenotype caused by inactivation of a gene in some cells through its abnormal juxtaposition beside heterochromatin.
Positional information	A theoretical feature of organisms which enables their cells to select courses of differentiation and development appropriate to their position in the body. According to this hypothesis, cells acquire a 'Positional value' as a result of exposure to an external influence, such as a particular concentration of a morphogen. This information is then interpreted by the cell with a resultant change in its determined or differentiated state.
Positional value	The status of the positional information deemed to be acquired by a cell in accordance with the theory of 'Positional information' (q.v.).
Positive control	Causation of biological activity by the presence, as distinct from the absence, of a specific controlling molecule. Usually used in relation to transcription of a prokaryote operon, in which a diffusible activator protein acts at an operator site on the DNA enabling transcription to take place. In inducible systems a diffusible inducer stimulates activator function; in repressible systems a diffusible co-repressor prevents activator function (*cf.* 'Negative control').
Potential pigment cells	A term used to describe the unpigmented precursors of pigment epithelial cells, especially those formed by transdifferentiation in cultures of neural retina.
Pribnow box	The consensus sequence TATAATG centred about 10 bases before the start of bacterial genes. It is part of the promoter.
Primary induction	The action of the primary organizer (in Amphibia the dorsal lip of the blastopore) in influencing the developmental fate of other tissues which come into its vicinity during the process of gastrulation. The inductive causation of neurulation.
Primary organizer (or 'organizer')	In amphibian embryos the dorsal lip of the blastopore. The region of an early vertebrate embryo which initiates gastrulation and defines the future major body axis.
Primary structure	Of a protein, the sequence of amino acids.
Processing RNA	Species of small nuclear RNA (snRNA) that are concerned with processing of mRNA by excision-splicing.
Proenzyme	The enzymically inactive precursor of an enzyme.
Progenesis	Paedomorphosis due to the precocious sexual maturation of an organism still in a morphologically juvenile stage.
Progesterone	A steroid hormone produced mainly by the corpus lutea of the ovary which prepares the uterus to receive the embryo and maintains the uterus during pregnancy.
Programmed cell death	Cell destruction, or necrosis, occurring as a normal part of ontogeny, as for example in the reduction of the tail in amphibian tadpoles.
Progress Zone Model	An explanation proposed by Summerbell, Lewis and Wolpert to explain determination of vertebrate limb parts along the proximo-distal axis. It is suggested that the final differentiative fate of a cell lineage depends on the number of mitoses its progenitor underwent in the 'progress zone' just behind the apical ectodermal ridge at the limb bud tip.
Prokaryote (procaryote)	Literally 'before a nucleus'; having the genetic material in the form of a simple filament of DNA within the cytoplasm (*cf.* 'Eukaryote').

Prolactin	A proteinaceous hormone secreted by the anterior pituitary, which stimulates milk secretion in mammals. In larval amphibians it acts antagonistically to thyroxine in inhibiting metamorphosis.
Promoter	The region of the DNA in a prokaryote operon to which RNA polymerase binds and initiates transcription (*see* 'Pribnow box', 'TATA box').
Pronucleus	Haploid nucleus derived from female or male parent, present in the ovum or zygote before nuclear fusion.
Prosthetic group	A non-protein accessory substance necessary for the proteinaceous part of an enzyme to work, which is strongly bound to the protein (*cf.* 'Co-enzyme').
Protein	A complex organic molecule or multimer composed largely of amino acids, but often complexed with other molecules, elaborated from one or more simple polypeptides.
Proteoglycan	A class of very large molecules formed by covalent linkage of glycosaminoglycan with protein; formerly called mucoprotein.
Prothorax	The most anterior of the three segments which make up the insect thorax, it bears a pair of walking legs, but no wings.
Protoeukaryote	The hypothetical ancestor of the eukaryotes; an anaerobically respiring phagocytic amoeboid cell, thought to have incorporated a species of aerobically respiring bacterium, which became the ancestor of the mitochondria.
Proximo-distal axis	With respect to a limb, the axis from the root of the limb to the tips of the digits.
Puffing	The formation of puffs on polytene chromosomes.
Punctuation code	The set of correspondences between nucleotide sequences in DNA or RNA and signals that delimit the extent of the coding sequence, or act as control points for gene expression, as distinct from those nucleotide sequences which define the amino acid composition of a polypeptide.
Pupa	The stage between the larva and the imago of a holometabolous insect during which larval structures break down and are replaced by adult ones, as for example by the development and differentiation of imaginal discs.
Purine	A type of nitrogenous base. The purine bases in DNA are adenine and guanine.
Pyrimidine	A type of nitrogenous base. The pyrimidine bases in DNA are cytosine and thymine.
Quaternary structure	Of a protein, its multimeric constitution.
Radial symmetry	Capable of being halved in two or more planes so that the halves are approximate mirror images of one another. Coelenterates and echinoderms show radial symmetry (*cf.* 'Bilateral symmetry').
Radioactive	Decomposing spontaneously with emission of α, β or γ rays.
Reading frame	One of the three possible ways of reading a nucleotide sequence as a series of triplets.
Read-through protein	An unusually extended protein produced by translation of the messenger beyond the normal 'stop' codon.
Re-annealment	Renaturation or reassociation of denatured DNA.
Recapitulation	The repetition of ancestral adult stages in embryonic or juvenile stages of descendants.
Recessive	In diploid organisms, an allele which is phenotypically manifest in the homozyous state, but is masked in the presence of its dominant allele.

Recognition site	A characteristic nucleotide sequence centred about 35 bases upstream from the transcription initiation site in prokaryotes. It has the consensus sequence TTGACA.
Recombination	Any process in a diploid or partially diploid cell that generates new gene or chromosomal combinations not previously found in that cell or its progenitors. The process that generates a haploid product of meiosis whose genotype differs from both the haploid genotypes of the parent.
Reference Points Theory	A theory which suggests that gradients of biochemical morphogens are established within the egg cytoplasm with respect to reference points such as the animal and vegetal poles.
Regulative development	Animal development of a type which allows reparative readjustments to take place following major disturbances, differentiation of tissues being demonstrably dependent on the influence of other tissues (*cf.* 'Mosaic development').
Regulator	The rate-controlling enzyme of a metabolic pathway which is subject to feedback inhibition by the end-product of the pathway.
Regulator gene	A gene which codes for an RNA or protein product that controls the rate of synthesis of the products of other non-adjacent genes.
Renaturation	The reassociation or re-annealment of denatured complementary single strands of a DNA double helix.
Repression	Inhibition of transcription (or translation) by binding of repressor protein to a specific site on the DNA (or mRNA).
Restriction endonuclease	An enzyme that recognizes a specific short sequence of DNA and cleaves it at that site or nearby.
Reticulocyte	A small erythrocyte or proerythrocyte which has a reticular (net-like) appearance when stained. The precursor of a mature red blood cell.
Retinoid	A substance biochemically closely related to retinene (retinal), the aldehyde of Vitamin A.
Reverse transcriptase	An enzyme produced by RNA viruses that promotes synthesis of DNA on an RNA template.
Rhodopsin	The retinal visual pigment of marine and land vertebrates derived from the aldehyde of Vitamin A_1.
Ribonucleoprotein	A complex of RNA and protein.
Ribosomal RNA (rRNA)	Species of RNA which contribute to the structure of the ribosomes. In eukaryotes this includes species of sedimentation coefficients 28S, 18S, 5.8S and 5S. rRNA is not translated.
Ribosome	The ribonucleoprotein particle which carries out the translation of messenger RNA into polypeptide. They are composed of two unequal subunits bound together by Mg^{2+} ions and may exist free in the cytoplasm or bound to the (rough) endoplasmic reticulum.
RNA (ribonucleic acid)	A long molecule consisting of a large number of nucleotides linked together in a strand one nucleotide wide. Each nucleotide contains the sugar ribose, plus either guanine, adenine, cytosine or uracil and phosphate. In some viruses the genetic material is RNA; in all species RNA assists with the interpretation of genetic information, as either messenger RNA, transfer RNA, ribosomal RNA or 'processing RNA'.
RNA polymerase	A class of enzymes that during transcription catalyse the formation of RNA from ribonucleotide triphosphates, using DNA as a template. There are three classes of the enzyme: Pol I or A

which synthesizes 18S, 5.8S and 28S rRNA; Pol II or B, which transcribes hnRNA and Pol III or C which produces small RNA species including tRNA.

RNA processing
The conversion of hnRNA into mRNA. It involves capping, excision-splicing and polyadenylation.

RNP
Ribonucleoprotein.

S
Svedberg unit. A unit of sedimentation velocity commonly used to describe molecular units of various sizes.

Saltation
An abrupt evolutionary change. An irreversible and inherited change in a cell or organism due to mutation or some other cause.

Satellite DNA
DNA with a sedimentation coefficient markedly different from that of the bulk of the DNA of that organism. It typically contains many tandem repeats of a short repeat unit, that in the chromosome are localized near the centromere.

Single-copy DNA
Gene or non-coding sequences that are present only once in the haploid genome.

Second messenger system
A system which permits transfer into the cell interior of a signal generated by binding of a ligand to the cell surface. The best-known examples are those that utilize cyclic-AMP and Ca^{2+} ions as second messengers.

Secondary induction
Inductive interaction between tissues following the action of the primary organizer. (In this text the term is taken to include all later inductions also.)

Secondary structure
Of a protein, the spiral or other arrangement of the polypeptide chain.

Segmentation
Also known as 'metamerism'. Repetition of a pattern of elements belonging to each of the main organ systems of the body along the animal's antero-posterior axis, or a comparable repetition along the axis of an appendage.

Selector
A sequence which defines where transcription should commence.

Sense strand
See 'Codogenic strand'.

Serial endosymbiotic theory (SET)
A theory which postulates the eukaryotes originating from a primitive anaerobic phagocyte, by sequential symbiotic incorporation of an aerobically respiring bacterium, a spirochaete-like, flagellated bacterium and ultimately a photosynthetic bacterium.

Sex chromatin
Chromatin that forms a condensed mass representing one of the X chromosomes in a normal female mammal. Also known as the Barr body.

Sex chromosome
A chromosome whose presence or absence is correlated with the sex of the bearer.

Shedding
Loss of cytoplasmic components, especially melanin granules during the dedifferentiation of pigment epithelium in Wolffian lens regeneration.

Signal sequence
The sequence of fat-soluble amino acids near the N-terminus of a secretary protein which, according to the 'Signal Hypothesis' aids the attachment of a polysome to the endoplasmic reticulum.

Sink
With respect to biochemical gradients, the site at which the component of the gradient is destroyed or eliminated.

Small nuclear ribonucleoprotein (snRNP)
A complex of snRNA and specific protein.

Small nuclear RNA (snRNA)
Small RNA species largely of unknown role that are confined to the nucleus.

Solenoid	A uniformly wound coil in the form of a cylinder having a length much greater than its diameter.
Somatic	Relating to the body (soma) as distinct from the gonads or germ cells.
Somatotrop(h)in	A protein hormone produced by the anterior lobe of the pituitary gland which promotes general body growth. Also known as 'growth hormone'.
Somites	Blocks of sequentially arranged tissue of mesodermal origin that arise pairwise on each side of the major body axis in chordates. From each somite is derived a block of striped muscle, the myotome, innervated by one ventral nerve root, and mesenchyme which forms connective tissue and the vertebral column.
Somitogenesis	The formation of sequentially arranged blocks of mesoderm on either side of the main body axis during development in chordates.
Specific Gene Amplification Theory	A theory which attempted to explain cytodifferentiation in terms of the tissue specific selective amplification of genes associated with specific functions.
Spindle	A structure formed within a cell at mitosis or meiosis which takes part in the distribution of chromatids to daughter nuclei. The chromosomes become arranged along its equator at metaphase.
Spiral cleavage	A type of cleavage which results in a characteristic spiral body form, in which blastomeres are distributed alternately clockwise and anticlockwise. This is shown for example by the annelids and the cephalopod molluscs.
Splicing	The joining together of exons during RNA processing, following excision of introns. The term is sometimes taken to include the excision stage also.
Start codon	The triplet codon AUG which signifies where translation of a mRNA strand should start.
Stop codons	Also called 'chain terminator' and 'nonsense' codons, they are triplet codons that indicate where on a mRNA molecule translation should cease. In genomic mRNA they are UAA, UAG and UGA, in mammalian mitochondrial transcripts AGA and AGG.
Structural gene	A gene coding for any RNA or protein product other than a regulator of the activity of another gene.
Structural protein	A protein which contributes to body structure, as distinct from enzymes, which catalyse chemical reactions.
Supercoiling	The coiling of a closed duplex DNA double helix so that it crosses over its own axis.
Switch site	A part of a chromosome containing repetitive nucleotide sequences which facilitates rearrangement of gene sequences, as for example among the immunoglobulin genes.
Synapsis	Close association of homologous chromosomes or chromatids.
Syncytium	A mass of cytoplasm enclosed within a single continuous plasma membrane and containing many nuclei. Examples are striped muscle and the trophoblast epithelium of the mammalian placenta.
TATA box	A conserved sequence 25–30 bases upstream of the cAp site of eukaryotic genes transcribed by RNA polymerase II. It acts as a 'selector' defining the site at which transcription should commence.
Tail	*See* 'Poly-A tail'.

Telomere	The natural end of a chromosome.
Termination	The third phase of translation of a mRNA molecule by a ribosome, during which polypeptide elongation is terminated following entry of a stop codon into the ribosome.
Tertiary induction	A term used by some authors to define inductive interactions that occur subsequent to 'Secondary inductions' (q.v.).
Tertiary structure	Of a protein, the folding or coiling of a secondary structure to form an often globular molecule.
Testosterone	An androgenic hormone produced mainly by the testis in vertebrates.
Tetraploid	Having four sets of chromosomes.
Thymidine	The nucleoside having thymine as its base.
Thymine	A pyrimidine base that in DNA pairs with adenine.
Thyroid-stimulating hormone (TSH)	A glycoprotein hormone secreted by the anterior pituitary which regulates growth of the thyroid gland and the formation and secretion of its hormones.
Thyroxine	A hormone derived from tyrosine and produced by the thyroid gland. In Amphibia it promotes metamorphosis, exerting diverse effects on different target cell types.
Tight junction	A type of inter-cellular linkage in which the two plasma membranes are closely apposed (*cf.* 'Gap junction').
Tissue	A part of the body consisting mainly of cells of similar differentiated type associated together and bound by inter-cellular material.
Topoisomerase	An enzyme which can control supercoiling of DNA, that is, can change the number of times a closed duplex of DNA crosses over itself.
Totipotency	The capacity of a cell or nucleus to give rise to any cell type in the body.
Trailer	A non-translated sequence at the 3′ end of mRNA, following the 'stop' codon.
Trans configuration	In repulsion; on the opposite member of a homologous pair of chromosomes (*cf.* '*Cis* configuration').
Transcription	The synthesis of RNA on a DNA template. This is carried out by the enzyme RNA polymerase.
Transcription initiator	The point in a gene sequence at which transcription begins, the cAp site.
Transcriptional unit	The stretch of DNA between sites of initiation and termination of transcription by RNA polymerase. One transcriptional unit may contain more than one gene.
Transdetermination	A change in the determined state of cells to a different state, such that pathways of differentiation become accessible leading to structures different from those originally specified. The classic example is provided by the imaginal disc cells of *Drosophila* which, after a prolonged period of proliferation, then exposure to ecdysone, will differentiate into adult organs different from those that would normally develop from those particular discs. An important feature of transdetermination is that the changes occur in cells which show evidence of determination, but which are not ostensibly differentiated.
Transdifferentiation	A switch from one differentiated cell type to another. The term was originally used by Selman and Kafatos in relation to cells of the labial glands of silk moths.

Transfer RNA (tRNA)	Any of at least 20 structurally similar species of RNA which can combine covalently with a specific amino acid and hydrogen bond with at least one mRNA triplet codon. Transfer RNA transfers amino acids onto the growing polypeptide during translation of mRNA within the ribosomes.
Translation	The synthesis of polypeptide on the mRNA template, which takes place with the agency of the ribosomes.
Transposable element	A mobile genetic element capable of producing an unstable mutant target gene; two types exist, the regulator and the receptor elements.
Tricarboxylic acid cycle (TCA cycle)	Also known as the citric acid or Krebs cycle. A very important metabolic cycle associated with aerobic respiration and mitochondria. It involves the interconversions of various acids resulting in reduction of NAD and FAD. The NADH$_2$ and FADH$_2$ produced are used for generation of ATP by oxidative phosphorylation.
Triplet codon	A sequence of three adjacent amino acids in mRNA that code for an amino acid or signify the beginning or end of the section to be translated.
tRNA	*See* 'transfer RNA'.
Turner's syndrome	A human condition caused by monosomy of the X chromosome. Patients are sterile females.
Tyrosinase	The key enzyme in the synthesis of melanin. It converts the amino acid tyrosine to DOPA and oxidizes this to DOPA-quinone, which undergoes spontaneous polymerization to form melanin.
U1	A species of small nuclear RNA concerned with processing of hnRNA by excision-splicing.
Ubiquitin	A widely distributed globular acidic protein with a high affinity for histones. It is associated with active genes in eukaryotes.
Upstream	Sequences located in the direction opposite to that followed by the RNA polymerase during transcription. The initiation codon is upstream of the coding region.
Uracil	A pyrimidine base that appears in RNA in place of the thymine found in DNA.
Uridine	The nucleoside having uracil as its base.
Vasopressin	A polypeptide hormone secreted by the posterior pituitary that stimulates unstriated muscle, causes constriction of arteries, raises blood pressure and has an anti-diuretic effect.
Vegetal pole	The point on the surface of an egg furthest from its nucleus, marking the highest concentration of the graded distribution of yolk (*cf.* 'Animal pole').
Wave-form Gradient Model	An explanation proposed by Wilby and Ede to explain the determination of elements of the vertebrate limb along its antero-posterior axis. The model proposes that alternating activities of synthesis and destruction of a morphogen become created across the limb bud, which establish patterns of mesenchyme condensates that then give rise to cartilage and bone.
Wild-type	The most frequently observed phenotype, or the one arbitrarily designated as 'normal'. Most wild-type features are coded by dominant alleles.
Wobble	The ability of the third base in a tRNA anticodon (5′ end) to hydrogen bond with any of two or three bases at the 3′ end of a codon. This allows a single tRNA species to recognize several different codons.

Wolffian lens regeneration	Regeneration of an eye lens from the iris epithelium following lentectomy. This phenomenon was in fact first reported by V. Collucci in 1891, who worked with the salamander *Triturus cristatus*. Without knowing of Colucci's work, G. Wolff discovered the same effect in *T. taeniatus* and devoted a series of papers to it. It has been reported during embryonic stages in several vertebrates, but occurs in adults only in members of the order Urodela.
Z-form DNA	A left-handed zig-zag spiral conformation of DNA (*cf.* 'B-Form').
Zone of polarizing activity (ZPA)	A mesodermal region at the posterior margin of the vertebrate limb bud presumed to regulate cytodifferentiation along the antero-posterior axis of the limb.
Zygote	The product of fusion of gametes of opposite types; a fertilized egg.
Zymogen	Proenzyme; the enzymically inactive protein precursor of an enzyme.

Index